GRAVITY FIELD, SEISMICITY AND TECTONICS OF THE INDIAN PENINSULA AND THE HIMALAYAS

SOLID EARTH SCIENCES LIBRARY

R. K. VERMA

Professor of Applied Geophysics,
Indian School of Mines, Dhanbad, Bihar, India

GRAVITY FIELD, SEISMICITY AND TECTONICS OF THE INDIAN PENINSULA AND THE HIMALAYAS

D. REIDEL PUBLISHING COMPANY

A MEMBER OF THE KLUWER ACADEMIC PUBLISHERS GROUP

DORDRECHT / BOSTON / LANCASTER

Library of Congress Cataloging in Publication Data

Verma, R. K.
Gravity field, seismicity, and tectonics of the Indian peninsula and the Himalayas.

(Solid earth sciences library)
Bibliography: p.
Includes index.
1. Gravity—India. 2. Seismology—India. 3. Geology, Structural. 4. Gravity—Himalaya Mountains Region. 5. Seismology—Himalaya Mountains Region.
I. Title. II. Series.
QB335.I4V47 1985 538'.7'0954 85-2248
ISBN 90-277-1864-4

Published by D. Reidel Publishing Company,
P.O. Box 17, 3300 AA Dordrecht, Holland.

Sold and distributed in the U.S.A. and Canada
by Kluwer Academic Publishers,
190 Old Derby Street, Hingham, MA 02043, U.S.A.

In all other countries, sold and distributed
by Kluwer Academic Publishers Group,
P.O. Box 322, 3300 AH Dordrecht, Holland.

Dedicated to
my father and mother

TABLE OF CONTENTS

FOREWORD

Recent developments in Science and Technology have removed the boundaries between the so-called 'fundamental' and 'applied' fields of research. This has been particularly evident in the rapidly expanding geophysical sciences with their far reaching applications. Fundamental geophysical studies of surface, subsurface and crustal structures where gravity investigations play an important role, are of immediate potential importance in defining major structural features or geological units which may control or influence the development of metallogenic provinces, petroleum potential areas or coal fields.

Gravity studies in India have a long history and tradition. The earliest gravity measurements were made with two brass pendulums loaned by the Royal Society of England, when Basevi and Heaviside established 30 gravity pendulum stations between Cape Comorin and Ladakh during 1865 to 1873. Over the subsequent years, large areas of the country have been systematically covered through gravimeters and a number of regional and detailed studies have been undertaken by various agencies like Survey of India, Geological Survey of India (GSI), Oil and Natural Gas Commission (ONGC), National Geophysical Research Institute (NGRI) and a number of universities. Assistance was provided by late Prof. G. P. Woollard and his former students Muckenfuss, Bonini, Shankaranarayan and Murali Manghnani, who have connected India to the international gravity net work.

Gravity studies carried out by the Survey of India have been of great value in geodetic studies. Gravity investigations by ONGC have provided potential areas for more expensive seismic surveys for defining suitable structures favourable for petroleum accumulation. In the GSI, many detailed mineral investigations have been based on gravity surveys and other geophysical and geological studies. The NGRI has collected all the gravity data in the country and published in 1978, a series of five maps on 1:5 million scale indicating the Station Locations, Bouguer gravity anomalies, Free-Air gravity anomalies, Isostatic gravity anomalies, and Bouguer gravity anomalies super-imposed on regional geology. NGRI has made a number of regional and detailed studies in various parts of the country and published papers on these investigations. By now the country as a whole has been covered fairly well to attempt systematic and indepth studies of the gravity field and its corelation with geology, tectonics and deepseated structural features.

I am pleased to see that Dr. R. K. Verma has taken pains to identify the various aspects of gravity investigations in India and has studied the major trends of Bouguer anomalies and prominent geological features of the country. He has given particular attention to gravity anomalies of Gondwana Basins, in the Dharwar and Bastar provinces of India, the Eastern Ghats, the Singhbum and adjoining areas and over the

Himalaya. If suitably handled, gravity data can be of immense value to study the surface geology, tectonics, the deepseated structures and the mass inhomogeneities within the crust and the upper mantle. Data reduction and computer oriented modelling techniques are available today to use gravity as a powerful tool for such studies. Dr. Verma has treated these aspects extremely well and has also discussed the concepts of isostasy and its applicability to the Indian Peninsula and the Himalaya. He has discussed gravity field and seismicity in the North-Eastern parts of India and Burma and has also discussed the relationship between gravity, seismicity and plate tectonics as applicable to the Himalayan region.

Being himself a very competent research worker and a teacher in Geophysics of long standing and having carried out detailed gravity investigations in several parts of the country, Dr. R. K. Verma is one of the most eminent persons in the country to write a book of this nature. While covering the various topics in fair detail, he has emphasized studies on one of the oldest global shields of Peninsular India and the youngest and the highest mountain ranges of the Himalaya. I feel confident, that this book will meet the growing demands of the students of Geology and Geophysics and of many research workers in Earth Sciences who are engaged in studies of interdisciplinary nature.

April 16, 1984
Hyderabad

HARI NARAIN
D. Phil (Alld), Ph.D. (Sydney), F.N.A.Sc., F.N.A.
Chief Project Coordinator, UNDP,
Director (Retd.), National Geophysical Research Institute,
Former Vice-Chancellor,
Banaras Hindu University, Varanasi

ACKNOWLEDGEMENT

I am sincerely thankful to a large number of eminent geophysicists and geologists who gave their valuable comments and suggestions for improvement of text of various chapters. They include Prof. G. D. Garland, University of Toronto; Prof. Murli Manghnani, University of Hawaii; Prof. R. H. Rapp, University of Ohio; Prof. S. N. Sarkar, Indian School of Mines; Prof. C. S. Pichamuthu (Retd.); Prof. A. K. Saha, Presidency College, Calcutta and Prof. J. C. Bhattacharya (Retd.), University of Roorkee.

I am deeply indebted to Dr. Hari Narain, Director (Retd.) National Geophysical Research Institute, Hyderabad, for agreeing to write Foreword for the book and many useful suggestions.

Dr. M. Mukhopadhyay and Dr. C. Subrahmanyam critically reviewed chapters on gravity in Peninsular India, Dharwars, Eastern Ghats, Singhbum and Northeastern India. Valuable suggestions were received from Dr. K. L. Kaila, Dr. B. K. Rastogi and Shri D. Sarkar for chapter on Seismicity of the Himalayas.

Thanks are due to Shri R. S. Chugh, Director (Retd.) of Geodetic and Research Branch, Survey of India, for permission to reproduce figures from his unpublished paper. Dr. D. V. Subba Rao permitted me to use some material and figures on Mahanadi Valley from his Ph.D. thesis. Permission to reproduce figures from published papers was obtained from Prof. C. M. Powell and N. J. Conaghan, Prof.C. N. Allègre, Prof. Peter Molnar, Prof. P. Tapponnier, Dr. R.A. Gibb, Dr. M. Thomas, Prof. K. S. Valdiya, Prof. L. Seeber and J. Armbruster, Dr. K. L. Kaila, Shri D. R. Nandy and Dr. W. E. K. Warsi.

Shri R. S. Kundu took considerable pains to draft figures for the book. Shri M.K. Roy, Shri K. K. Bose, and Shri A. N. Singh typed the manuscript.

PREFACE

Gravity field has been studied in India since early 1830's when Col. George Everest, the Surveyor General of India, carried out precise measurement of the Great Arc established around 77° 30′ E longitude and discovered difference between geodetic and astronomic measurements of latitude between Kaliana and Kalianpur. Interpretation of these differences gave birth to the ideas of Isostasy which were put forward by John Henry Pratt and George Airy around 1854-1855 in terms of two well known hypotheses of isostatic compensation. Gravity observations in India using pendulums had started in 1865. Since then there has been considerable progress in the study of gravity field in India. The Survey of India has established nearly 6500 gravity stations in India. The National Geophysical Research Institute, Hyderabad and Oil and Natural Gas Commission (ONGC) have contributed appreciably by establishing over 13 000 and 10 000 observations respectively. Several other organisations including the Hawaii Institute of Geophysics, and various University Departments have added several thousand more. Most of the values obtained by ONGC and the University Department have not yet been published. The overall gravity coverage in India is fairly adequate to carry out analysis of regional gravity field in India, in order to throw some light on the nature of underlying rock formations and tectonics of the subcontinent.

The peninsular shield of India as well as parts of the Himalayas have been evolved through different geological processes which have taken place from Archaean to Neogene times. During this period large parts of the crust have undergone processes such as sedimentation, folding, faulting, uplift, erosion, metamorphism etc. and experienced large scale intrusions of basic and igenous rocks. The upper part of the lithosphere has been actively involved in these processes which have left their imprint on the gravity field of the Indian subcontinent. Analysis of gravity anomalies over several mobile belts such as Dharwars, Eastern Ghats, Aravallis, Singhbhum and the Gondwana grabens has been discussed in terms of their geological history. In order to give to the reader sufficient background to the subject, basic concepts of gravity field measurements and processing of data have been explained in the very beginning. A brief introduction to the Geology of India has also been given in order to provide sufficient background to the discussion of gravity anomalies over different geological/tectonic provinces of India.

Since Isostasy is a major phenomenon which is of world-wide interest and can be studied through analysis of gravity field, this subject has been dealt with in detail. Nature of isostatic compensation prevailing over mountain ranges, plateaus and hill masses over Southern Peninsula of India has been discussed.

Gravity field over the Himalayas has been dealt with to the extent possible and ideas concerning the state of its isostatic compensation put forth. The latest concepts concerning isostatic compensation of various types of loads have been outlined in terms of characteristics of the underlying lithosphere.

In order to give an overall view of gravity field over Indian Peninsula, the Himalayas and the surrounding regions such as the Indian Ocean, Tibetan plateau and the Hindu Kush mountains, the geoid undulation and the free-air anomaly maps prepared from satellite gravity data have been presented and interpreted.

It is now well established that the drift of the Indian landmass from the southern latitudes to its present position and its collision with the Eurasian plate has resulted in the formation of the Himalayas through large scale thrust movements which were directed from north towards south. The forces which brought about this collision are still active as is evidenced by considerable seismicity observed all along the Himalayas, the Arakan-Yoma, the Kirthar and Suleiman Ranges of Pakistan. The Himalayas being the outcome of continental-continental collision are of considerable interest. Their seismicity has been analysed in detail and problems concerning its interpretation in terms of plate-tectonics model as applicable to the Himalayas, Northeast India and Northern Burma have been discussed. The latest models concerning their evolution have been presented.

Throughout its geological history the nature has created major mountain ranges, plateaus, sedimentary basins and orogenic belts through processes which are not well understood. The book is a humble attempt to probe into the mysteries of nature through analysis of gravity field and seismicity over Indian Peninsula and the Himalayas.

R. K. VERMA

CHAPTER 1

MEASUREMENT OF GRAVITY FOR STUDY OF FIGURE OF THE EARTH

1.1. Basic Concepts

Ever since 1687 when Sir Issac Newton gave his famous law of gravitation scientists all over the world have used it to study the figure of the earth. The law can be written as follows,

$$F = G \cdot \frac{m_1 m_2}{r^2} \tag{1.1}$$

Where F is the force of attraction between two point masses m_1 and m_2 and r is the distance between them, G is the universal gravitational constant. The value of G has been determined by several workers including Henry Cavendish in 1798 and its latest value adopted is $(6.673 \pm 0.003) \times 10^{-8}$ c.g.s. units (IUGG, 1967). In Equation (1.1) if we substitute $m_1 = 1$ unit, $m_2 = M$, the mass of the earth, R–the distance from the centre of gravity of the earth to its surface, we can write,

$$g = \frac{GM}{R^2} \tag{1.2}$$

Where g is the acceleration due to gravity. It is expressed in ft sec^{-2} or cm sec^{-2}. Geophysicists generally use a unit of 'gal' which is defined as an acceleration of 1 cm sec^{-2}. One milligal is 1/1000 of a gal.

Equation (1.2) can be used to find mass of the earth by using suitable values for g, radius of the earth and the gravitational constant. Thus, with

$$R = 6.37 \times 10^3 \text{ km}$$
$$G = 6.67 \times 10^{-8} \text{ c.g.s. units}$$
$$g = 980 \text{ cm sec}^{-2}$$

we get, $M = 6.14 \times 10^{27}$ gm

Using this value for mass of the earth, and dividing by the volume of the earth ($4/3\ \pi R^3$), we get mean density of the earth $\rho = 5.32$ gm cm^{-3} approximately. This value is appreciably greater than the mean density of rock formations on the surface of the earth which is approximately 2.67 gm cm^{-3}. Application of Newton's law therefore suggests that the overall density of materials composing the earth is appreciably higher than that of the surface rocks.

Although it had been known for a long time that the earth had a spherical shape, accurate measurements of the size of the earth started around the 17th century. First estimate of departure of the shape of the earth from a sphere was made by Newton in the Principie. Assuming that the earth was a liquid sphere, he argued that due to its rotation, it should assume the shape of an oblate spheroid. He imagined canals to be extending from the pole to its center and from the center to its equator. Water in both the canals would be attracted towards its center. He concluded that the equatorial canal would be longer than the polar canal by 1/230, on the basis of gravitational attraction. Newton also predicted that the value of gravity increased with latitude on account of the figure of the earth.

At any latitude ϕ, the acceleration acting on a particle of mass consists of two parts. 1. Acceleration due to gravity, denoted by g, and 2. Centrifugal acceleration on account of angular rotation of the earth. Vertical component (f) due to this acceleration is $f = \omega^2 a \cos \phi$, where a is the equatorial radius of the earth and ω is its angular acceleration. This component is maximum at the equator and minimum at the poles. It has long been known that g at the poles is greater than g at the equator by 1/189 or 5180 mgal approximately. The variation due to centrifugal force accounts for 3460 mgal, while the effect due to difference in equatorial and polar radii is 1720 mgal, the total being 5180 mgal.

A major advance in the study of figure of the earth was possible with the development of instrumentation for measuring the angle between the horizontal plane and the zenith over the pole as marked by the Pole Star. It has now been possible to measure exactly the latitude of a place astronomically. Over a spherical earth, the angle θ_a at any point A between the horizontal plane and the pole star defines the latitude of the place, as shown in Figure 1.1.

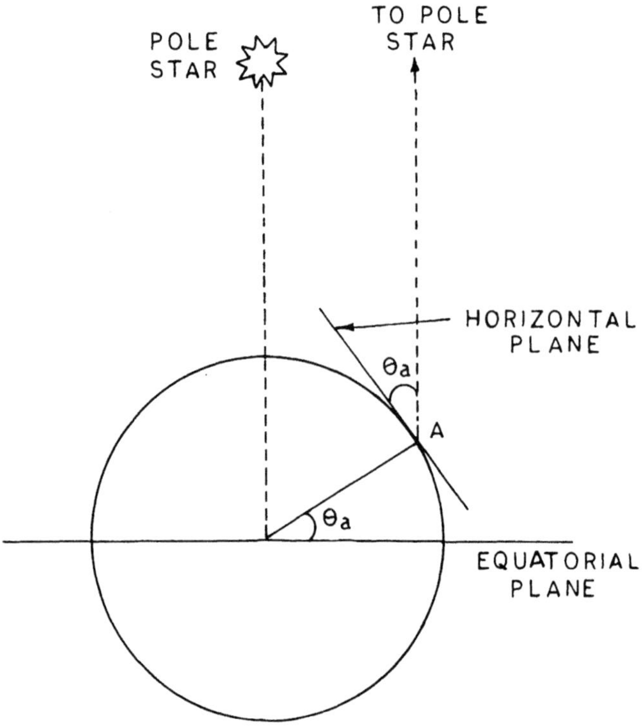

Fig. 1.1. Latitude of a site is nearly equal to the angle θ_a between the horizontal plane at the site and site to the pole star or axis of rotation of the earth.

Near about the 17th century, attempts were initiated to measure circumference of the earth accurately. Willebrord Snell in Holland, around 1617, made measurement of one degree of latitude and obtained the length of one degree of meridian as 69 miles (110.4 km). Piccard in Paris also made a similar effort around 1671.

Newton as well as Huygens, had predicted that the earth was probably an oblate spheroid instead of a perfect sphere. In order to verify this result, the French Academy in Paris sent expenditions, one to Lapland in North Finland and the other to Peru, in South America. The objective was to measure length of 1° of latitude on the surface of the earth. The results verified Newton's hypothesis and indicated a flattening of 1/218. Snell was the first to use a chain of interconnecting triangles as a substitute for direct measurement of distances. This system has now been much developed and is known as triangulation. In order to determine latitude and longitude of a place by triangulation, four quantities are needed. These are: (1) Latitude and longitude of an initial point, (2) Azimuth direction of a line through the point, (3) Two parameters (the axes) specifying the terrestrial form on which the triangles are to be fitted.

In India the geodetic datum used for this purpose is situated at Kalianpur (latitude 24°07′10.97″, longitude 77°39′17.57″) in Madhya Pradesh near Bhopal. The reference surface on which the triangulations are carried out is the Everest Spheroid, which has an equitorial radius '*a*' of 6 377 299 m, polar radius '*b*' of 6 356 075 m, or a flattening

$$\left(f=\frac{a-b}{a}\right) \text{of } 1/300.8.$$

Knowing the coordinates of the geodetic datum, side length of each triangle, latitude and longitude of the vertex of each triangle, azimuth of each side can be determined without the astronomical observations. However, the geodetically determined coordinates differ from the astronomical ones. These differences do not represent errors in measurement but are due to a phenonemon, known as the 'deviation of the vertical'. This results from local undulations of the equipotential surface, which depends upon the actual gravitational field.

The gravitational field largely depends upon the distribution of masses in the crust and the upper mantle and hence is much affected by local geology, resulting in mass excess at some places and mass deficiency elsewhere. A bubble in a spirit level aligns itself in the direction of the equipotential surface and the plumb line is perpendicular to it. It is therefore understandable that astronomical observations will be affected by local distribution of masses (or local gravity field). Errors of 30″ from this cause are not uncommon, and errors up to 1′ are known. The latter would result in a difference of about 1 mile in 60 miles.

For practical purposes of gravitational field computations geoid is taken as a reference surface. It is an equipotential surface of the earth's gravitation and rotation and on an average coincides with the mean sea level in the open sea. It cannot be a regular surface as it depends upon the visible irregularities of the physical surface of the earth and the variable densities inside the earth. Earth's mathematical surface defined by '*a*' and '*b*' values, is known as the ellipsoid. Had the solid earth itself resembled a perfect mathematical model, e.g., an ellipsoid of revolution with no mass anomalies inside, the geoid would have taken nearly the same form. But the irregularities in shape and mass distribution of the real earth cause the geoid to depart from a perfect mathematical form and develope into an irregular surface as shown in Figure 1.2. It

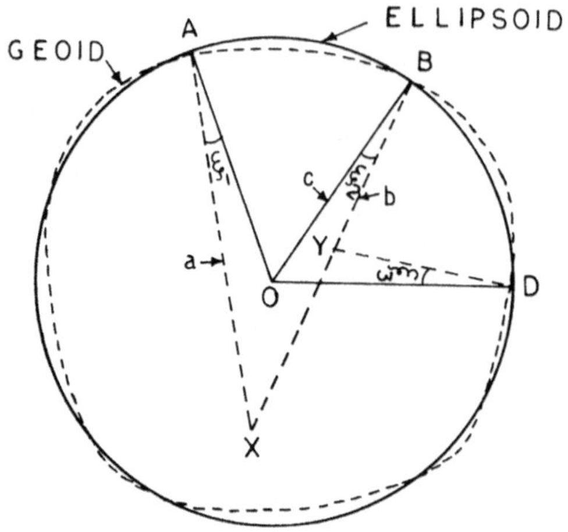

Fig. 1.2. Relationship between the ellipsoid and geoid. While ellipsoid is a mathematical surface, the geoid is an undulating surface. *a* and *b*: normal to the geoid, *c*: normal to the ellipsoid. ξ_1, ξ_2, ξ_3 are deflections of the vertical at *A, B,* and *D* as explained in text, (after Heiskanen and Vening Meinesz, 1958; Reproduced with the permission of McGraw Hill Book Co. Inc., Copyright, 1958).

can be seen from the figure that where the geoid is flatter than the reference spheroid, as between *A* and *B*, the verticals will meet at point *X*, more distant than the true center of the earth at 0. Where the geoid bulges as between *B* and *D*, the verticals will meet at point *Y*, indicating a smaller radius as compared to the spheroid.

Deviation of the vertical is determined in terms of ξ and η (measured along the meridian and at right angles to it) defined as,

$$\xi = \phi_a - \phi_g, \quad \text{and} \quad \eta = (\lambda_a - \lambda_g) \cos \phi_g.$$

Where, ϕ and λ denote latitude and longitude of the place of determination and suffix '*a*' and '*g*' stand for astronomical and geodetic measurements. The undulation of geoid or geoidal height (or depression) '*N*' can be deduced from measurements of ξ and η. Figure 1.3 shows how the mass surplus of a mountain and mass deficiency under oceanic areas causes the deviation of the vertical at point *A* towards the mountain, positive undulation of the geoid under land areas and negative under oceanic areas. Invisible sub-surface anomalous masses also produce the same effect, mass surpluses causing the positive undulations of the geoid and vice versa. On account of undulation of the geoid the problem of determining the exact shape of the earth is still to be fully resolved.

In India triangulation work was started around early 1800 by the Survey of India. From measurements of distances, latitudes and longitudes of various places Sir George Everest had determined the value for equatorial radius of the earth, assuming it to be a spheroid. He obtained a value of $a = 6\ 377\ 276$ m and $f = 1/300.8$ for the earth.

Historically, the values of '*a*' and '*b*' the equatorial and polar radii of the earth have been determined by measurements of arcs by several workers. Bessel obtained these values in Germany in 1841, Pratt in India in 1863, Clarke in England in 1880, Hayford in U.S.A. in 1906. Heiskanen gave values for '*a*' and '*b*' in 1926 which were

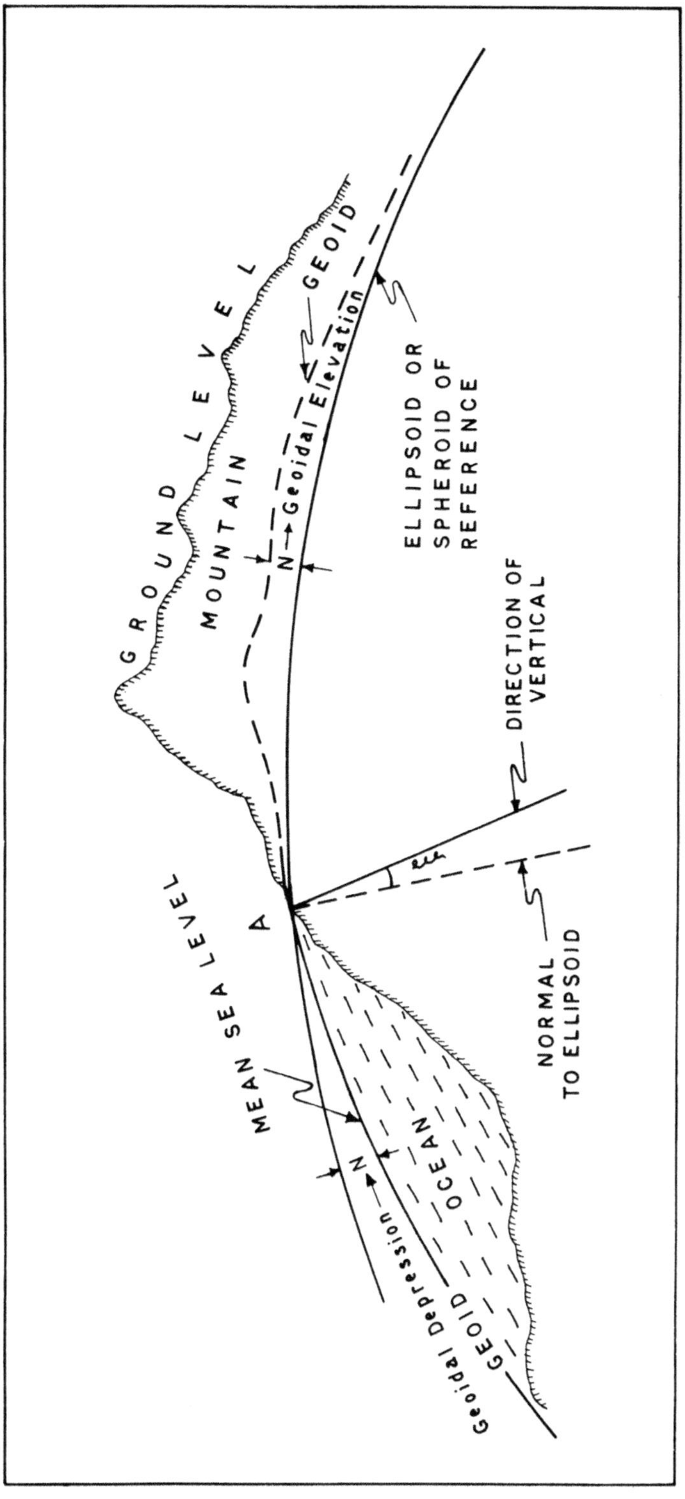

Fig. 1.3. Relationship between geoid and spheroid over mountainous and oceanic areas. Deviation of the vertical ξ and geoid height (N) are shown in the figure.

TABLE 1.1

Values for earth's parameters 'a' and 'b' deduced by various workers. Part of the data from Daly, 1969.

(Reproduced with permission of M/s. Prentice-Hall Inc., U.S.A., Copyright 1940.

Investigator and year	Equatorial radius 'a'	Polar radius 'b'	$e = \frac{a}{a-b}$
Everest, 1830	6 377 253	—	300.8
Bassel, 1841	6 377 397	6 356 079	299.2
Pratt, 1863	6 378 245	6 356 643	295.5
Clarke, 1880	6 378 249	6 356 515	293.5
Hayford, 1909	6 378 388	6 356 909	297.0
Heiskanen, 1926	6 378 397	6 356 921	297.0
International Ellipsoid, 1924	6 378 388	6 256 912	297.0
International Ellipsoid, 1967 (IUGG, Zurich)	6 378 140	6 356 774	298.257

internationally accepted. The parameters for the earth given by various workers are shown in Table 1.1.

The earth's polar and equatorial radii being unequal, it is natural to expect variations in the value of g from place to place, since the distance of a place at sea level from the center of the earth varies with latitude. However, in addition to distance from the center of the earth, the gravitational acceleration is also affected by centrifugal force acting on a body. Clairaut, a French mathematician obtained an expression for the acceleration due to gravity as a function of latitude, taking into account the flattening of the spheroid, as well as the angular velocity of rotation of the earth. The relationship is of the form,

$$\gamma_0 = g_0\ [1 + b_1 \sin^2\phi - b_2 \sin^2 2\phi + \ldots.]. \qquad (1.3)$$

Where, b_1 and b_2 are constants, and ϕ is the latitude of a site, g_0 is the value of 'g' at the equator at sea level and γ_0 is the theoretical value of gravity at the surface of the spheroidal earth at that latitude. The International Union of Geodesy and Geophysics (IUGG) accepted the following values for g_0, b_1, and b_2 in 1924.

$$g_0 = 978.049 \text{ gals}$$
$$b_1 = 0.0052884$$
$$b_2 = -0.0000059.$$

Using these values, Equation (1.3) can be written as,

$$\gamma_0 = 978.049\ (1+0.0052884 \sin^2\phi - 0.0000059 \sin^2 2\phi)$$

This equation is known as the International Gravity Formula.

The above mentioned value of 'g' at the equator (978.049 gals) is based upon the value of 981.274 gal at the international base station at Potsdam and flattening $f = 1/297$. These values have since been revised on the basis of precise terrestrial gravity observations. New International Gravity Formula adopted by IUGG (1967), corresponding to the new Geodetic Reference System (GRS-67) is,

$$g = 978.03185\ (1+0.005278\,895\ \sin^2\phi+0.000023462\ \sin^4\phi) \tag{1.5}$$

To an accuracy of 0.1 mgal, the new formula may be written as,

$$g = 978.0318\ (1+0.0053024\ \sin^2\phi - 0.0000059\ \sin^2 2\phi \tag{1.6}$$

The difference between the 1924 and 1967 formula is,

$$g_{1967} - g_{1924} = (-17.2+13.6\ \sin^2\phi)\ \text{gal}. \tag{1.7}$$

1.2. Reduction of Gravity Data

The theoretical value of gravity (γ_0) is valid only at the surface of the spheroid. In practice, gravity observations are taken over the surface of the earth and often under sea water. The gravity anomaly is obtained by deducting the theoretical value for the ideal earth from the actual gravity value observed (g) and reduced to the same level.

Let A be a point on the surface of the earth (as shown in Figure 1.4) where gravity observation is taken, and B a point vertically below A on the surface of the geoid, while B is a point on the surface of the spheroid. Theoretical value of gravity is valid at B'. In order to obtain the gravity anomaly, it is necessary to reduce the observed value to the surface of the ellipsoid. This operation is known as the reduction of gravity values.

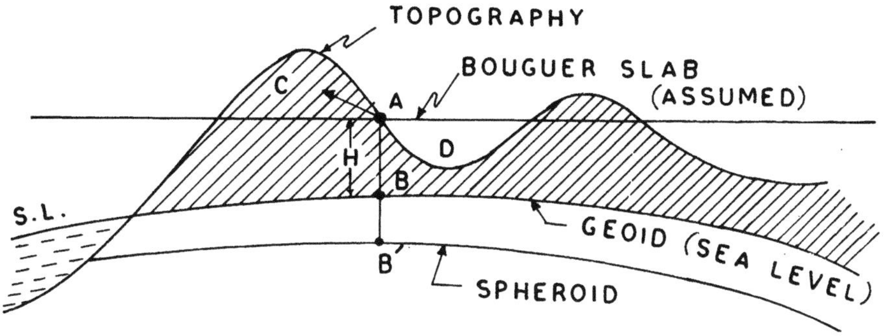

Fig. 1.4. Relationship between the topography, the geoid and the spheroid. A is an observation station, B is vertically below A on the geoid, while B' is vertically below A on the spheroid.

In general, the geoid and the spheroid surfaces do not coincide and height of the geoid above (or below) the spheroid is not known. In common practice therefore, gravity observations are reduced to the surface of geoid.

On land the corrections made to the gravity observations are the following:

a = Free-air correction
b = Bouguer correction
c = Terrain correction
d = Curvature correction
e = Geologic correction

These are discussed below.

1.2.1. FREE-AIR CORRECTION

The variation in the value of 'g' due to change in the distance from the center of the earth (R) is obtained by differentiating equation (1.2),

$$-\frac{\partial g}{\partial R} = \frac{2GM}{R^3} = \frac{2g}{R}. \tag{1.8}$$

Here, negative derivative is used, because elevation 'h' is positive outward, while normal gravity is positive downward.

Using a value of mean radius of the earth (R) as 6.367×10^8 cm, and theoretical value of gravity at sea level at 45° latitude, $g = 980.629$ gal, then at sea level Equation (1.8) gives the vertical variation of gravity as −0.3086 mgal/m or −0.09406 mgal/ft. For most purposes this value may be used anywhere on the earth. A more complete expression, taking into account the earth's spheroidal shape gives,

$$\frac{\partial g}{\partial R} = -\ 0.30855 - 0.00022 \cos 2\phi + 0.000144\ h \tag{1.9}$$

in mgal/m, where ϕ is geocentric latitude of site, and 'h' is the height of the station above sea level in meters.

The second order term is only important at high elevations and high latitudes. The magnitude of second order term at various elevations and latitudes is given in Table 1.2. By adding the free-air correction to the observed value of gravity at the station, we can reduce the value at the station to the value at geoid, assuming that the space between the station and the geoid is filled with air. Subtracting from this value the theoretical value at the spheroid (γ_0), we get free-air anomaly which is defined as,

$$\Delta g_F = (g_{obs} + 0.3086\ h) - \gamma_0$$

where, 'h' is the elevation in meters.

The assumption of free-air between the station and the geoid implies that the

TABLE 1.2

Magnitude of second order term of Equation (1.9) (values in mgal at various elevations)

Latitude	Elevation in meters					
	0	1000	2000	3000	4000	5000
0°	0.0	−0.2	−0.2	−0.0	+0.2	+0.6
10°	0.0	−0.2	−0.2	−0.0	+0.3	+0.7
20°	0.0	−0.2	−0.1	−0.0	+0.4	+0.8
30°	0.0	−0.1	0.0	+0.2	+0.6	+1.0
40°	0.0	−0.0	+0.1	+0.5	+1.0	+1.5
50°	0.0	+0.1	+0.3	+0.7	+1.2	+1.8
60°	0.0	+0.1	+0.4	+0.9	+1.5	+2.2
70°	0.0	0.2	+0.5	+1.1	+1.7	+2.5
80°	0.0	0.2	+0.6	+1.2	+1.9	+2.7
90°	0.0	0.2	+0.7	+1.3	+2.0	+2.8

topographic masses above sea level are virtually 'condensed' to form an infinitesimally thin layer of surface mass which is placed vertically below the observation station, just underneath the geoid as explained by Wilcox (1974). This unquestionably represents a gross excess of mass with respect to the normal sea level earth, which has no masses above sea level. Consequently there is a strong direct correlation between elevation and the free-air gravity anomaly as observed in continental areas. On a regional basis a mild correlation between average elevation and average free-air anomaly exists. For broad topographic features such as plateaus, average free-air value tends to near zero.

The existence of a strong local correlation between free-air anomaly and elevation suggests that local topographic variations, and hence local density variations of any type are either poorly compensated or not compensated at all (Wilcox, 1974).

1.2.2. BOUGUER CORRECTION

In reducing the value of observed gravity to geoid, it is not correct to assume that the space between the geoid and the observation station is filled with air. In general, the space is filled with some rock masses whose density may vary from place to place. If the gravity observation were to be taken at the geoid, this overlying rock mass would exert gravitational attraction in opposite direction to that of the earth. Correction for this rock mass is therefore taken to be negative. For applying this correction, it is assumed that the entire topography from the geoid to the observation station is composed of rock mass of density ρ and it is a flat plate, infinite in two dimensions. Gravitational attraction of such a two dimensional body with elevation 'h' above sea level is given by,

$$\Delta g = 2\pi G\rho h.$$

Where, G is the gravitational constant. This is known as the Bouguer correction, whose magnitude is 0.04185 ρh, were 'h' is in meters and 0.01276 ρh, where 'h' is in feet. Reducing the observed value of gravity to the geoid and applying this correction, we get simple Bouguer anomaly as,

$$\Delta g_B = (g_{\text{obs}} + 0.3086\ h - 0.04185\ \rho h) - \gamma_0$$

for 'h' in meters, and

$$\Delta g_B = (g_{\text{obs}} + 0.094\ h - 0.1276\ \rho h) - \gamma_0$$

for 'h' in feet.

It is common practice to use the value of $\rho = 2.67$ gm cm $^{-3}$ as representative density value for rocks above sea level. Using this value, the above equations can be written as,

$$\Delta g_B = (g_{\text{obs}} + 0.3086\ h - 0.1119\ h) - \gamma_0, \quad \text{for '}h\text{' in meters,}$$

and

$$\Delta g_B = (g_{\text{obs}} + 0.094\ h - 0.03407\ h) - \gamma_0, \quad \text{for '}h\text{' in feet.}$$

It may be mentioned here that a correction term taking into account the departure of the geoid from the spheroid is often omitted in the calcuation of Bouguer anomalies for non-geodetic purposes.

1.2.3. TERRAIN CORRECTION

As mentioned earlier in applying Bouguer correction it is assumed that topography around the station is a horizontal plane of height equal to the elevation of the observation station. This is not the case in reality, since there are places with higher elevations (such as hills) and also of lower elevation, (depressions) around the station.

In order to apply terrain correction two cases are to be considered. (1) Elevated areas above the Bouguer plate such as *C* in Figure 1.4, and (2) areas depressed below the level of the Bouguer plate, such as *D* shown in the same figure. When the attraction of Bouguer plate ($2\pi G\rho h$) is subtracted from the observed gravity, too little mass is subtracted for areas such as *C* and too much for areas such as *D*. The resulting Bouguer anomaly is therefore not free from topographic effects.

The terrain correction when applied removes these discrepancies and takes into account the gravitational effect of actual topography. The gravitational effect of masses above the Bouguer plate as well as of depressions is to reduce the value of observed gravity at the station. Since we assume that the topography around the station is flat and this effect is subtracted from the observed gravity, the corrections for topographic effects discussed above is always positive.

The correction for topography can be conveniently estimated by dividing the surrounding topography into different zones around the station, each zone into several compartments and determining the correction for each compartment as well as zone. Paticulars of various zones modified after Hayford and Bowie are given in Table 1.3. For actually applying topographic correction, a template with radii of various zones on the scale of the topographic map to be used is constructed. Various zones are divided into compartments as shown in Table 1.3. A typical template used for maps of

TABLE 1.3

Dimensions of various zones for applying topographic isostatic correction, modified after Hayford and Bowie, as used by the Survey of India

Designation of internal zone	No. of compartments	Outer radii of zone (ft.)	Designation of No. zone	No. of compartments	Outer raddii of zones
A	1	7	18		1°41′13″
B	5	223	17		1°54′52″
C	10	755	16		2°11′53″
D	10	1936	15		2°33′46″
E	10	4199	14		3°03′05″
F	10	7513	13	16	4°19′13″
G	10	11549	12	10	5°46′34″
H	20	17192	11	8	7°51′30″
I	20	27690	10	6	10°44′
J	20	40682	9	4	14°09′
K	20	61680	8	4	20°41′
L	20	94488	7	2	26°41′
M	20	192913	6	18	35°58′
N	20	324803	5	16	51°04′
O_1	20	435860	4	12	72°13′
O_2	20	547034	3	10	105°48′
			2	6	150°56′
			1	1	180°

1″ = 1 mile scale is shown in Figure 1.5. For local gravity surveys, where Bouguer anomaly is to be determined with high precision, a terrain correction chart given by Hammer (1939) is quite convenient. If ΔT is the terrain correction for a station, the terrain-corrected Bouguer anomaly $\Delta g_B(T)$ can be written as,

$$\Delta g_B\ (T) = (g_{obs} + 0.3086\,h - 0.04185\,\rho h + \Delta T) - \gamma_0$$

where, 'h' is the elevation in meters.

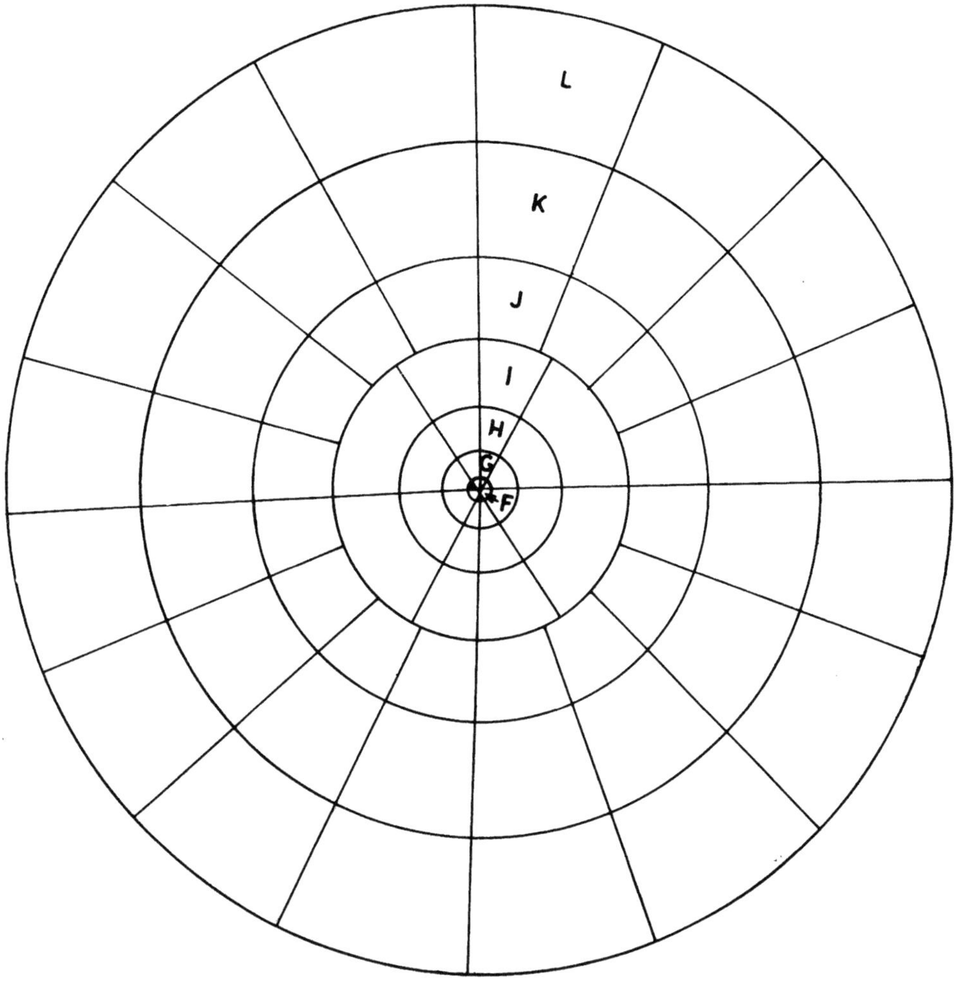

Fig. 1.5. An example of template used for applying topographic correction using Hammer (1939) chart.

Topographic correction for Hammer zones B to I, using density of 2.67 gm cm^{-3} for masses above sea level (after Telford *et al.*, 1976) is given in Table 1.4. An example of terrain correction computation for a station is shown in Table 1.5.

TABLE 1.4

Terrain corrections chart for zones B to I, based on average density, = 2.67 $g\,cm^{-3}$, from Hammer's Tables. (After Telford *et al.*, 1976). Z = avg. sector elevation-station elevation.

(Reproduced with permission of W. M. Telford (1976) and Cambridge University Press, England).

Zone B		Zone C		Zone D		Zone E		Zone F		Zone G		Zone H		Zone I	
4 sectors		6 sectors		6 sectors		8 sectors		8 sectors		12 sectors		12 sectors		12 sectors	
6.56′-54.6′		54.6′-175′		175′-558′		558′-1280′		1280′-2936′		2936′-5018′		5018′-8578′		8578′-14612′	
$\pm z$	dg_T	$\pm z$	dg_T	$\pm z$	dg_T	$\pm z$	dg_T	$\pm z$	dg_T	$\pm z$	dg_T	$\pm z$	dg_T	$\pm z$	dg_T
0.0–1.1	0.00000	0.0– 4.3	0.00000	0.0– 7.7	0.00000	0–18	0.00000	0–27	0.00000	0–58	0.00000	0–75	0.00000	0–99	0.00000
1.1–1.9	0.00133	4.3– 7.5	0.00133	7.7–13.4	0.00133	18–30	0.00133	27–46	0.00133	58–100	0,00133	75–131	0.00133	99–171	0.00133
1.9–2.5	0.00267	7.5– 9.7	0.00267	13.4–17.3	0.00267	30–39	0.00267	46–60	0.00267	100–129	0.00267	131–169	0.00267	171–220	0.00267
2.5–2.9	0.0040	9.7–11.5	0.0040	17.3–20.5	0.0040	39–47	0.0040	60–71	0.0040	129–153	0.0040	169–200	0.0040	220–261	0.0040
2.9–3.4	0.0053	11.5–13.1	0.0053	20.5–23.2	0.0053	47–53	0.0053	71–80	0.0053	153–173	0.0053	200–226	0.0053	261–296	0.0053
3.4–3.7	0.0067	13.1–14.5	0.0067	23.2–25.7	0.0067	53–58	0.0067	80–88	0.0067	171–191	0.0067	226–250	0.0067	296–327	0.0067
3.7–7	0.0133	14.5–24	0.0133	25.7–43	0.0133	58–97	0.0133	88–146	0.0133	191–317	0.0133	250–414	0.0133	327–540	0.0133
7–9	0.0267	24–32	0.0267	43–56	0.0267	97–126	0.0267	146–189	0.0267	317–410	0.0267	414–535	0.0267	540–698	0.0267
9–12	0.040	32–39	0.040	56–66	0.040	126–148	0.040	189–224	0.040	410–486	0.040	535–633	0.040	698–827	0.040
12–14	0.053	39–45	0.053	66–76	0.053	148–170	0.053	224–255	0.053	486–552	0.053	633–719	0.053	827–938	0.053
14–16	0.067	45–51	0.067	76–84	0.067	170–189	0.067	255–282	0.067	552–611	0.067	719–796	0.067	938–1038	0.067
16–19	0.080	51–57	0.080	84–92	0.080	189–206	0.080	282–308	0.080	611–666	0.080	796–866	0.080	1038–1129	0.080
19–21	0.0935	57–63	0.0935	92–100	0.0935	206–222	0.0935	308–331	0.0935	666–716	0.0935	866–931	0.935		
21–24	0.107	63–68	0.107	100–107	0.107	222–238	0.107	331–353	0.107	716–764	0.107	931–992	0.107		
24–27	0.120	68–74	0.120	107–114	0.120	238–252	0.120	353–374	0.120	764–809	0.120	992–1050	0.120		
27–30	0.133	74–80	0.133	114–120	0.133	252–266	0.133	374–394	0.133	809–852	0.133	1050–1105	0.133		
		80–86	0.147	120–127	0.147	266–280	0.147	394–413	0.147	852–894	0.147				
		86–91	0.160	127–133	0.160	280–293	0.160	413–431	0.160	894–933	0.160				
		91–97	0.174	133–140	0.174	293–306	0.174	431–449	0.174	933–972	0.174				
		97–104	0.187	140–146	0.187	306–318	0.187	449–466	0.187	972–1009	0.187				
		104–110	0.200	146–152	0.200	318–331	0.200	466–483	0.200	1009–1046	0.200				

TABLE 1.5

An example of terrain correction for a station (M.13) up to zone I - Average elevation and difference between station and compartment given in Ft., correction in mgal

Station - M.13 Elevation - 1676 ft.

Zone	Compartments	1	2	3	4	5	6	7	8	9	10	11	12	Total (mgal)
B	Avg. Elev.	1672	1672	1672	1672									
	Diff.	4	4	4	4									
	Corr.	0.0133	0.0133	0.0133	0.0122									
C.	Avg. Elev.	1666	1666	1666	1666	1666	1666							0.053
	Diff.	8	8	8	8	8	8							
	Corr.	0.00267	0.00267	0.00267	0.00267	0.00267	0.00267							
D	Avg. Elev.	1650	1650	1650	1550	1600	0.016							0.016
	Diff.	26	26	026	126	76	76							
	Corr.	0.0133	0.0133	0.0133	0.147	0.0133	0.0133							0.2135
E	Avg. Elev.	1700	1650	1600	1550	1550	1650	1575	1700					
	Diff.	24	26	126	126	126	26	101	24					
	Corr.	0.00133	0.00133	0.00133	0.0267	0.0267	0.00133	0.0267	0.00133					0,0099
F	Avg. Elev.	1875	1625	1550	1700	1600	1650	1750	1650					
	Diff.	199	51	26	24	76	26	74	26					
	Corr.	0.04	0.0013			0.0053	-	0.0053						0.052
G	Avg. Elev.	2200	2050	1600	1550	1600	2050	1875	1650	1800	1800	1600	1650	
	Diff.	524	374	76	126	76	374	199	026	124	124	076	026	
	Corr.	0.053	0.0267	0.00133	0.00267	0.00133	0.0267	0 0133	-	0.0267	0.0267	0.00133		.0.1317
H	Avg. Elev.	1500	2400	1600	1525	1625	2200	2500	2250	2050	2000	2000	1750	
	Diff.	824	724	76	151	51	524	824	574	374	324	324	74	
	Corr.	0.08	0.067	0.00133	0.00267	-	0.0267	0.08	0.04	0.0133	0.0133	0.0133		0.338
I	Avg. Elev.	2500	2600	1875	1750	1525	2500	2250	2300	2250	2375	1950	2250	
	Diff.	824	924	199	74	151	824	574	624	574	699	274	574	
	Corr.	0.04	0.053	0.00267	-	0.00133	0.04	0.0267	0.0267	0.0267	0.04	0.0053	0.0267	0.289
											Total (A-I)			1.1922

1.2.4. CURVATURE CORRECTION

This correction is applied in order to take into consideration the curvature of the earth's surface. For applying Bouguer correction, the Bouguer slab is assumed to be a plane, infinite in two dimensions. In reality the surface of the earth is curved. This factor is accounted for by the Bullard correction. The correction is positive up to an elevations of about 4000 m above sea level, and is negative above this. The magnitude of the correction is given in Table 1.6.

Over rugged terrain such as the Himalaya, the magnitude of terrain correction is considerable, sometimes as high as 80 to 100 mgal. It is therefore highly essential to

TABLE 1.6

Magnitude of curvature correction for different elevations

Elevation (m)	Correction (mgals)	Elevation (m)	Correction (mgals)
0	0.0	1000	1.2
100	0.2	1200	1.3
200	0.3	1500	1.5
300	0.4	2000	1.7
400	0.6	2500	1.7
500	0.7	3000	1.5
600	0.8	3500	1.1
700	0.9	4000	0.6
800	1.0	4500	−0.1
900	1.1	5000	−1.0

apply this correction to such stations and provide terrain corrected values of Bouguer anomaly. The magnitude of curvature correction is very small as compared to that of terrain correction.

1.2.5. GEOLOGIC CORRECTION

This correction is generally applied to obtain some insight into local lateral density variations in the upper part of the crust, especially in sedimentary areas. In general, the density of masses above sea level is taken to be 2.67 gm cm^{-3}. We consider the application of Bouguer correction to areas shown in Figure 1.6.

The area A corresponds to a normal granitic crust of density 2.67 gm cm^{-3}, B, a basic intrusive with average density 2.9 gm cm^{-3} and C, a sedimentary basin with average density of 2.4 gm cm^{-3}. While applying Bouguer correction for the area A, the correct amount of gravitational effect for masses above sea level is subtracted. Within area B an insufficient mass has been subtracted, since the density of mass above sea level is more than 2.67 gm cm^{-3}. Within the area C, reverse is the case. The result is a gravity profile such as shown in Figure 1.6.

In areas characterised by large thickness of sediments such as the Indo-Gangetic basin or the Assam Valley, geologic correction for sediments can be considerable. This correction can be computed for a known geologic section, considering the proper density contrast between the sediments and the basement rocks. Details of such cases have been discussed in Chapter 11.

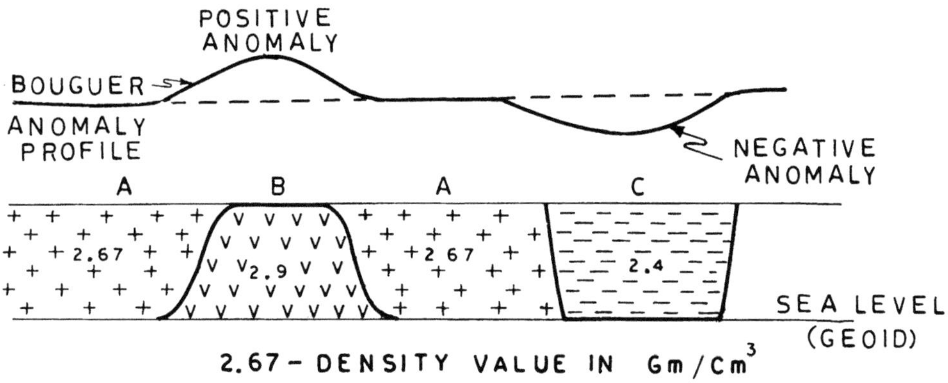

Fig. 1.6. A schematic diagram showing the nature of gravity anomalies caused by near surface variation of density.

1.3. General Remarks

Gravity field over land areas is strongly influenced by underlying geology, since different geological formations are characterised by different rock densities that typify the underlying mass distributions. If rock masses are of different density, the average density for the upper crustal rocks may differ considerably from 2.67 gm cm^{-3}. We may therefore get gravity highs or lows, depending upon whether the rock masses underlying an area have densities higher or lower than normal values. For example, the gravity anomalies over alluvium covered areas of the Indo-Gangetic basin are strongly negative, being of the order of -50 to -150 mgal, on account of mass deficiency created by large thickness of alluvium and Upper Tertiary sediments, whose density is of the order of 2.0 to 2.2 gm cm^{-3}. Over areas underlain by high density metamorphics/metasediments such as Dharwars, granulites or basic igneous rocks, Bouguer anomalies are less negative or positive with respect to the surrounding areas.

A simple definition of Bouguer anomaly is that it is the difference between the observed gravity value reduced to the geoid and the theoretical value at the spheroid. If the absolute magnitude of the Bouguer anomaly is negative (and this is the case over most land areas), it indicates that the value of the observed gravity field is lesser than the theoretical value expected at the level of the spheroid and vice-versa. In areas underlain by high density rock masses, such at the Shillong Plateau, the Eastern Ghats, Offshore Bombay region, Bouguer anomalies are found to have positive values.

Over continental plateau areas Bouguer anomalies are moderately negative, while over mountainous areas such as the Himalaya, Bouguer anomalies are appreciably negative. Over the latter type free-air anomalies are usually dominantly positive, though Bouguer anomalies bear a strong inverse correlation with regional elevation. This cannot be related to near surface lateral density variations, since these anomalies are much more regional in extent. The only possible explanation for this is that negative Bouguer anomalies are caused by regional mass deficiency under the elevated areas. The mass deficiency, in general, is expected to be proportional to the regional elevation of topography. This mass deficiency, resulting due to the compensation of the topographic load is called 'compensation' and the phenomenon giving rise to it is called 'isostasy'. This is discussed in detail in Chapters 8 and 9.

TABLE 1.7

Typical values of observed gravity, free-air, Bouguer, Hayford and Airy Isostatic Anomalies in India

Station Location	Elevation (ft.)	Latitude	Longitude	g_{obs} (gal)	Theor. value	Free-air (mgal)	Bouguer (mgal)	Hayford (mgal)	Airy (mgal)
Near Sea Level									
Calcutta	19	22° 34′48″	88° 20′24″	978.80867	978.80867	− 6.2	− 6.8	-	-
Madras	20	13° 04′08″	88° 14′54″	978.2837	978.31236	−26.8	−26.4	−73.5	−59.6
Cuttack	86	20° 29′05″	85° 52′01″	978.660	978.67000	−12	−15	−12	-
Barrackpur	21	22° 45′59″	88° 20′50″	978.810	978.81968	− 8.5	− 9.2	3.0	−5.0
Plains									
Lucknow	385	26° 49′54″	80° 55′06″	978.9813	979.09904	−81.6	−93.7	−54.2	−62
Hosiarpur	995	31° 31′52″	75° 54′50″	979.2357	979.45899	−129.6	−163.9	−65.2	−76.8
Plateaus									
Mysore	2501	12° 18′52″	76° 40′20″	978.0482	978.28326	− 5	− 89	−42	-
Hyderabad	2000	17° 33′24″	78° 31′48″	978.3369	978.51772	+ 7.3	− 61.7	−22.1	−14.5
Ranchi	2167	23° 23′05″	85° 19′ 0″	978.691	978.86076	+ 33	− 40	+12	-
Mountains									
Simla	7043	31° 06′19″	77° 09′55″	978.8410	979.42489	+787	−150.3	+30.1	+30.5
Pangu	11231	27° 34′24″	86° 44′57″	978.2065	979.15334	+109.7	−233.9	+51.6	+67.1
Deosai	12805	35° 02′04″	75° 23′47″	978.627	979.74850	+ 83.2	−353.4	+41.6	+60.8
Pahalgam	7200	34° 01′20″	75° 19′12″	978.9467	979.66323	− 39.2	−261.6	+54.0	+73.8
Chatarpur	4801	27° 37′23″	86° 45′52″	977.9548	979.15703	+190.3	−225.3	+86.8	+97.8

Regional Bouguer anomalies serve as a good indicator of the degree of compensation. If they are more negative than expected for any given regional elevation level, a state of overcompensation (excess root formation) is normally suggested. That is, the gravatational effect of the mass deficit at depth excels that of the topographic mass. In some areas, a state of undercompensation (opposite of overcompensation) can similarly exist. However, this simple picture can be quite complicated in the case of real earth due to abnormalities in crustal or upper mantle configurations. For example, an abnormally dense crustal block can be in complete isostatic equilibrium, and yet generate relatively positive anomalies. This otherwise would suggest a state of undercompensation (Woollard, 1969).

In areas of limited regional extent (say less than $1° \times 1°$), Bouguer anomalies may exhibit inverse relationship to topography. This does not necessarily suggest lack of compensation because crust has sufficient strength and flexural rigidity to withstand mass inequalities of limited areal extent.

Some typical values of various anomalies observed over mountainous areas, plateaus, and near sea level in India are given in Table 1.7 as examples.

CHAPTER 2

GEOLOGY AND TECTONICS OF INDIAN PENINSULA

2.1. Introduction

For geological considerations the Indian subcontinent can be divided into three different regions: (1) Peninsular shield, (2) the Indo-Gangetic basin and (3) the Himalaya. For understanding the nature of the gravity field in peninsular India, it is essential to have a good understanding of the geology and tectonic history of various parts of the peninsular shield.

A large number of geological events have left their imprint on the topography of the Indian subcontinent. These events are responsible for major mountain trends, such as the Eastern Ghats, Vindhyans, Satpuras and the Aravallis; plateaus, such as the Mysore plateau, Bihar plateaus (Chotanagpur and Hazaribag) and Shillong plateau, as well as valley-like structures i.e., Godavari valley and Mahanadi valley in the peninsula.

The shield of India has witnessed a great deal of geosynclinal activity since earliest known times, in several different parts. The major geosynclines include the Dharwar geosyncline in the southwestern part, the Iron-Ore geosyncline in northeastern India, and the Aravalli–Delhi and Vindhyan geosyncline in the northwest. Major geological or tectonic trends have been established as a result of these geosynclinal and tectonic activities. The Dharwars, the Eastern Ghats, the Satpuras, and the Aravallisform some of the major orogenic trends in Peninsular India.

The peninsular shield of India may further be divided into two parts; the northern shield north of the Satpura mountains, and the southern, south of these.

The shield has had a very complex geological history ranging in age from Archaean to Recent. The major rock formations in the area are the following (a) Sediments ranging in age from Tertiary to Recent, (2) Coastal sedimentary formations ranging in age from Cretaceous to Tertiary, (3) Deccan basalts, mostly Paleocene to Eocene in age, (4) Gondwana formations, at present mostly confined to Godavari, Mahanadi, and Damodar valleys as well as Satpura basin, (6) Cuddapah–Kurnool, Vindhvan and Chhatisgarh formations belonging to the respective basins, (7) The Iron Ore Group of sediments, (8) Metasedimentary and metavolcanic sediments including the Dharwar Supergroup, Bijawar Group, Sausar and Sakoli Groups, Aravalli and Delhi Groups, (9) Granitic batholiths such as Bundelkhand and Singhbhum granites, (10) Charnockite intrusives, and (11) Peninsular Gneisses.

2.2. Precambrians of Southern Peninsula

Practically all the rocks mentioned above except the Deccan lavas, the Tertiary sediments and the Gondwanas, are of Precambrian age. The geological history of the Precambrian rocks from the southern shield has been revealed by the efforts of a large number of geologists (Rama Rao, 1945; Pichamuthu, 1953; Radhakrishna, 1956, 1967) from the Mysore Geological Survey as well as radiometric age determination by a large number of workers (Holmes, 1955; Vinogradov and Tugarinov, 1968; Ramamurthy and Sadashivaiah, 1967; Crawford, 1969; Venkatasubramanian, 1974a, b; Sarkar, 1980, 1983).

The Basement Complex in south India appears to be older than 3200 m.y. (Sarkar, 1980). Sargur Group in Karnataka is believed to be the oldest rock formation of S. India (Radha Krishna and Vasudev, 1977). Some of the charnockites in south India are believed to be nearly about 3000 m.y. old (Crawford, 1969). The Peninsular Gneiss Complex (which includes banded gneisses, granitic gneisses and granites) which covers large parts of the southern shield of India, from Mysore up to Hyderabad has ages varying from 2950 m.y. to 2585 m.y. (Crawford 1969; Sarkar 1980). It is believed by many geologists that the Peninsular Gneiss was the basement over which rocks belonging to the Dharwar Group were deposited. This gneissic complex is the most wide spread rock formation in Peninsular India. The oldest gneisses from Gorur-Hassan districts are 3360 m.y. old (Beckinsale *et al.*, 1980).

The charnockites are believed to be of metamorphic origin (Rama Rao, 1945). They have been studied extensively by Pichamuthu (1953) and are believed to be of different ages. They have acidic, intermediate, basic as well as ultrabasic composition. Radiometrically their ages are found to vary from 3000–1300 m.y. (Crawford, 1969). A major event in the geologic history of south India was the deposition of rocks belonging to the Dharwar Group. This Group consists of basic volcanics, rhyolites, banded magnetite quartzites, manganiferous cherts, various types of schists, limestone, dolomites, quartzites and conglomerates. The Dharwars are also characterised by mafic, ultramafic as well as granitic intrusives. The Dharwar sedimentation probably started around 2600 m.y. and culminated with the formation of Chitradurga granite around 2450 m.y. (Sarkar, 1980).

The Chitradurga, Shimoga, Kudremukh and Bababudan form the major structural belts of these provinces. The general trend of Dharwars varies from NW–SE in Karnataka State, to approximately N–S and then tends to SW and WSW aligning with the trend of Eastern Ghats belts of Salem and Nilgiris.

According to Rama Rao (1962) as well as Pichamuthu (1967), the Peninsular Gneiss is believed to be younger than the Dharwars. However, recent radiometric ages on these rocks do not support this contention.

The granitic intrusives, including Closepet and Chitradurga granites occupy large parts of the southern shield. These are believed to be younger than the oldest Dharwars ranging in age from about 2450 to 2000 m.y. (Crawford, 1969). Possibly the granitic intrusives have been formed during the later stages of geological history of the Dharwar geosyncline. It may be mentioned here that several large granitic masses of the batholithic dimensions such as the Closepet granite, the Singhbhum granite, the Bundelkhand granite have had their origin as a part of the geological history of various geosynclines. Figure 2.1 shows a simplified geological map of India.

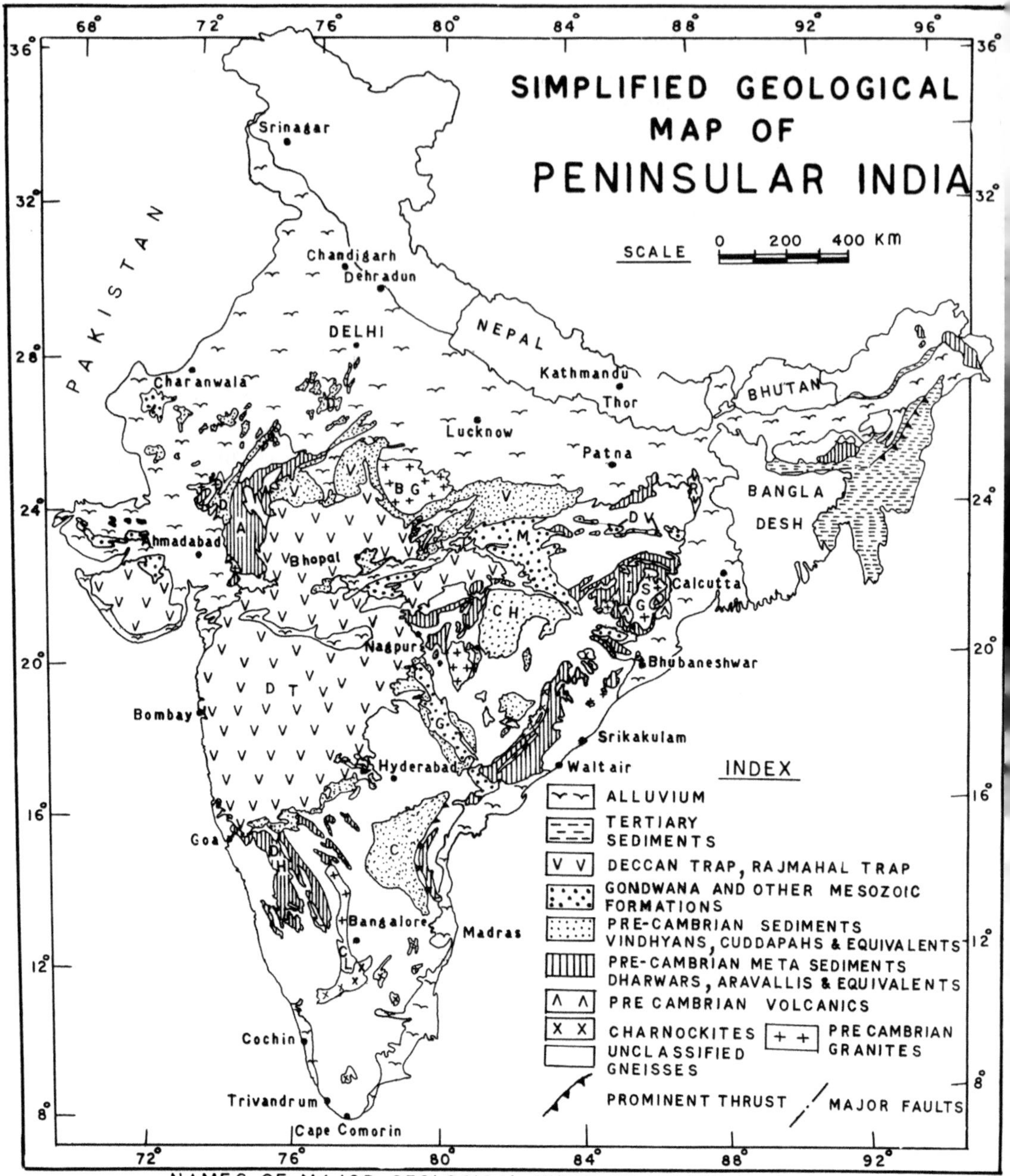

NAMES OF MAJOR GEOLOGICAL UNITS OF PENINSULAR INDIA

A — ARAVALLI, BG — BUNDELKHAND GRANITE, C — CUDDAPAH BASIN, CH — CHHATISGARH BASIN, CL — CLOSEPET GRANITE, D — DELHI SERIES, DH — DHARWARS, DT — DECCAN TRAP, DV — DAMODAR VALLEY GRABENS, G — GODAVARI VALLEY, M — MAHANADI VALLEY, I — IRON ORE BASIN, R — RAJMAHAL VOLCANICS, SG — SINGHBHUM GRANITE, V — VINDHYAN BASIN.

Fig. 2.1. Simplified geological map of India (based on G.S.I. map).

2.3. Eastern Ghats

The Eastern Ghats mountain belt is the longest Precambrian mountain system of Peninsular India. It extends from the Niligiri Hills in the south to near Cuttack in Orissa State. It has its maximum width in Cuttack–Ganjam region of Orissa. The belt consists mostly of subparallel alternating layers of khondalites (Garnet-sillimanite-graphite schists and gneisses, and their variations), charnockites, pyroxene granulites, kodurites, as well as granites and granitic gneisses. According to Holland (1900) the charnockites are classified into acid, intermediate, basic and ultrabasic varieties based on their silica content. Pichamuthu (1953) has put forward his concept of charnockitisation of Peninsular Gneisses by metamorphic processes. Subramaniam (1967) reviewing the geological data has suggested that while some of the charnockites are of magmatic origin, others are of metasomatic origin. The charnockite-khondalite suite of rocks are engulfed in a sea of granulite facies rocks, gneisses and migmatites, which form the next most important lithological unit in the Eastern Ghats belt. These high grade gneisses are often clubbed with the gneisses of south India and commonly called the Peninsular Gneisses. The predominance of high grade metamorphics is an important feature of this belt. The predominant trend of the mountain system is NE–SW in Tamil Nadu as well as in the northern part of Andhra Pradesh and Orissa. The trend follows the outline of the Cuddapah Basin to its east. Further south they follow the trend of Nallamalai and Shevaroy hills and finally merge into the Niligris (See Figure 2.3 for location).

According to Subramaniam (1967), the charnockite suite of rocks have been emplaced as sheets and lenses in a gently folded basement sequence of khondalites with interstratified pyroxene granulites and were later involved in regional metamorphism. Radiometric ages determined by Crawford (1974), Vinogradov *et al.* (1968), and Sarkar, Saha and Miller (1969) for several rocks and minerals have thrown some light on the history of evolution of this belt. On the basis of these, it has been suggested that the emplacement of the charnockites of Madras and Nilgiris, and the granulite facies metamorphism of these (Phase I) as well as of khondalites, took place between 2900 and 2600 m.y. period. There is indirect evidence of a basement gneissic complex older than 3100 m.y. (Sarkar, 1980). A second phase of metamorphism of the belt accompanied by granitic and pegmatitic activity with culmination of Eastern Ghats Orogeny closed in around 1600 m.y. Emplacement of charnockites and granites in certain parts of the belt took place around 1520 to 1300 m.y. (Aswathanarayan, 1964). The emplacement of massive anorthosite complex at Chilka Lake took palce around 1312 m.y. A much younger phase of metamorphism (phase III) and mountain building took place around 700–450 m.y. (Sarkar, 1968, 1980). These events support the ideas of Fermor (1930), who suggested that the Eastern Ghats north of Krishna river have been uplifted during Precambrian or later times. This has been supported by Wadia (1975, p. 20) who maintains that these mountains have been block uplifted through epeirogenic forces. The most prominent examples of these at present are the Nilgiri Hills and their continuation, the Palni and Cardamom Hills in the south.

Figure 2.2. shows the radiometric ages of major geological formations, metamorphic and orogenic events that took place in Peninsular India.

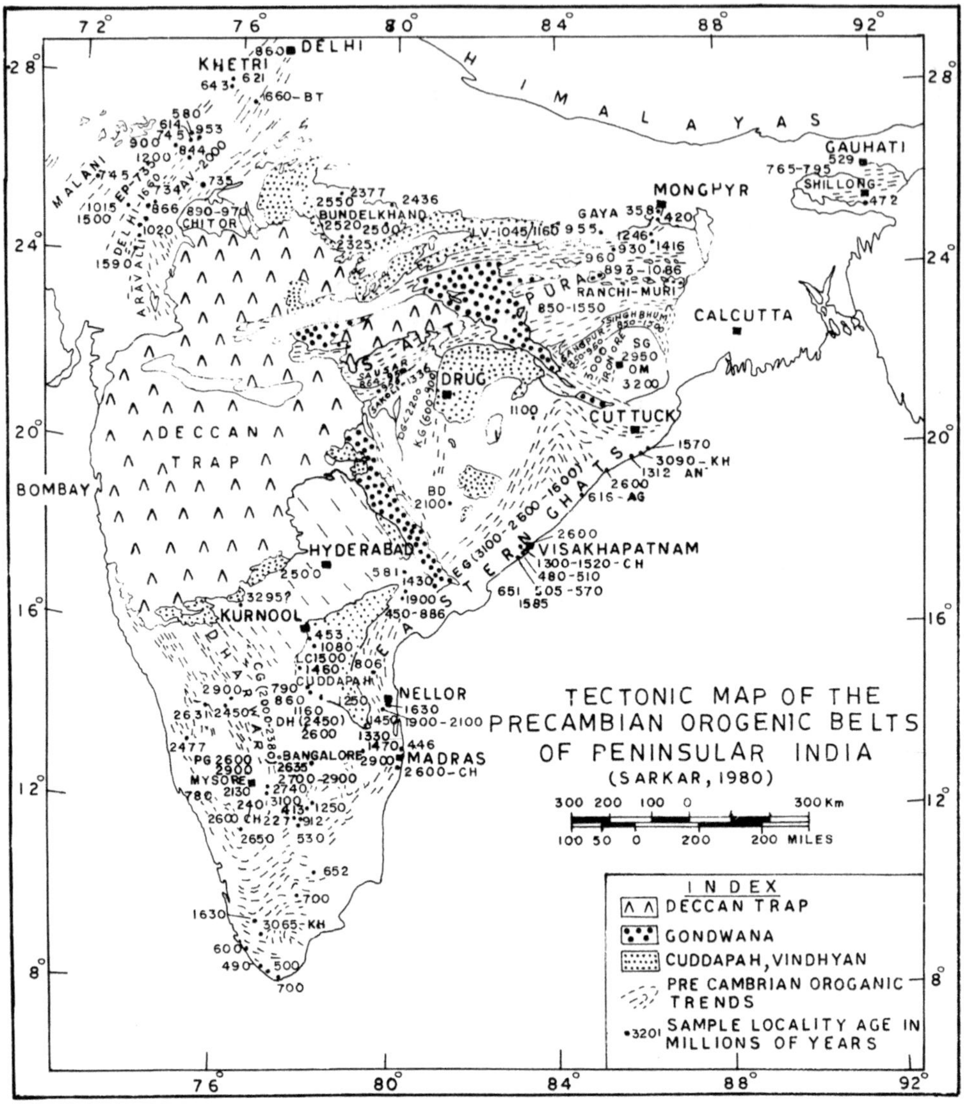

Fig. 2.2. Tectonic map of Precambrian orogenic belts of Peninsular India (after Sarkar, 1980) showing radiometric ages of major events. Reproduced by permission of S. N. Sarkar.

2.4. Cuddapah Basin

The Cuddapah Basin is the most important sedimentary basin in the southern part of Peninsular India. It is crescent shaped, convex to the west, and occupies an area of about 35 000 km ^{2} in Andhra Pradesh. Much of the basin is a dissected plateau with parallel ridges and more rugged topography towards the east. The basin is filled with Precambrian (Cuddapah) sediments consisting of quartzites, sandstones, shales, slates, limestones and Kurnool formations. The Kurnool rocks are confined to two areas, the western part of the basin and the area in NE, known as the Palnad basin. These rocks are separated from the Cuddapah sediments by an unconformity. Significant igneous activity has been witnessed in the form of basic sills in the western margin of the basin. The sediments rest on the denuded surface of the older gneisses which have been dated around 2590 ± 40 m.y. Along the concave eastern margin, they have been folded along axes parallel to the Eastern Ghats orogenic trend.

On the basis of radiometric age determinations (Crawford and Compston, 1973) it has been surmised that the Cuddapah sedimentation started around 1700 m.y. (after the culmination of Eastern Ghats orogeny, phase II) and lasted till about 1225 m.y. The radiometric data suggests that the Cuddapah sediments, at least in part, are older than the Vindhyan sediments. The Kurnool sediments are possibly younger than 870 m.y. The sediments belonging to the Cuddapah and Kurnool systems were later affected during the Indian Ocean Orogenic Cycle, closing around 450 m.y. in the Eastern Ghats region (Sarkar, 1968).

2.5. Singhbhum in Bihar and Orissa

The rock formations in Bihar and Orissa primarily belong to two different geological environments and history, geosynclinal sediments belonging to the Iron-Ore Province and metamorphics belonging to the Satpura orogenic cycle.

The Iron Ore Group of rocks have had their origin as far back as 3200 m.y. after the end of the Older Metamorphic orogenic cycle, which resulted in the formation of Older Metamorphic Group (OMG) Tonalite Gneiss. The OMG tonalite was formed c. 3800 m.y ago (Basu *et al.*, 1981) and is the oldest rock so far found in India, equivalent in age to the oldest rocks found anywhere in the world. During the earlier phase of its history, a huge thickness (estimated at 10 000 m) of Iron Ore Group (IOG) of rocks consisting of conglomerates, sandstones, banded hematite quartzites and jaspers (BHQ/BHJ) along with basic lavas were formed. The end of this cycle (around 2900–3000 m.y.) witnessed the large scale intrusives, now called the Singhbhum granite batholith covering about 10 000 km^2 area. The batholith was probably formed as a result of partial meleting of the older granitic basement. Then followed a long period of erosion and stabilisation of the Iron Ore orogenic belt. During the second phase of its history (later than 2300 m.y.), a large thickness of sandstones, shales and conglomerates (about 10 km thick) were deposited along the northwestern, northern and eastern edges of the stabilised belt. This was followed by a large scale eruption of basic volcanics (around 2100–2200 m.y.) belonging to Dalma, Dhanjori and Jagannathpur lavas along the entire above mentioned belt. The eruption of Jagannathpur lavas was followed by the deposition of the Kolhan group of sediments (c. 2100 m.y.) in two intracratonic basins (Sarkar, 1980; Sarkar and Saha, 1983). The Newer Dolerite intrusives continued intermittently up to 1000 m.y. ago. Folding, regional metamorphism and cross folding of Gangpur and Singhbhum Groups of rocks took place in several phase between the period 1500 to 850 m.y. The Singhbhum orogenic cycle ended around 850 m.y. Towards the later part of the cycle, the arcuate Singhbhum thrust developed nearly along the southern limit of the Singhbhum Group of rocks against the Iron-Ore Craton. During the thrust movements high grade metamorphics in the north were thrust over low grade metamorphics in the south. The thrust movements were accompanied by granitization leading to the development of biotite granite gneisses, soda granite, granophyre, Chakradharpur granite gneiss, etc.

2.6. Satpura Orogenic Belt

The Satpura Orogenic cycle affected large areas of northern India in Madhya Pradesh as well as Bihar states. The metamoprhic rocks affected by this cycle are generally called the Chotanagpur granite gneisses. These include Precambrian quartzites, quartze-felspathic gneisses, and amphibolites. Rocks in this belt have been metamorphosed at places to high grade amphibolite-granulite facies with the development of anorthosite at Bankura. On the basis of radiometric age determinations by Holmes (1950), Vinogradov and Tugarinov (1968), Sarkar (1968, 1980, 1983), Sarkar, Saha and Miller (1969), the major metamorphic events in this belt are considered to have taken place around 960 m.y. The Satpura mountains were probably formed or uplifted during this period. The strike of the Satpura moutains is nearly ENE–WSW.

2.7. Aravallis and Vindhyans

In northwestern India, large areas are covered by rocks belonging to the Aravallis and Vindhayan geosynclines. These rocks have been deposited over the Precambrian basement complex which may be as old as 3500 m.y. The Bundelkhand granite consists of gneisses, schists and amphibolites whose age is estimated to be more than 2600 m.y. (Sarkar, 1968). The sediments belonging to the Aravalli System now consist of phyllites, limestones, quartzites and composite gneisses. According to Crawford (1969) and Sarkar (1980, 1983), Aravalli sedimentation began around 2500 m.y. and lasted till about 2000 m.y. The sediments were metamorphosed during the Aravalli cycle which closed around 2000 m.y. The Aravallis were overlain by sediments beloging to the Raialo and Delhi Systems. These sediments were later metamorphosed during Delhi cycle (c. 1600 m.y.) and consist of garnetiferous and biotite schists, marbles, quartzites, calc schists and gneisses. Towards the end of the Delhi cycle, a number of acidic intrusions took place which are evidenced by the formation of Erinpura granite (740 m.y.) and eruption of Malani suite of rhyolites (during 783–745 m.y.) on the western side of the Aravalli axis. A later phase of metamorphism which took place during 643–621 m.y. has been identified along the northern end of the Aravalli mountains and is termed as the Khetri phase. The Aravalli mountains have been formed as a result of folding of the sediments belonging to the Aravalli and Delhi Systems.

Rocks belonging to the Vindhyan System (named after Vindhya Mountain) occupy a large area (about 100 000 km^2) in northern India, north of Satpura mountains. A considerable thickness of these is at present covered by the Deccan traps. The Vindhyans are divided into two categories, Lower and Upper Vindhyans. While the Lower Vindhyans are mostly a margin sequence consisting of limestones and shales and have about 1000 m thickness, the Upper Vindhyans are mostly sandstones, gently overlying the Lower Vindhyans and are of considerable thickness. Within the Vindhyan sediments there are unconformities. The Vindhyan basin extends from

Dehri-on-Sone to Hoshangabad and from Chitorgarh to Agra and Gwalior surrounding the Bundelkhand granite batholith practically on all sides. The total thickness of the sediments is estimated to be about 4000 m. Radiometric ages on several glauconite samples from the formation give an average age of about 1110 ± 60 m.y. for the Lower Vindhyans and about 923 ± 30 m.y. for the Upper Vindhyans. Recent determinations of ages for the Vindhyans shows that sedimentation extended for a long period from about 1200 m.y. to about 550 m.y. Basic intrusives have taken place into the Vindhyans as evidenced by the presence of diamond bearing Kimberlite bodies near Panna, dated to be around 1100 m.y. Along the Great Boundary Fault of Rajasthan, highly folded Aravallis have overthrusted against undisturbed Bhander Series belonging to the Upper Vidhyans. There is little doubt that the Vindhyan rocks continue to the north under the Ganagetic alluvium as shown on the ONGC tectonic map.

The Vindhyans of Rajasthan are invaded by the Malani group of igneous rocks which include Idar, Jalor and Siwana granites and granophyres.

2.8. Precambrian of Madhya Pradesh

The Precambrian of M.P. comprises of large thickness of sediments and intrusives which has been named as the Dongargarh Belt by Sarkar (1968). These consist of quartzite, sericite schists, feldspathic quartzites, hornblende biotite quartzite, hornblende schists and amphibolites belonging to the Amgaon Group orogeny and metamorphism, closing earlier than c. 2300 m.y. (Sarkar *et al.*, 1981). These are overlain by rhyolites, andesites, sandstones, limestones and shales belonging to Nandgaon, Khairagarh and Chhattisgarh formations. The Dongargarh granite (less than 2200 m.y.) is intrusive into the sediments and rhyolites. The age of younger (Khairgarh) Group is estimated to be less than 1680 m.y. The sedimentary rocks occupy a prominent exposure in northeastern India in the Chhatisgarh basin (less than 900 m.y.) north of Godavari valley.

2.9. Gondwanas of Peninsular India

After the deposition of the Vindhyan sediments and their uplift in the form of a landmass, there was a long period of inactivity over the Indian peninsula till the end of Palaeozoic Era i.e. towards the Upper Carboniferous. During this period there existed a landmass, comprising the present day continents of India, Australia, Antarctica, Africa, Madagascar and South America, called the Gondwanaland. In this subcontinent similar climates prevailed from Upper Carboniferous to Jurassic times as evidenced from fossils of plants and animals from those continents.

Several parts of the Indian peninsula (as in other continents) experienced tensions resulting in fracturing and faulting on the earth's crust leading to the deposition of fresh water or lacustrine sediments in grabens or basins. In the Indian peninsula, these sediments were brought by the rivers Godavari, Mahanadi, Damodar, Narmada, Son and their tributaries. The Gondwana sediments are found at present in basins/grabens formed during this period. Besides the Indian peninsula the Gondwanas are also found along the foothills of Himalaya in Nepal, Bhutan and Assam.

The basins of deposition might have been shallow and sinking gradually, as we see each cycle of deposition starting from coarse sandstones and proceeding through shales and coal seams. There must have been an abundance of plants during this period whose remains formed the coal seams upon deposition and burial under high temperature and pressure conditions. All the present day coalfields of India are located in the Gondwana basins.

2.10. Earth Movements During Cretaceous and Eocene Times

The Cretaceous was a period during which good deal of earth movements took place all over Europe, Mediterranean region, the Baluchistan Arc, the Makran coast, Himalaya, Assam as well as the Arakan Yoma in Burma. In the Indian peninsula, these movements were confined to few areas such as the Ahmednagar area in Gujarat, Wadhawan in Kathiawar, near Bagh in Gwalior, Khasi Garo and Jaintia Hills in Assam and coastal areas near Madras, where perhaps the maximum deposition of sediments took place. In practically all these areas marine transgressions resulted in the deposition of marine sediments including sandstones, claystones, shales and limestones.

2.11. Deccan Lavas

The Deccan lavas (commonly called traps) are perhaps the most prominent feature on the geological map of Peninsular India (see Figure 2.1). They occupy an area of about 500 000 km^2 in several states including Maharashtra, Gujarat, Kathiawar, Madhya Pradesh, Andhra Pradesh and adjoining areas. They are found as far as Belgaum in the south, Rajahmundry in southeast, Amarkantak in the east and Kutch in northwest. Over most of the area the traps overlie older sedimentary formations or metamorphics.

Most of the trap country is characterised by flat-topped hills and step-like terraces. The topography is mostly a result of the variations in the thickness of the traps and their hardness. The hard and compact traps lie mostly on top of the terraces and plateaus.

Recent radiometric ages on the traps show that their ages vary from about 65 to 59 m.y. (Wellman and McElhinny, 1970). The maximum thickness of trap has not so far been ascertained with certainty. However, recent D.S.S. results from Koyna and the west coast region shows that their thickness may be of the order 1 to 2 km.

2.12. Formation of Indo-Gangetic Basin

The rise of the Himalaya took place in a series of earth movements beginning from Upper Cretaceous through Eocene and Middle Miocene times, the last one probably being the most powerful one. It was perhaps during the Middle Miocene times that the Himalaya achieved their major features (Krishnan, 1960). At the same time a long narrow trough extending from Assam to Kashmir, was formed between the Himalaya and the Peninsular landmass. This trough has generally been referred to as the foredeep. In this trough a large thickness of sediments estimated to be 5 to 6 km thick were deposited by the rivers Indus, Ganges, Yamuna, Brahmaputra and their tributaries. Most of these rivers brought sediments from the rising Himalaya to the foredeep. The basin so formed is known as the Indo-Gangetic basin. These sediments constitute the Siwalik System of the Himalayan foothills. Their counterparts in Pakistan and Burma are known as the Manchar and Irrawaddy formations.

Another area in which large thicknesses of Teritiary sediments were deposited is the Bengal basin and the Assam Valley.

2.13. Mountain Ranges and Plateaus of Peninsular India

The gravity field of Peninsular India is considerably influenced by its mountain ranges as well as plateaus. It is therefore essential to discuss, something about their geological history. The major mountain ranges of Peninsular India are; the Western Ghats, the Eastern Ghats, the Satpuras, the Vindhyans and the Aravallis. Prominent plateaus are, the Mysore plateau and Deccan plateau (formed by Deccan basalts) in the south, Bihar plateaus (Chotanagpur and Hazaribag) in the east, and Shillong plateau in eastern India. Locations of major mountains and plateaus are shown in Figure 2.3.

2.13.1 THE WESTERN GHATS

The Western Ghats (nearly 1600 km long) form a well marked feature along the west coast of India from Tapti valley in the north up to Cape Comorin in the south. Their northern part is composed of the Deccan Volcanics, while its southern part consists of Archaean gneisses, schists and charnockites. Their average elevation is 1100 to 1200 m but many peaks rise up to 2200 to 2600 m. Near the southern end, the Western Ghats merge with the Eastern Ghats in the Nilgiri Hills forming a mountain knot whose highest point is Doddabetta (elevation, 2637 m). South of the Nilgiri Hills is a prominent pass known as the Palghat gap. South of this gap, the Western Ghats rise again to form the Anaimalai Hills whose highest peak is Anaimudi (2694 m). The Western Ghats form a real watershed of the Indian peninsula, as many of the important rivers including Godavari, Krishna, Cauvary and Tungabhadra originate from these Ghats and flow towards east in the bay of Bengal (Krishnan, 1960, pp. 5–7).

2.13.2. THE EASTERN GHATS

The Eastern Ghats is a major topographic feature in south-eastern India. The geology and tectonic history of these Ghats have been discussed already in Section 2.3.

2.13.3. SATPURA AND VINDHYAN MOUNTAINS

The Vindhyan Mountains lie to the north of Narmada river and separate northern peninsula from the southern peninsula. They extend from Jobat (22° 77′ N, 74° 35′ E) in Gujarat to Sasaram (24° 57′ N, 84° 02′ E) in Bihar, through Indore, Bhopal and Bundelkhand. They have a general ENE–WSW strike. Their elevation is nearly 500–600 m. The majority of the hill ranges of Vindhyans are composed of sandstones and quartzites, belonging to the Vindhyan System. The western part of the Vindhyan mountains consists mainly of Deccan traps, which may be underlain by Vindhyans or older formations.

The Satpura mountains have nearly the same strike as that of the Vindhyans. Their

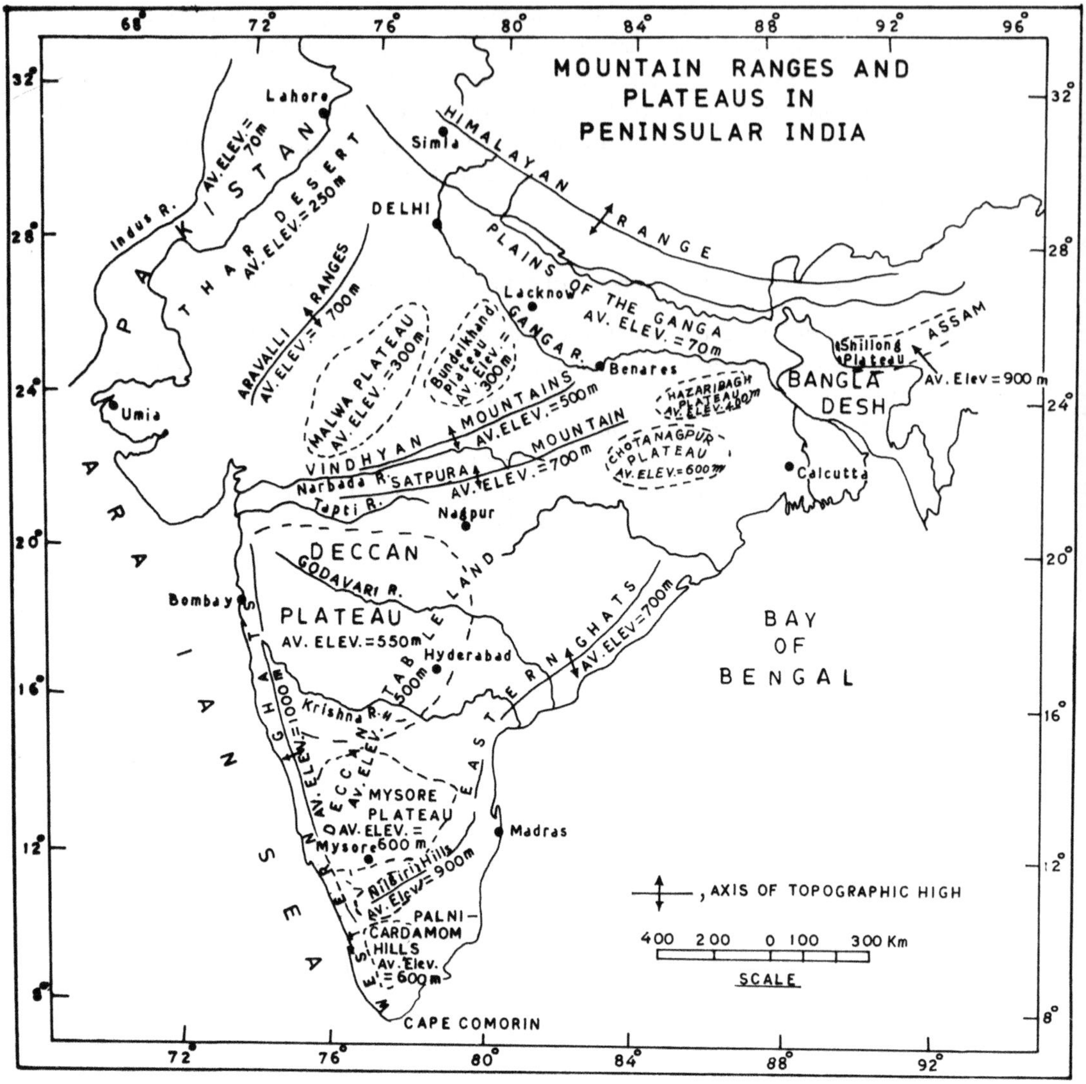

Fig. 2.3. Mountain ranges and plateaus in Peninsular India.

western termination is in the Rajpipla hills in Gujarat, while in the east they extend up to Hazaribag through Pachmarhi hills, Sarguja and Ranchi. Their average elevation is about 500 m, while some peaks such as Pachmarhi reach up to 1100 m.

In the eastern part, the Satpuras are composed of Gondwanas and Archaean gneisses, while in the west they consist mostly of Deccan trap lava flows. The Vindhyans as well as Satpuras form a watershed for Central India and many important rivers such as the Narmada, Chambal, Betwa, Son and others originate from these mountains.

2.13.4. THE ARAVALLI MOUNTAINS

These mountains consistute one of the finest examples of a true tectonic range. They were formed during Precambrian (Post-Delhi) times and were probably uplifted again in post-Vindhyan times (Krishnan, 1960). The last movement may have been block uplift, as suggested by Fermor (1930).

The mountains are composed of rocks belonging mostly to the Delhi and Aravalli Systems. The mountains have a consistent NE–SW strike and extend from near Sirohi (25° N, 73° E) in the south to near Delhi (29° N; 77° E) in the north. The average elevation of the mountains is around 700 m, the highest peak Mt. Abu being 1722.5 m. Although the mountains terminate in Gujarat in the south and near Delhi in the north, there are good indications that they extend in the north up to the Garhwal Himalaya and in the south near Cambay basin. They form a major watershed, separating the drainage system of the Ganges river from that of the Indus.

CHAPTER 3

DESCRIPTION OF BOUGUER AND FREE-AIR ANOMALY MAPS OF PENINSULAR INDIA

3.1.1. Bouguer Anomaly Map

The gravity field prevailing over the Indian peninsula and the Himalayas is best reflected on the Bouguer anomaly map shown in Figure 3.1. Bouguer anomalies are most sensitive to the mass distributions and hence are influenced by the nature of underlying geological formations. In areas underlain by masses with relatively higher density, Bouguer anomalies are reflected as gravity highs and vice versa. The calculation of Bouguer anomaly involves no assumption regarding the root formation under the mountains, plateaus or other elevated areas. For this reason Bouguer anomalies are negative over elevated areas, showing inverse relationship to topography i.e., higher the elevation, more negative is the Bouguer anomaly. The negative values under mountains/elevated areas therefore give an indirect evidence in support of the root formation. However, in areas where the topography is not locally but regionally compensated or the mountains are underlain by high density rocks, there may be departures from linear relationships between topography and Bouguer anomalies.

A general view of the Bouguer anomaly map (Figure 3.1) of India shows that the entire area may be divided into three major parts on the basis of character of these anomalies. These are, (a) The Indo-Gangetic basin, Assam valley and the Himalaya. This area is characterised by large Bouguer anomalies ranging from −70 to −270 mgal or more. (b) The northern peninsula, north of lat. 20° N. This is on the whole an area of gravity high, Bouguer anomalies ranging from 0 to −70 mgal. (c) The southern peninsula south of 20° N latitude. This is on the whole an area of low gravity as compared to the northern peninsula, the Bouguer anomalies ranging from −50 mgal to −130 mgal.

The geology and tectonics of the Indian peninsula has a strong influence on the gravity field which is well reflected by major trends (highs and lows) on the Bouguer anomaly map.

3.2. Major Trends on Bouguer Anomaly Map

The major trends on this map are shown in Figure 3.2 and their relationship to tectonic features are discussed below.

3.2.1. EASTERN GHATS HIGH

In southern peninsula of India, a major trend in a general NE–SW direction with gravity high (H–1) corresponds with that of the Eastern Ghats belt. The high starts

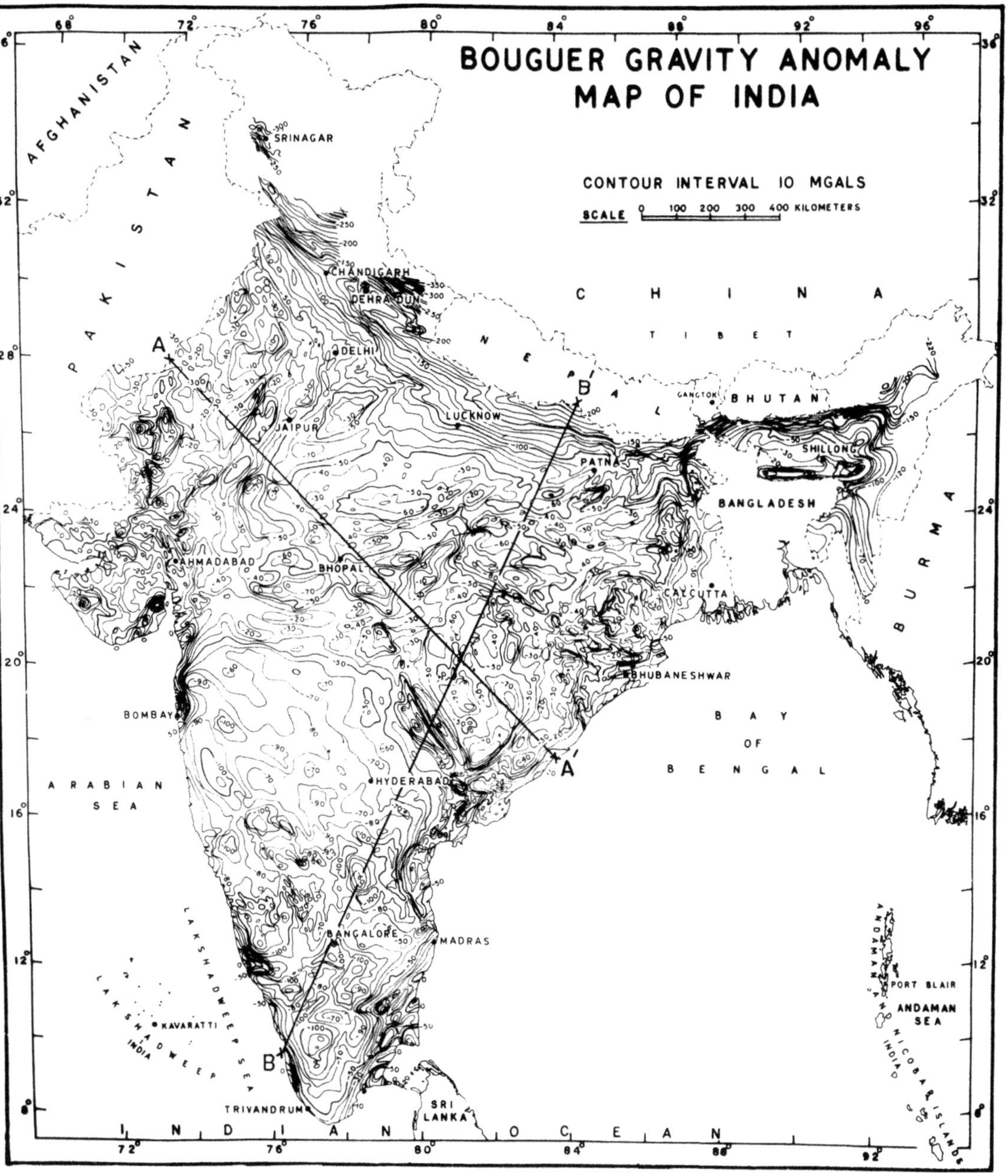

Fig. 3.1. Bouguer anomaly map of India published by NGRI, Hyderabad (Reproduced with permission of Director, NGRI).

from near Tiruchchirapalli (11° N, 79° E) in the south, trends towards Madras, follows to the east of the Cuddapah basin in an arcuate fashion and continues northwards along the east coast of India. Although the Eastern Ghats terminate near Mahanadi Valley, the trend of gravity high continues further towards north with a

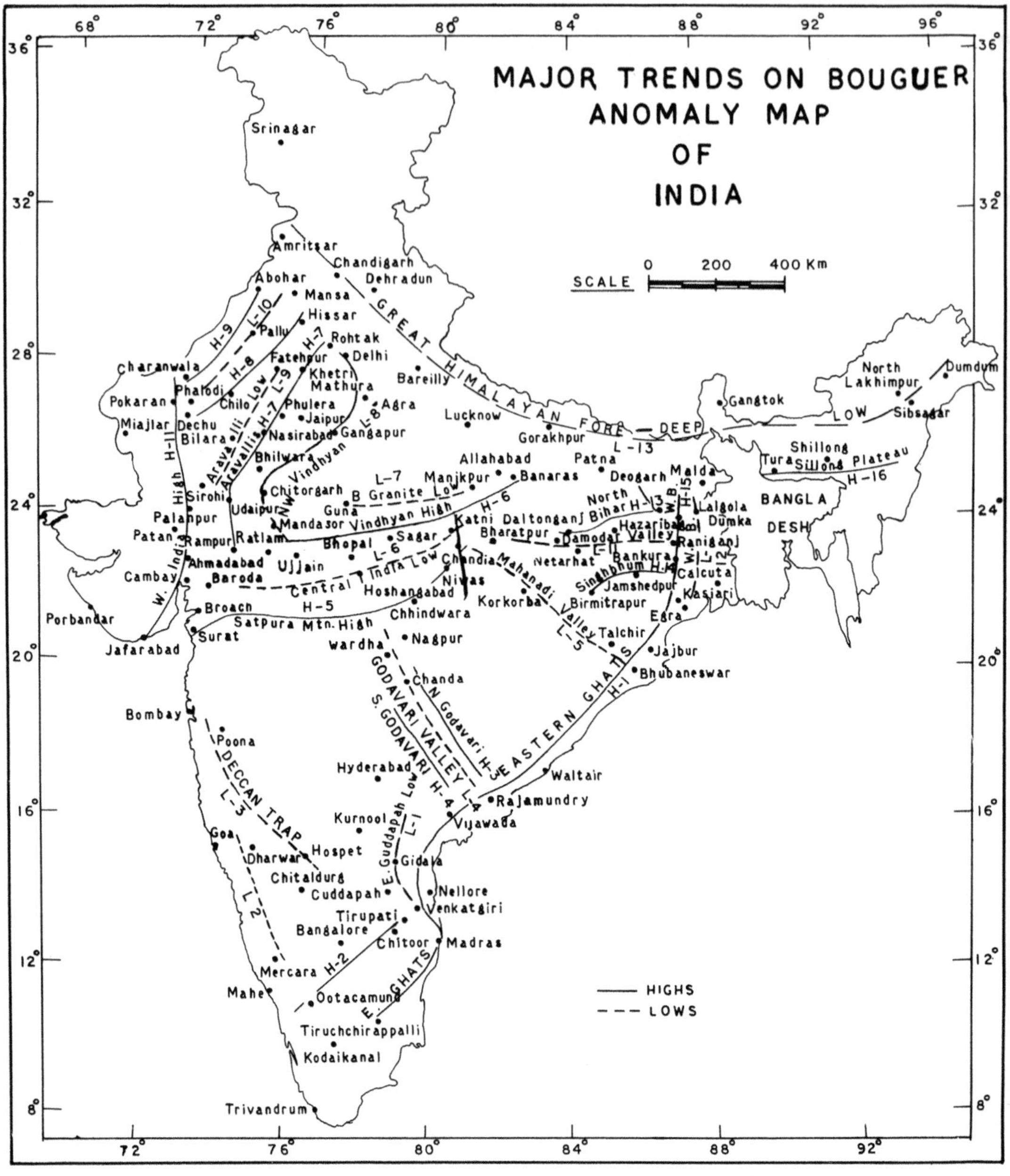

Fig. 3.2. Major trends on the Bouguer anomaly map of India.

north-south strike as H–15 and ends north of Dumka (24° 15′ N, 77° 30′ E) over Rajmahal Hills. The Bouguer anomalies along the trend range from −20 to +20 mgal.

The entire belt of the Eastern Ghats is characterised by high grade metamorphism with intrusives of charnockites, gabbros, anorthosites and similar rocks. Detailed analysis of gravity field along this belt is discussed in Chapter 6.

3.2.2. TRENDS ASSOCIATED WITH DHARWARS, CUDDAPAHS AND GNEISSES

3.2.2.1. *East Cuddapah Low*(L–1)

The East Cuddapah Low (L–1) lies to the west of the Eastern Ghats along the eastern part of the Cuddapah Basin, conforming to its arcuate shape. The low extends from Venkatagiri in the south to the northern age of the Cuddapah Basin in the north. A steep gradient of Bouguer anomaly in the eastern part of the basin suggests a faulted contact, between the crystallines in the east and the basin on the west. The western part of the Cuddapah basin shows a gravity high with respect to the eastern part.

The lowest value of Bouguer gravity along this trend is −110 mgal. The nature of the low itself suggests that it is related to the tectonics of the Cuddapah basin. Recent results of Deep Seismic Soundings (D.S.S.) carried out by the National Geolphysical Research Institute, Hyderabad, in the Cuddapah basin suggest that this low is attributable to the thickening of underlying crust in the eastern part of the basin (Kaila and Bhatia, 1981).

3.2.2.2. *Oottacamund-Chittoor High* (H–2)

The southern shield of India (south of 16° N) is characterised by large scale occurrences of a wide variety of rock types such as greenstones, gneisses of various types, migmatites, granites and charnockites of acidic, basic and intermediate varieties. These rocks when occurring over large areas influence the gravity field considerably. One example of this is the NE–SW trending gravity high (H–2), which extends from near Oottacamund over Nilgiri Hills in the south to near Chittoor in the east. The Bouguer anomaly along this trend is about 20 to 30 mgal higher with respect to the surroundings. Over a large part of its length the high is associated with outcrops of charnockites.

3.2.2.3. *Dharwar-Mercara Low* (L–2)

A conspicuous gravity low in the southern shield of India is the Dharwar-Mercara Low. This low extends from near Mercara in the south to near Dharwar in the north in a general NNW–SSE direction. Minimum gravity along this low is −110 mgal. This low occurs partly over the gneisses (which have a density around 2.85 gm cm^{-3}) of the southern shield and suggests the presence of granitic rocks underlying the greenstones/migmatites and the gneisses. The nature of gravity field over Dharwar Craton is discussed in detail in Chapter 5.

3.2.2.4. *Deccan Trap Low* (L–3)

A prominent gravity low (L–3) trending in a general NW–SE direction extends from near Poona in the north to Hospet in the south. The low occurs partly over the gneisses and migmatites of the southern shield and partly over the Deccan traps. Since, the Deccan traps are high density rocks (about 2.90, gm cm^{-3}), the gravity low suggests the presence of granitic rocks underlying the gneisses and the traps.

3.2.3. TRENDS ASSOCIATED WITH GODAVARI, MAHANADI AND DAMODAR VALLEYS

The Godavari, Mahanadi and Damodar Valleys form the major Gondwana basins of India. In these grabens appreciable thickness of Gondwana sediments was deposited during Carboniferous to Cretaceous times. The tectonics of Godavari Valley is well reflected on the Bouguer anomaly map. The valley is characterised by a gravity low (L-4) trending in NW-SE direction, with −90 mgal Bouguer anomaly in the central part of the valley as compared to −40 mgal over the surroundings. The low is attributable to large thickness of Gondwana sediments which have a lower density as compared to the surrounding metasediments (see Chapter 4 for density of Gondwana sediments).

Over the northeastern as well the southwestern flanks of the valley, gravity highs (H-3) and H-4) trend in the same direction as the low, with an amplitude of +20 to +40 mgal. These highs appear to be due to the presence of high density metasediments, known as Pakhals in the area.

The Mahanadi Valley low (L-5) extends from near Bhubaneswar in the southeast to near Chandia in the northwest in a general northwesterly direction, for nearly 500 km following the trend of the Mahanadi river. The low is clearly attributable to low density Gondwana sediments that were deposited by the Mahanadi river and its tributaries, as in the case of Godavari Valley.

The gravity low L-11 lies over Damodar Valley Gondwana basins of Auranga, Hutar, Karanpura, Bokaro, Jharia and Raniganj Coalfields and extends in east-west direction from Ramanujganj in the west to Raniganj in th east. Detailed gravity surveys over these basins show typical graben-like structure (see Chapter 4).

The Godavari, Mahanadi and Damodar Valley lows suggest the presence of major lineaments in the peninsular shield of India along which the earth's crust fractured during Lower Paleozoic times resulting in the formation of grabens and their subsequent filling with sediments belonging to the Gondwana System.

3.2.4. TRENDS ASSOCIATED WITH SATPURA AND VINDHYAN MOUNTAINS

The gravity field in the area lying between 20° to 24° N and 74° to 84° E is predominantly influenced by the tectonics associated with the Satpura and Vindhyan mountains as well as the Vindhyan sediments which have been deposited over the Precambrian crystalline basement. The area is associated with several highs and lows trending in a general ENE-WSW direction. These are discussed below.

3.2.4.1. *Satpura Mountain High* (H-5)

A major trend characterised by gravity high extends from near Surat in the west to near Chandia in the east. This high has a general east-west strike from Surat to near Chindwara on the east beyond which it takes a north-easterly turn towards Chandia. The high lies over the Deccan trap along most of the way. Along the eastern side, the Satpura high is interrupted by the low overlying the Mahanadi valley (L-5) and appears to continue further as North Bihar high (H-13) in a general east-west direction over Hazaribag plateau of Bihar. The metamorphics in this area have been affected by Satpura orogenic cycle (Sarkar, 1968).

3.2.4.2. *Central India Low* (L-6)

North of Satpura mountains lies a large area underlain by Vindhyan sediments, the Deccan trap volcanics, and the Bundelkhand granite. The central India low lies north of the Satpura high and extends from near Baroda in the west to near Katni in the east, via Hoshangabad, almost following the course of the Narmada river. The low lies partly over the Deccan trap (in the western parts) and partly over the Alluvium, Gondwana and Vindhyan sediments. Minimum gravity (Bouguer) along the low is −70 mgal. The nature of the low suggests that it is possibly due to the relatively low density Vindhyan and Gondwana sediments which appear to underlie large parts of the Deccan traps in the northern part of the peninsula.

3.2.4.3. *Vindhyan High* (H–6)

North of the Central India low lies a well defined gravity high which extends from near Ratlam in the west to near Banaras in the east. The high has an amplitude of nearly +25 to +30 mgal relative to the surroundings. It lies partly over the Deccan trap and partly over the Vindhyan sediments. The high appears to be related to the structure of the Vindhyan basin, and could possibly indicate a basement high underneath the sediments.

3.2.4.4. *Bundelkhand Granite Low* (L–7)

North of the Vindhyans high is a well defined low termed as the Bundelkhand granite low with Bouguer anomaly of −60 mgal. This lies mostly over the Bundelkhand granite and its surroudning Vindhyan rocks. The low suggests that the granitic batholith has a lower density as compared to the surrounding metamorphics/Vindhyans.

3.2.4.5. *NW Vindhyans Low* (L–8)

The Vindhyan sediments take a sharp turn in an arcuate manner against the Aravalli sediments on the northwestern side. The northwest Vindhyans low conforms to the shape of the Vindhyans basin and extends from Mandasor in the south to near Gangapur in the north passing close to Chitorgarh. It has an amplitude of −30 to −40 mgal with respect to the surroundings. Further north the low trends towards Delhi in north westerly direction.

3.2.5. TRENDS ASSOCIATED WITH THE ARAVALLIS AND WESTERN INDIA

The area to the west of the Vindhyan Low (L–8) up to the India–Pakistan border is dominated by structures associated with history of Aravalli geosyncline. Gravity field reveals the presence of a number of highs and lows trending in a general NE–SW direction. A possible explanation of these is discussed below.

A very prominent gravity high (H–7) is associated with the Aravalli mountains which are composed of sediments belonging to the Delhi and Aravalli Systems of Precambrian age. The associated high is termed as the Aravallis high and extends from near Rampur in the south to near Rohtak in the north in a general NE–SW direction passing through Udaipur, Nasirabad, Phulera, and Khetri. The gravity high has an amplitude of about +60 to +70 mgal with respect to the surroundings, which are characterised by gravity low. Nature of the gravity field suggests that the Aravallis are underlain by high density rocks and supports the idea of a horst type structure for these mountains.

West of the Aravalli mountains are the acidic rocks belonging to the Erinpura granite complex, and alluvium as well as Vindhyan sediments. The Aravallis low (L–9) is associated with these rocks and extends from near Sirohi in the south to near Fatehpur in the north, having the same trend as the Aravallis high, discussed earlier.

The gravityfield to the west of Aravalli mountains is high as compared to thaton its east. The highs H–8, H–9 and low L–9 appear to be related to the nature of rocks underlying the alluvium to the west of Aravallis. It is possible that the basement is faulted and fractured giving rise to highs and lows.

In western India the gravity high H–11 extends over a large distance from Jafarabad in the south to Charanwala in the north. The high has an amplitude of 30–40 mgal and lies partly over Deccan trap and partly over the Cambay-Mehsana basin. It appears to be related to the tectonics of this basin and could be partly attributed to the large thickness of traps underlying this basin (Kailasam and Qureshy, 1964).

3.2.6. TRENDS IN EASTERN INDIA

Gravity field in Eastern India has been influenced by the geological events which have taken place in this part of the country. These include the formation of Iron Ore geosyncline (3000–1500 m.y.), Satpura orogenic belt (1100–900 m.y.), the Gondwana sedimentation, Mesozoic volcanism, formation of Bengal basin during Tertiary times and post-Mesozoic tectonic movements.

The Damodar Valley low (L–11), North Bihar high (H–13) and West Bengal high (H–15) have been discussed earlier. In Singhbhum region of Bihar (South of Damodar Valley grabens), several highs and lows are found which are related to the Iron Ore Group of rocks, basic volcanics and granitic intrusives as discussed in detail in Chapter 7. One of the prominent highs is H–14 which extends from near Birmitrapur in the west to near Kuilpal in the east. The high is related to Dalma volcanics which are intrusive into Singhbhum Group of rocks.

To the east of Singhbhum and Hazaribag region lies the W. Bengal low, L–12, which is attributable to large thickness of Tertiary sediments of Bengal basin.

A very prominent gravity high trending nearly in an east-west direction lies over Shillong Plateau. The high extends from near Tura in the west to near Haflong in the east and covers almost whole of Shillong Plateau. The amplitude of gravity high is nearly +100 mgal with respect to the Bengal basin to the south. The gravity high suggests the presence of high density rocks underlying this plateau (see Chapter 11 for details). Steep gradient to the south of the plateau is due to the presence of a major fault, known as the Dauki fault.

3.2.7. TRENDS ASSOCIATED WITH HIMALAYA

3.2.7.1. *Great Himalayan Foredeep Low* (L–13)

The most prominent gravity low on the Bouguer anomaly map is the Great Himalayan Foredeep Low (L–13). It extends all the way from northwest to northeast following the trend of the Himalayan foothills and the Assam Valley. Gravity low (Bouguer) along the foothills is of the order of −140 mgals. The low is mostly due to the large thickness of uconsolidated sediments belonging to the Siwalik formation (density, about 2.2 to 2.3 gm cm^{-3} of Upper Tertiary age, which have been deposited in the foredeep, created by the uplift of the Himalaya in the north. General elevation of the plains underlain by the *Indo-Gangetic* alluvium and Siwalik formation is of the order of 200 to 300 m. Bouguer anomalies in the plains range from about −50 mgal in the areas close to the shield to about −140 mgal near the foothills, indicating progressive increase in the thickness of sediments towards the foothills. The largest Bouguer anomaly found in the foothills is −250 mgal near North Lakhimpur in Assam.

North of Himalayan foothills the Bouguer anomalies become progressively more negative, as the elevation increases. The lowest value of Bouguer anomaly observed in the Himalayas is of the order of −500 mgal. This increase is attributable to the effect of root formation underneath the Himalaya and give an indirect evidence in support of the phenomenon of isostasy (discussed in detail in Chapter 9).

3.3. Free-Air Anomalies in Peninsular India

Free-Air anomaly map of Peninsular India is shown in figure 3.3. It may be recalled here that free-air correction is +0.3086 mgal m^{-1}, while Bouguer correction is 0.1119 mgal m^{-1} for a density of 2.67 gm cm^{-3}, for masses above sea level. As explained in Chapter 1, the free-air anomaly tends to follow the elevation over mountainous areas, while Bouguer anomaly bears an inverse relationship to topography. Over areas of low elevation free-air correction is very small and hence the anomaly generally follows the Bouguer anomaly and can be explained in terms of geological effect of underlying rock masses.

Over mountain ranges of relatively small dimensions which are locally uncompensated (but regionally compensated), the free-air anomalies are found to be highly positive. Over continental areas, regional compensation is found to be a rule rather than an exception. Over large mountain ranges, such as the Rocky Mountains of U.S.A., the free-air anomaly tends to be nearly zero on an average.

Major trends of free-air anomaly over Peninsular India are shown in Figure 3.4. Their correlation with tectonic features is briefly discussed below.

Examples of large free-air anomalies over mountain ranges, hills and plateaus are:

H-1 over the Eastern Ghats with a N–S to NE–SW trend along the east coast of India. Here maximum anomaly is nearly +110 mgal. The high correlates with Bouguer anomaly high H–1 over the same area (Figure 3.2).

H–12 over Aravalli mountains. The high correlates well with H–7 of Bouguer anomaly.

H–9 over Satpura mountains. The high has a maximum value of +70 mgal and continues in northeasterly directions as H–10 over Hazaribag plateau of Bihar.

H–11 correlates well with the strike of Vindhyan mountains.

H–2, H–3, and H–4 are located over Mysore plateau, Nilgiri and Palni-Cardamom hills respectively. All the hill masses are characterised by large free-air anomalies, the maximum value reaching +200 mgal over Nilgiri Hills.

H–6, H–7 and H–8 are located over Deccan volcanics and are related to the topography of Deccan plateau.

H–14 is a major anomaly (+170 mgal maximum) as associated with Shillong plateau (See Chapter 11 for details).

An example of free-air anomaly related to the underlying formations (Pakhals) is H–5 which is located north of Godavari valley. The high correlates with H–3 of Figure 3.3. A similar high (H–13) west of Aravalli mountains appears to be related to the underlying formations.

Lows over free-air anomaly map are located at low elevation areas or over

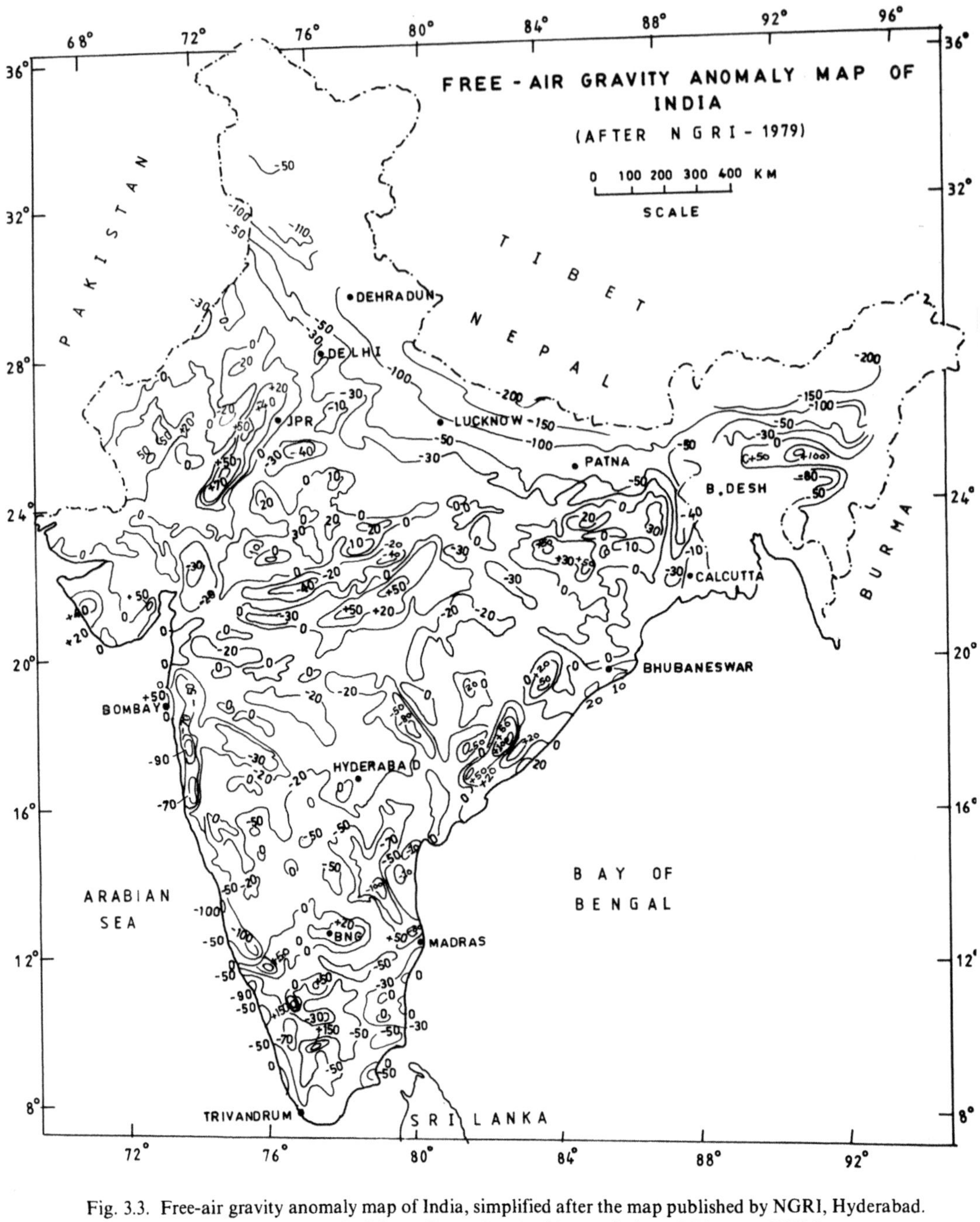

Fig. 3.3. Free-air gravity anomaly map of India, simplified after the map published by NGRI, Hyderabad. BNG-Bangalore, JPR-Jaipur (Reproduced with permission of Director, NGRI).

sedimentary basins. L–1 is located along the West Coast, while L–2 follows the arcuate shape of Cuddapah basin and is correlatable with L–1 of Figure 3.2. L–3, L–4, and L–6 correspond to gravity lows over Godavari, Mahanadi and Damodar Valley grabens. L–5 follows the Satpura trend and is sandwitched between the Vindhyan and the

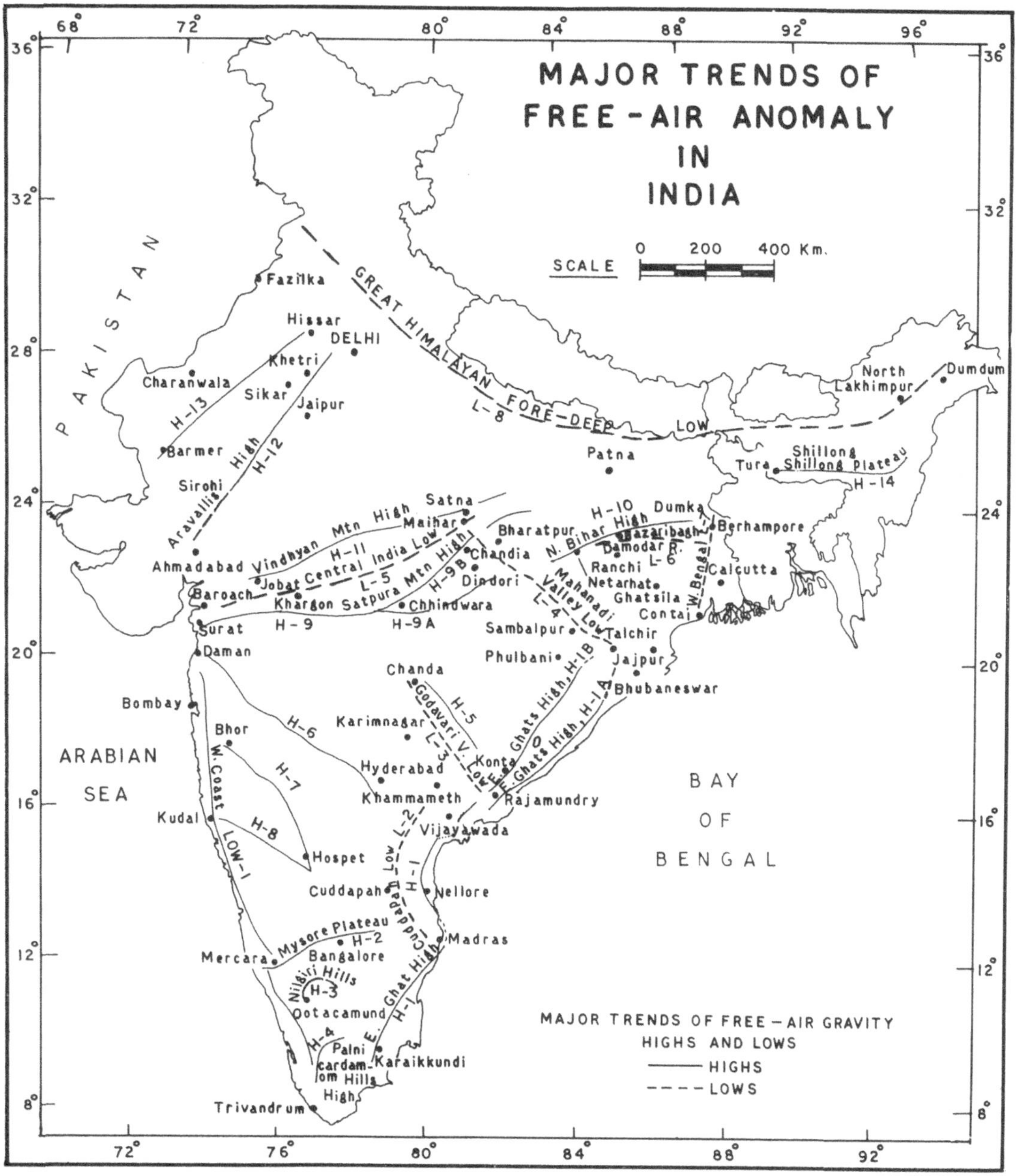

Fig. 3.4. Major trends of free-air anomaly in India.

Satpura mountains. A major low L–8 lies along the foothills of the Himalayas. Here minimum free-air anomaly is −250 mgal, near North Lakhimpur.

3.4. Inter-Relationship of Various Anomalies

The inter-relationship of various gravity anomalies and their correlation to various geological and tectonic features has been illustrated along two profiles AA′ and BB′ passing through all the major geological units of Peninsular India. These are discussed below.

3.4.1. PROFILE AA′

The profiles trends in a general north-westernly direction and extends from near Charnwala in the northwest to near Srikakulam in the south-east. Location of the profile is shown in Figure 3.1. Starting from the alluvial covered areas in the north-west, the profile passes over the Aravalli mountains, the Vindhyan basin, the Vindhyan mountains, the Satpura Gondwana basin, the Satpura mountains, the Precambrian metasediments and granites belonging to Archaean complex, unclassified gneisses, charnockites and high grade metamorphics belonging to the Eastern Ghats, ending at east coast of India.

It is interesting to note that all the three anomalies, (free-air, Bouguer and isostatic) indicate gravity highs over the major mountain systems of Peninsular India i.e. the Aravallis, the Vindhyan mountains, Satpuras as well as the Eastern Ghats. The anomalies rise sharply over the mountains with respect to the surrouding areas, thus indicating their horst type structure. As expected, the Bouguer anomalies are more negative as compared to the free-air anomalies, which are in general positive over these mountains. Isostatic anomalies are also in general positive but their magnitude is lesser than that of the free-air anomalies.

Over the Vindhyan basin and the Vindhyan mountains (underlain by sediments as well as the Deccan trap), the anomalies are undulating, the lower elevation areas being characterised by more negative values as compared to the areas with higher elevation.

Over the areas underlain by alluvium as well as Gondwanas, the Bouguer as well as free-air anomalies show strongly negative values, although isostatic anomalies are only moderate. Over the crystalline areas, south-east of the Satpura mountains, the anomalies are undulating, till they rise uniformly over the Eastern Ghats and thereafter tend to come close to zero near the east coast.

3.4.2. PROFILE BB′

The nature of various gravity anomalies along this profile is shown in Figure 3.6. This profile starts from the west coast of India near Cochin, passes over the Nilgiri Hills, the Mysore Plateau, Cuddapah basin, the Chhatisgarh basin, the Godavari Valley, Mahanadi Valley, Vindhyan sediments, Ganga basin and ends at Thor over Himalaya near India-Nepal border.

It may be noted that the Bouguer anomaly is appreciably negative over the west coast, becomes less negative over the Nilgiri Hills and shows undulating nature over the Mysore Plateau. A very strong correlation between the free-air anomalies and the elevation over the Nilgiri Hills as well as the Mysore plateau is noteworthy. The

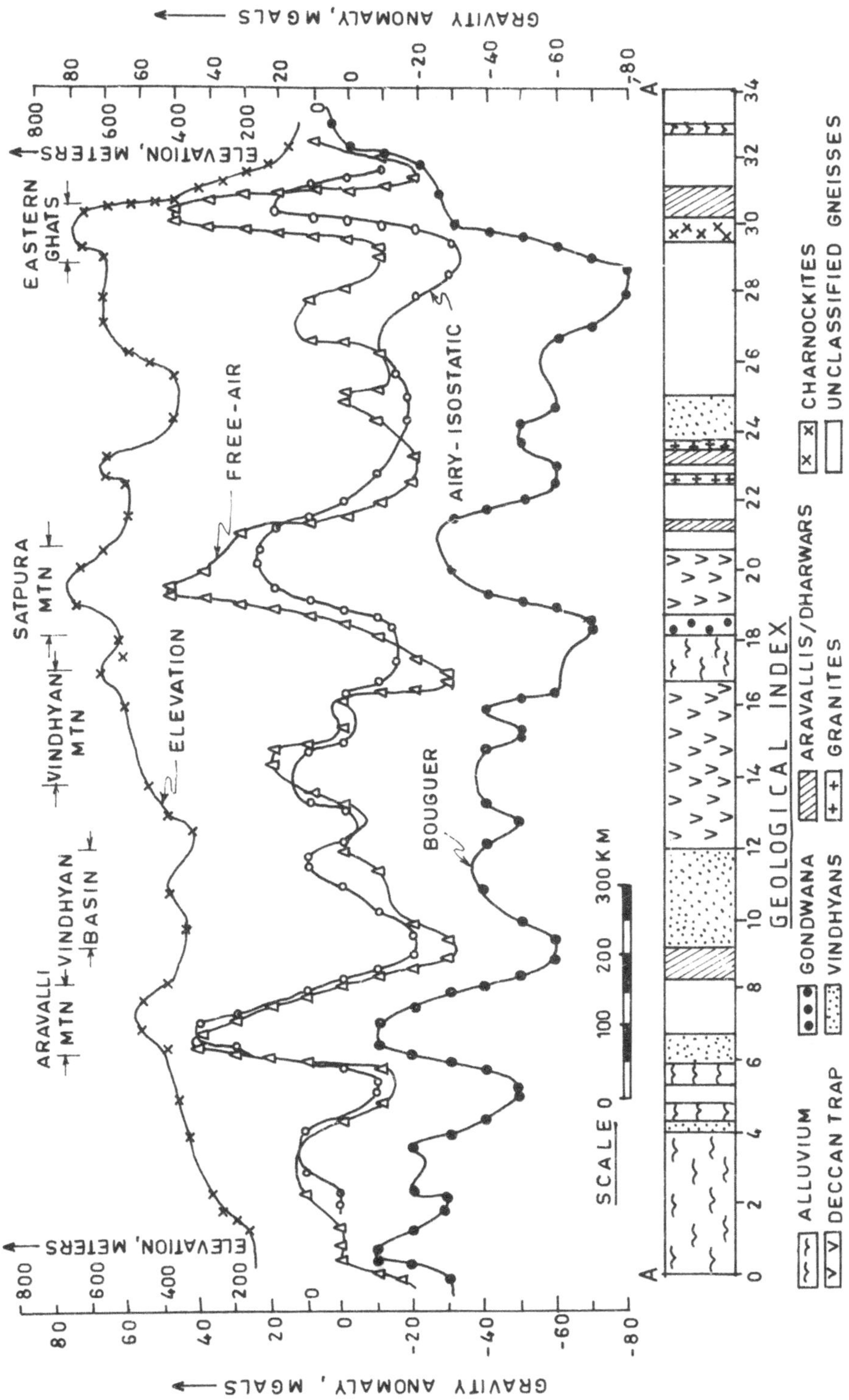

Fig. 3.5. Profile AA′ crossing the Aravalli mountains, the Vindhyan basin, the Vindhyan mountains, the Satpura moutains and the Eastern Ghats in a general NW–SE direction across the Indian Peninsula. The nature of various gravity anomalies, the underlying geology and the elevation are shown in the figure.

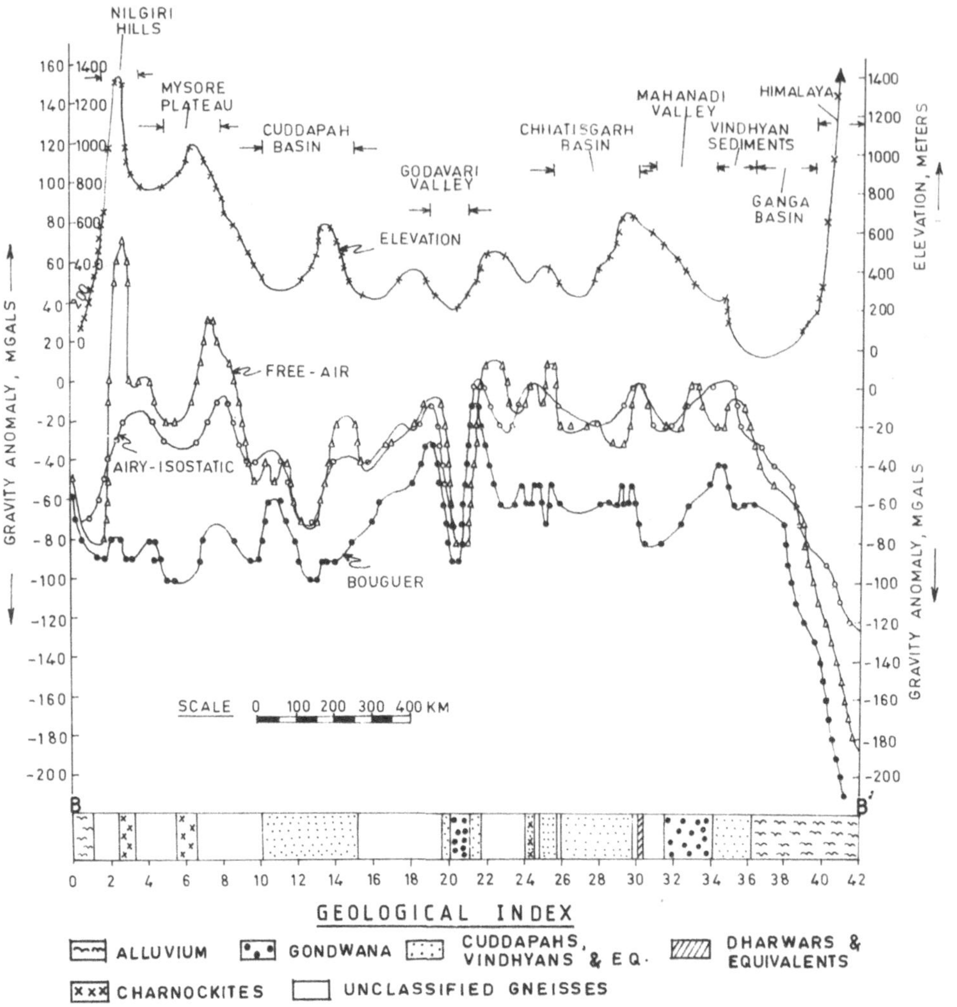

Fig. 3.6. The nature of various gravity anomalies, the underlying geology and elevation along profile BB′ across the Indian peninsula from south to north. The profile passes through several topographic and geological units such as the Nilgiri Hills, the Mysore Plateau, Cuddapah basin, Godavari Valley, Chhatisgarh basin, Mahanadi Valley, Vindhyan basin, the Gangetic alluvium and ends in Himalaya.

isostatic anomalies over these areas are much less negative compared to Bouguer anomalies, and tend to come close to zero, thus suggesting near isostatic conditions prevailing in the area as a whole. Further north, the profile passes over the south-western part of the Cuddapah basin which is an area of gravity high, as compared to the north-eastern part as seen on the Bouguer anomaly map. The gravity high on the southeastern part is partly due to the large thickness of basaltic sills, which have intruded into the sediments in this part of the basin.

All the three gravity anomalies show conspicuous low over the Godavari Valley as compared to the surrouding areas. These lows suggest a graben type structure for the valley, as discussed earlier. Northeast of Godavari Valley, the profile passes over the unclassified gneisses and Precambrian sediments belonging to the Chhatisgarh basin. The fluctuating anomalies over the crystallines seem to be due to different densities of rock masses, while the Chhatisgarh basin shows a gravity low, about 10 to 15 mgal in magnitude with respect to the surroundings.

Gravity low over the Mahanadi Valley is well reflected in all the three gravity anomalies along the profile. The gravity field rises over the Vindhyan sediments and anomalies fall steeply further north, as we cross the Vindhyan sediments into the area underlain by Gangetic alluvium. All the three gravity anomalies show a similar trend over this area, on account of the predominant influence of the low density sediments on the gravity field. Crosssing the Ganga basin, we enter into the Himalaya, where the Bouguer anomaly continues to show a negative trend. The inverse relationship between the elevation and the Bouguer anomaly is generally maintained over the Himalaya.

It may be recalled here that the free-air and isostatic anomalies reverse their trend after crossing the Ganga basin and become positive further north. The behaviour of free-air and isostatic anomalies over the Himalayas and their explanation is an important problem and is discussed in detail in Chapter 9.

CHAPTER 4

GRAVITY FIELD AND TECTONICS OF GONDWANA BASINS OF PENINSULAR INDIA

4.1. Introduction

The Godavari, Mahanadi, Damodar, Narmada and Son are the major rivers along which Gondwana sediments were deposited during Carboniferous to Cretaceous times along well formed valleys, basins or rift valley structures. The present-day Gondwana occurrences in Peninsular India are shown in Figure 4.1. These include, the Godavari Valley in the south-east, the Narmada-Son-Mahanadi basin in Central India, Damodar valley basins including the Raniganj, Jharia, Bokaro, Karanpura and Ramgarh coalfields in eastern India. Besides these, Gondwanas are found in Palamau and Deoghar districts in Bihar, the Rajmahal Hills, the foot-hills of Himalaya and along the east coast of India.

The Gondwana basins represent major features on the Geological Map of India and are believed to have been formed as a result of faulting in the earth's crust during Upper Carboniferous to Jurassic times. Various theories have been proposed from time to time regarding the tectonic evolution of these basins. The main controversy centers around the age of different faults relative to the age of sedimentation. Fox (1931) has been of the opinion that the Gondwana sediments were laid down in rift valleys and faulting continued throughout the period of deposition. Gee (1932), and few others consider that the Gondwanas were laid down in a large unified basin and later preserved by faults in the down-thrown blocks and therefore the present day basins are but the remnants of more widespread ones.

Chatterji and Ghosh (1970), in their recent analysis of the evidences of the development of sedimentary and structural elements of the Gondwana basins, expressed the view that these basins neither fit in with the rift-valley concept of Fox (1931), nor are compatible with the ideas of Gee (1932) that they are faulted remnants of more widespread ones. They conclude that the tectonics associated with the Gondwana basins is an example of 'resurgent tectonics', based on the principle that the activation of the pre-existing weak features of the basement gave rise to the subsequent tectonism and sedimentation, within a new set of dynamic forces.

4.2. Gravity Anomalies and Gondwana Tectonics

Gravity studies throw a considerable light on the nature of tectonics of the Gondwana basins. Since their structure has been linked to that of rift valleys, we may first discuss typical rift valley structures.

The term 'rift valley' was used by Gregory to mean a long strip of terrain let down between normal faults or parallel series of step faults, as if a fractured arch has been

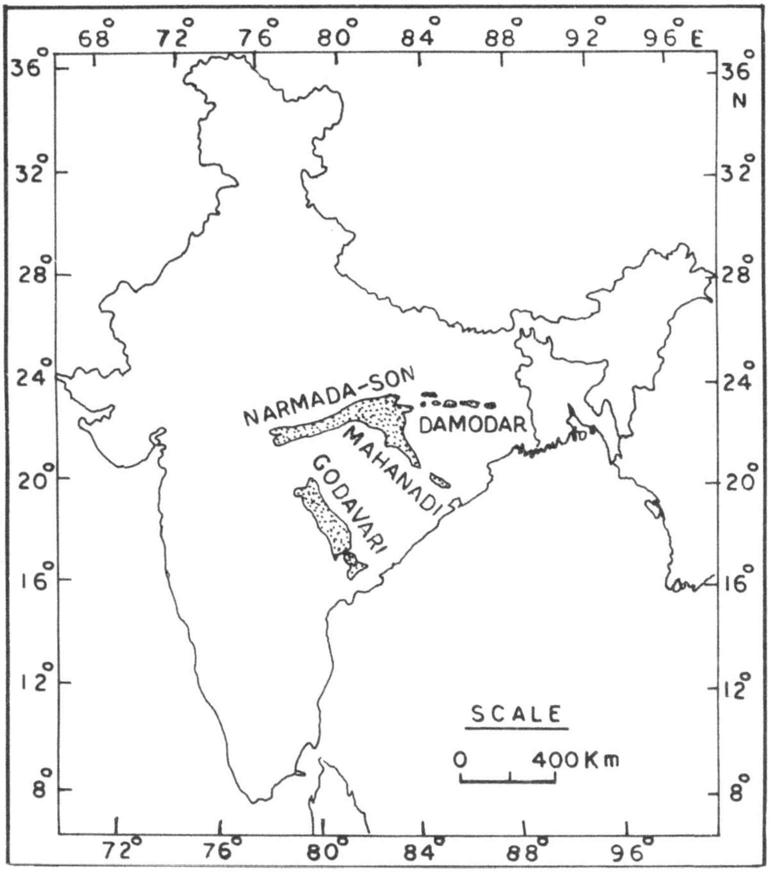

Fig. 4.1. Present day occurrence of major Gondwana basins of Peninsular India.

pulled apart by tension, so that the keystone dropped in block or in strips (Holmes, 1965). Brock (1952) in discussing the origin of African rift valleys suggests that the rift valley structure is a manifestation of vertical tectonics. He discards both the tension and compression theories and argues that the processes of vertical tectonics have been active since the Precambrian and the Precambrian faults have been active till now. Vening Meinesz and Heiskanen (1958) consider tension to be the cause of graben formation.

Typical rift valleys are found in Africa. Here Girdler (1964) has shown that the gravity field is consistent with normal faulting on both sides of the rift valley. Searle (1970) and others, have carried out detailed gravity studies over the African rift valley of Kenya. They have shown that the gravity field is characterised by a regional low over the valleys, while highs in the central part are superimposed on this low. The broad gravity low (about 1000 km wide) has been attributed to the presence of a lower density asthenosphere under the rift valleys, (Girdler *et al.*, 1969).

The results of gravity studied carried over different Gondwana basins are discussed below.

4.3. Geology of Gondwana Basins – General Remarks

The Gondwana formations in general, overlie either the crystalline metamorphics or the older sedimentary formations considered to be equivalents of Cuddapahs or Vindhyans in Peninsular India. The Gondwana rocks are divided into Lower and Upper Gondwanas (see Table 4.1). The sediments are mostly lacusterine or fluviatile in nature. The Talchir formation is known to be the oldest (Upper Carboniferous) and is supposed to be of glacial origin. This is overlain by Barakar formation consisting mostly of sandstones, shales, their intercalations and coal seams. The formation owes its importance to the coal bearing strata. The rest of the formations consist of sandstones, shales and clay beds. The uppermost Gondwanas are Upper Jurassic or Lower Cretaceous in age. The stratigraphic sequence in Godavari and Mahanadi valleys and Damodar Valley coalfields is given in Table 4.1.

TABLE 4.1

Stratigraphic Sequence in Godavari, Mahanadi, Damodar Valley Gondwana basins (Krishnan, 1961)

		Godavari Valley	Mahanadi Valley	Damodar Valley Coalfieds		
				Raniganj	Jharia	Bokaro
Upper Gond-wana	Cretaceous	Chikiala,	–	Dolerite dykes	Dolerite dykes	Dolerite dykes
	Jurassic	Chikiala. Kota	Athgarh	Mica perido-tite dykes and sills	Mica periodo-tite dykes and sills	–
	Upper Triassic	Maleri	Mahadevas	Supra-Panchets	–	Mahadevas
		UNCONFORMITY				
	Lower Triassic	Mangli	Panchet Series	Panchet Series	–	Panchet Series
Lower Gond-wana	Upper Permian	Kamthi	Hingir	Raniganj Series	Raniganj Series	–
	Middle Permian	–	–	Ironstone Shales	Barren Measures	Barren Measures
	Lower Permian	Barakar Series	Barakar Series	Barakar Series	Barakar Series	Series
	Upper Carboniferous	Talchir Formation	Talchir Boulder Bed	Talchir Boulder Bed	Talchir Boulder Bed	Talchir Boulder Bed
		UNCONFORMITY				
	Precambrian	Pakhal Series	Crystalline Metamorphics			

4.4. Density of Gondwana Sediments

The Gondwana rocks consist mostly of sandstones, sandy shales and shales. Histograms of density values for the sandstones, sandy shales and shales collected from parts of Jharia, Raniganj, Bokaro and Sonhat Coalfields are shown in Figure 4.2 (Rao, 1976).

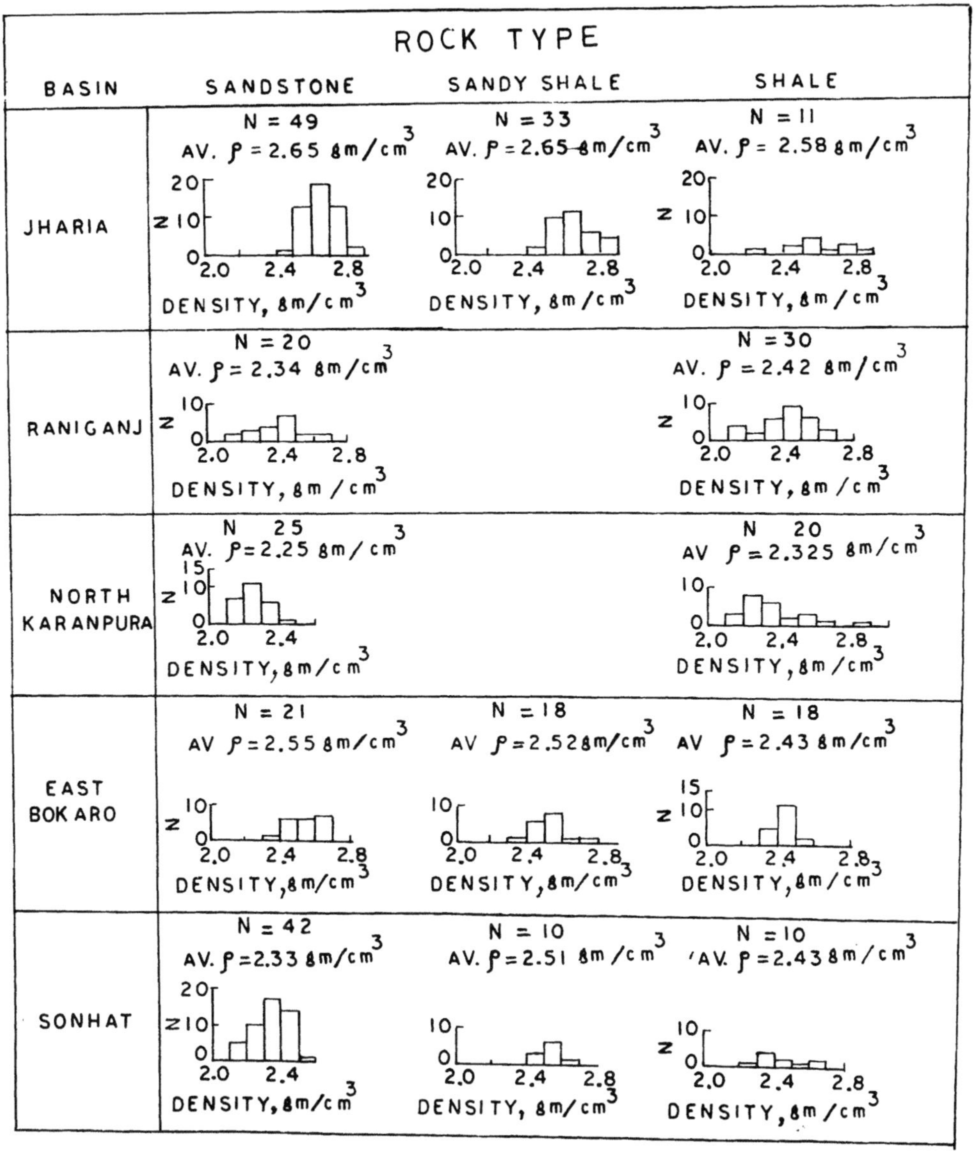

Fig. 4.2. Histograms of density values of Gondwana sediments from various coalfields (after Rao, 1976).

It has been observed that the density values vary considerably from one coalfield to another, perhaps depending upon their degree of compaction and composition. The average for sandstones varies from 2.33 gm cm^{-3} in Sonhat coalfield to 2.65 gm cm^{-3} in Jharia coalfield. The range of values for various rock types is given in Table 4.2.

In order to determine the representative density value of Gondwana rocks for a particular coalfield, the density values for various rock types need to be weighed according to their stratigraphic thickness. This may differ from one coalfield to another. The metamorphic rocks surrounding Gondwana basins have highly variable

TABLE 4.2

Density of rocks from Damodar Valley Gondwana basins of India. Some of the data is obtained from Rao (1976).

Rock type	No. of samples	Density values (g cm^{-3})	
		Range	Mean
Sandstone	213	2.10-2.90	2.50
Sandy shale	107	2.15-2.85	2.52
Shale	135	2.10-2.80	2.43

density as discussed in the next. Interpretation of gravity field over different Gondwana basins is discussed below.

4.5. Godavari and Narmada-Son Valleys

Gravity anomaly map of the Godavari valley and its surroundings prepared by Qureshy *et al.* (1968) is shown in Figure 4.3. The figure shows that the valley is characterised by a gravity low of the order of 50 mgal trending in NW-SE direction conforming to the trend of the geological formations in the area. The low starts from Masulipatnam in the south-east and extends beyond Indore up to 23° N, through Wardha-Godavari valley as several closed minima in a line. It has an average width of about 50 km. The anomaly correlates well with the Godavari basin. The eastern margin of the anomaly coincides with the Precambrian-Gondwana boundary. An interesting aspect of the Bouguer anomaly map is the presence of small gravity 'highs' flanking the 'lows'. These 'highs' are significant in view of the uplifting that should occur over the boundaries of the graben, had they been formed according to the tensional theory of garben formation of Vening-Meinesz and Heiskanen (1958).

Further northwest a similar low is encountered west of Nagpur which is correlatable with Kamthi coalfield. It appears that Godavari valley Gondwanas may extend underneath the Deccan trap even beyond Nagpur. Fig. 4.3 shows that a prominent low of 50 mgal amplitude is located near Sohagpur where the contours have ENE-WSW trend of Narmada-Son valley. This low is also attributable to the Satpura Gondwana basin, which extends in northeasterly direction and merges with the Mahanadi valley Gondwanas (see Section 4.6).

In order to gain some idea of the basement configuration and the nature of faulting present in the valley, sections along profiles AA′ and BB′ (locations shown in Figure 4.3) have been interpreted by Qureshy *et al.* (1968) assuming two-dimensional models. A density contrast of -0.4 gm cm^{-3} has been assumed between the Archaean rocks and the Gondwana sediments. The interpreted sections are shown in Figure 4.4. The figure shows that the Godavari valley is characterised by steeply dipping faults on both sides, the fault along the northern side being steeper than on the southern side. The maximum thickness of sediments is estimated to be of the order of 4 to 5 km in Godavari valley, while over Satpura basin the estimated maximum thickness is about 3.0 km.

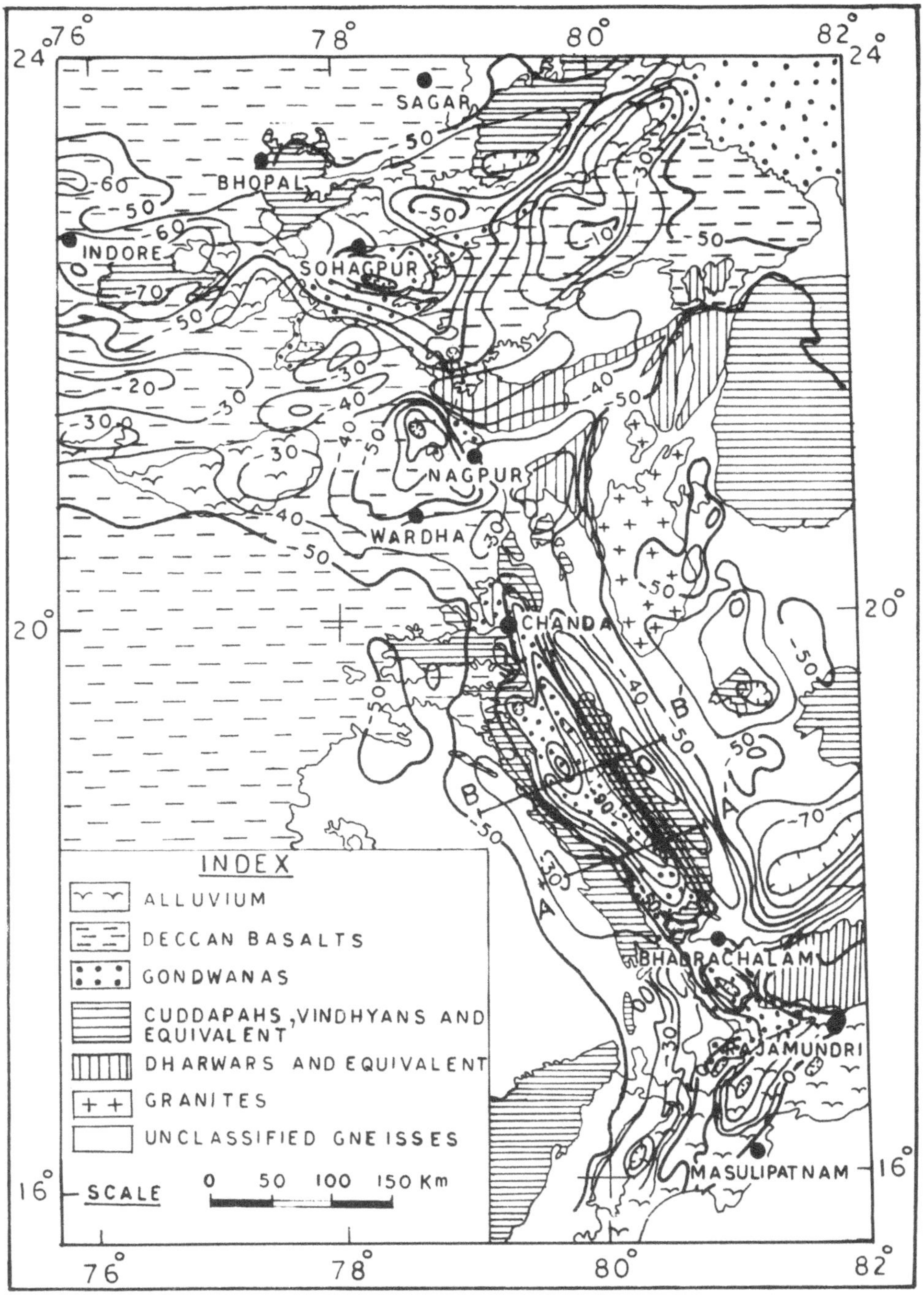

Fig. 4.3. Bouguer gravity anomaly map of Godavari Valley and the adjoining areas, suggesting the possible extension of Godavari Valley towards north-west underneath the traps (Qureshy *et al.*, 1968).

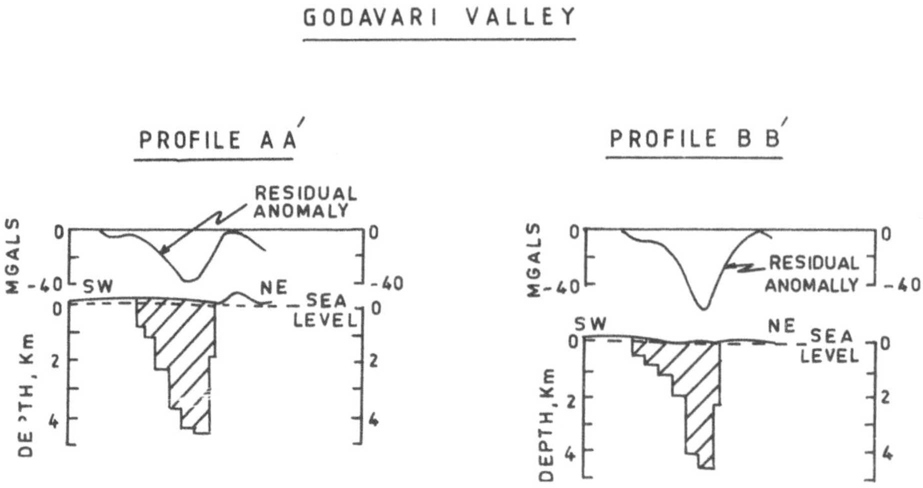

Fig. 4.4. Interpreted sections along two profiles AA′ and BB′ (location shown in Figure 4.3) across the Godavari Valley, based on two dimensional model, after Qureshy *et al.* (1968). The figure shows that the valley is bounded by steeply dipping normal faults.

According to Qureshy *et al.* (1968) the Godavari basin appears to be a rift valley with an average width of about 50 km and total length of about 800 km, extending from the vicinity of Masulipatnam in the SE to near Wardha and beyond in northwesterly direction, part of the valley at present being under the cover of the Deccan traps.

4.6. Mahanadi Valley

Gondwanas belonging to the Mahanadi valley occupy a very large area in eastern India extending from latitude 21½° N to 24° N and longitude 81° E to 84° E, roughly in a triangular fashion. These rocks are bounded by Bijawar Series and Vindhyans (Precambrian to Cambrian age) towards north, Deccan traps (Eocene) towards west, Peninsular Gneisses (Precambrian) towards south and east. In the southeastern part near Raigarh, rocks belonging to the Chhatisgarh basin abutt against the Lower Gondwanas. The area occupied by these rocks is approximately 40 000 km^2. The basin is narrow in the southeastern part (about 40–50 km), but widens to nearly 250–270 km towards north. The northern parts of the area is called the south Rewa or Son Basin, while the southern part is called the Mahanadi Basin. Figure 4.5 shows the geology of the area superimposed on Bouguer anomaly map.

4.6.1. BOUGUER ANOMALY MAP

The Bouguer anomaly map (after Subba Rao, 1977) is based on 700 gravity stations occupied in the area during the course of regional gravity surveys by NGRI parties. The station interval is 4 to 6 km along major roads in the area. It may be seen that a major gravity low extends from near Hingir in the south to near Umaria (23°30′ N, 80° 50′ E) via Chokan and Shahdol in northwesterly direction. The low correlates well with the Lower Gondwana formations of the Mahanadi Valley and has an amplitude

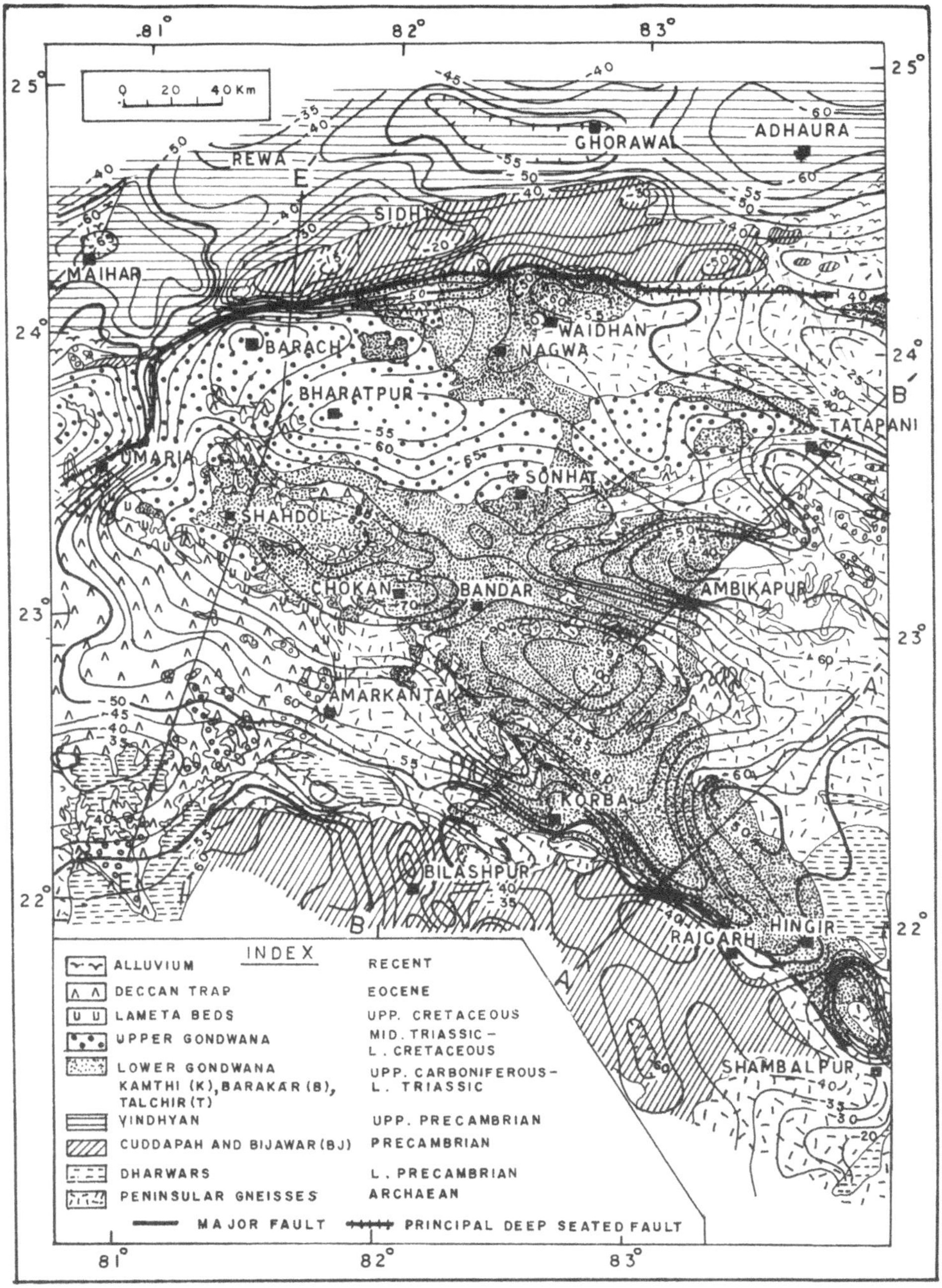

Fig. 4.5. Bouguer anomaly map of Mahanadi Valley (after Subba Rao, 1977) with 5 mgal contour interval.

of −20 to −30 mgal. However, it is interesting to note that part of the low lies over the crystalline metamorphics and Deccan trap particularly south of Ambikapur (23° 10′ N, 83° E) and west of Amarkantak (24°41′ N, 81°48′ E). Presence of steep gravity gradient in the southern part between Korba and Raigarh and near Ambikapur suggests faulted nature of the Mahanadi basin against the older crystalline and meta-sedimentary formations. A prominent gravity high of 20 mgal amplitude occurs over Bijawar formation north of Barach.

Interpretation of gravity field in this area in terms of basement configuration is a bit complicated, since the gravity low observed over Mahanadi Valley is not attributable to Gondwana sediments alone. A major part of the low is observed over the crystalline gneisses and other formations surrounding the Gondwanas as seen along profile BB′ (Figure 4.6) passing through the central gravity low in SW–NE direction. The causes for the low over gneisses are not easily understandable. The low may be due to deeper causes related to the formation of Mahanadi Valley graben (Subba Rao, 1977). This could be a rift cushion as suggested by Illies (1970), a low density sialic crust as underneath Rhine graben (Mueller, 1974) or a thinning of lithosphere as advocated by Girdler *et al.* (1969) underneath African rift valleys.

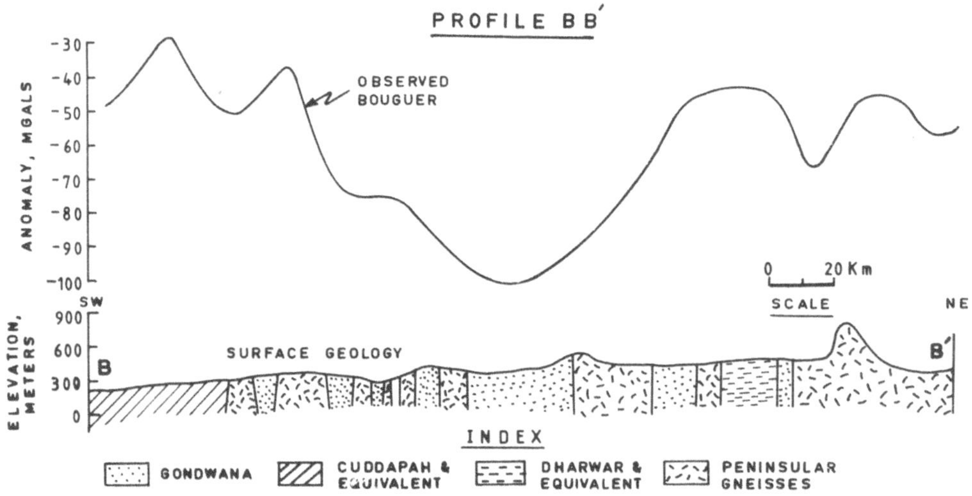

Fig. 4.6. Nature of observed Bouguer anomaly along profile BB′ (location shown in Figure 4.5), across Mahanadi Basin. The profile shows that a major part of the gravity low lies over the Peninsular gneisses surrounding the Gondwanas.

Subba Rao (1977) has interpreted few profiles across the Mahanadi Valley Gondwanas on the basis of residual gravity field obtained by using graphical method. A density value of 2.35 gm cm^{-3} obtained from field investigations was taken to be representative for the Gondwana sediments. For the surrounding formations (mostly Peninsular Gneisses), a representative value of 2.7 gm cm^{-3} was taken, based on laboratory determinations (Subba Rao, 1977). Interpretation of two profiles (AA′ and EE′; locations shown in Figure 4.5) after Subbha Rao (1977) in terms of depth-to-the-basement is shown in Figure 4.7. It may be seen from the figure (Profile AA′) that

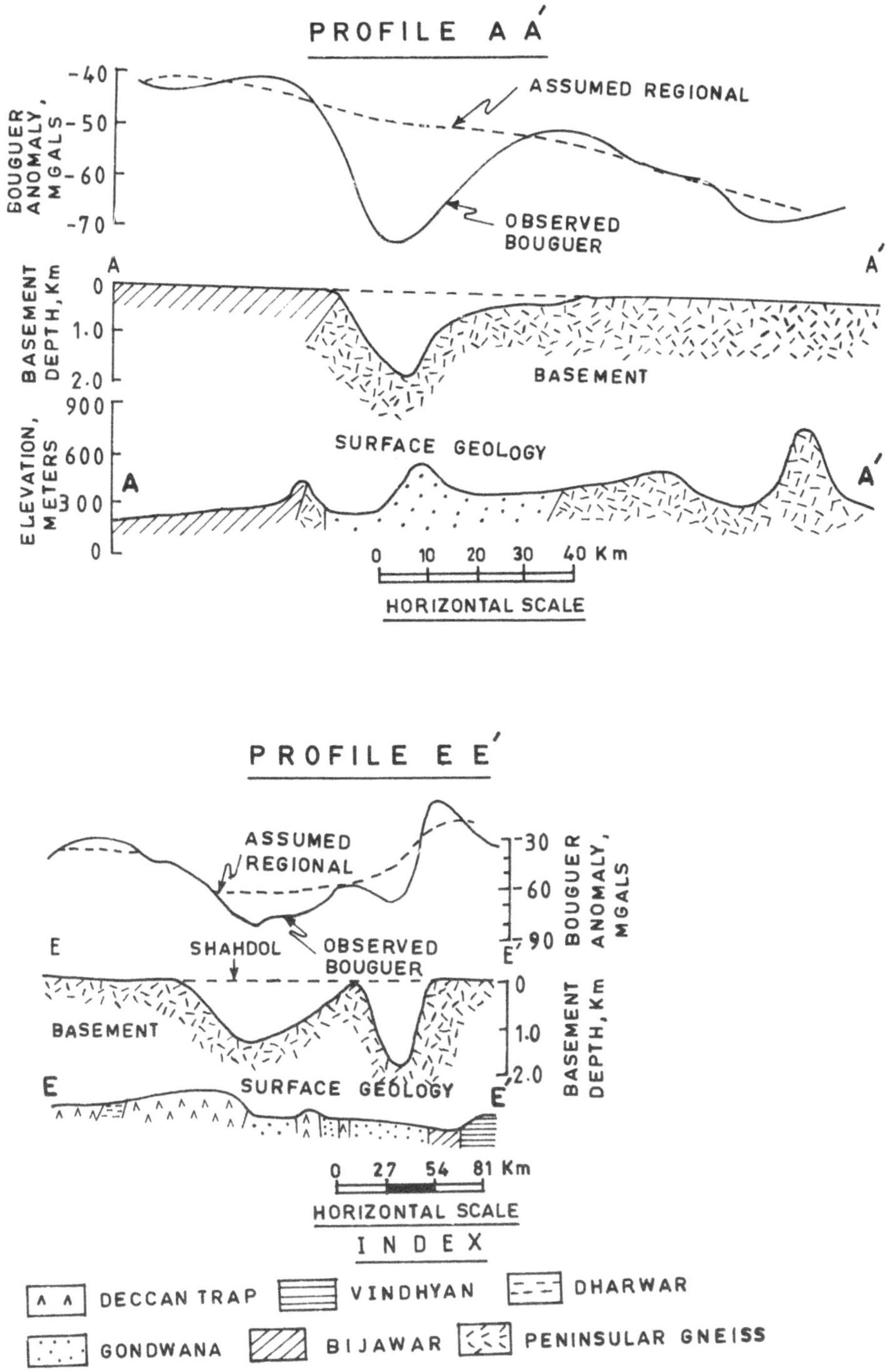

Fig. 4.7. Interpretation of residual gravity anomaly along profiles AA' and EE' across Mahanadi and South Rewa basins, in terms of basement structure (after Subba Rao, 1977).

Gondwanas are steeply dipping against the Older Precambrian formations in the southwestern part, reaching a maximum thickness of about 1.8 km along the profile. Also profile EE′ (Figure 4.7) shows that Gondwanas occur in two separate basins in the northwestern part of the Mahanadi Valley, one near Barach and the other near Shahdol, reaching a maximum thickness of about 1.8 and 1.4 km respectively in the two basins. They are steeply dipping against the Bijawar formations north of Barach and may underlie part of the Deccan traps towards south.

4.7. Damodar Valley Coalfields

Detailed gravity studies over Raniganj, Jharia and Bokaro coalfields belonging to Damodar Valley have been carried out by Verma and Ghosh (1974, 1977) and Verma *et al.* (1976, 1980). Gravity anomaly map of the area with 5 mgal contour interval is shown in Figure 4.8. The figure shows that prominent gravity lows of the order of −15 to −25 mgal are associated with the basins. The results obtained in different coalfields are outlined below.

4.7.1. RANIGANJ COALFIELD

The Raniganj coalfield (No. 1 in Figure 4.8) is the eastern-most coalfield of the Damodar Valley and is located approximately between 23° 35′ N to 23° 50′ N and 86° 40′ E to 87° 15′ E. The Lower as well as Upper Gondwana rocks are fully represented in the coalfield. It is surrounded by Archaean metamorphics on the western, northern and the southern sides, while towards the east the Gondwana rocks merge with the sediments of the Bengal Basin.

The Bouguer anomaly map of the Raniganj coalfield shows that the central part of the basin is oval shaped. The deepest part of the basin is located few km southwest of Asansol (23° 40′ N, 87° E), and has a residual anomaly of −32 mgal (Verma *et al.*, 1979). The western part of the coalfield forms a separate basin, while the Gondwana sediments extend towards southeast up to 87° 25′ E, merging with the sediments of the Bengal basin. The southern boundary of the coalfield is associated with steep gravity gradients.

Verma *et al.* (1980) have prepared a residual Bouguer anomaly map of the coalfield and interpreted the residual gravity field in terms of a three dimensional model of the basin using a density contrast of −0.37 gm cm^{-3} between the sediments and the surrounding metamorphics. One of the interpreted profiles (AA′) located in the central part of the coalfield (see Figure 4.8 for location) is shown in Figure 4.9. The section shows that the thickness of sediments increases towards the central part of the basin and reaches a maximum value of 2.5 km. The coal seams belonging to the Raniganj formation dip gently towards the south from the northern side. The southern part of the coalfield is bounded by a steeply dipping fault, known as the Southern Boudnary Fault. It appears that faulting in the basement is mostly responsible for deposition of sediments in the basin.

4.7.2. JHARIA COALFIELD

The coalfield (No. II in Figure 4.8) is roughly sickle shaped, being about 19 km wide in N–S direction, 38 km in E–W direction. The sedimentary strata mainly consist of the

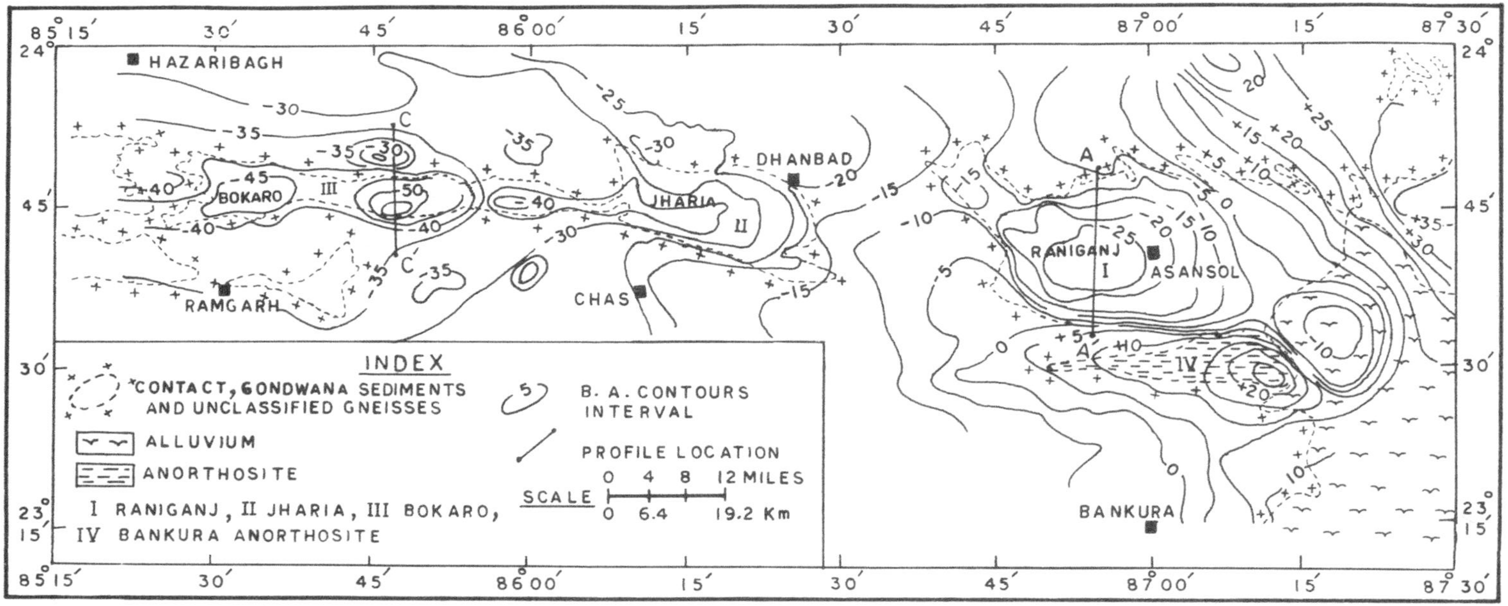

Fig. 4.8. Bouguer anomaly map of Raniganj, Jharia and Bokaro coalfields belonging to the Damodar Valley. It may be noted that all the coalfields are characterised by prominent gravity lows.

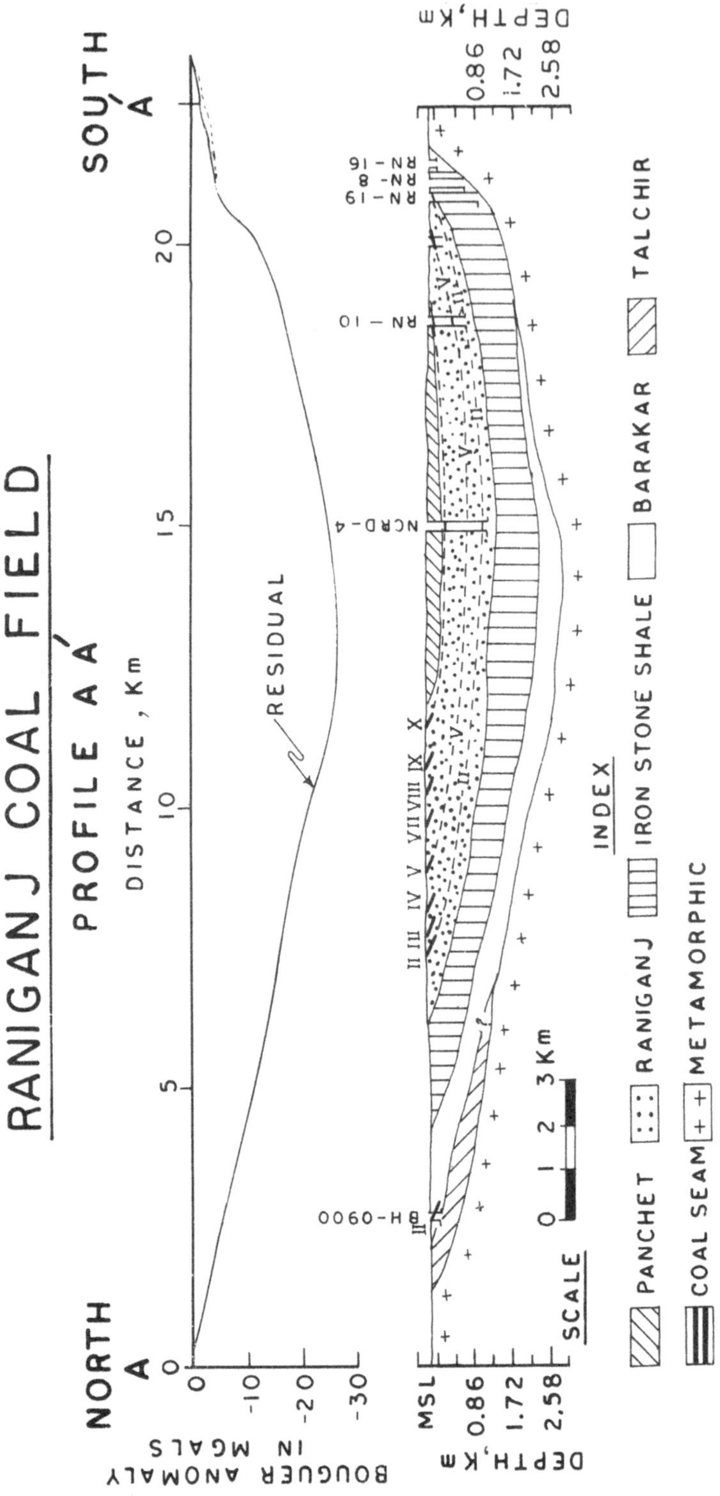

Fig. 4.9. Intepreted geological section across the central part of the Raniganj coalfield along profile AA′ (location shown in Figure 4.8).

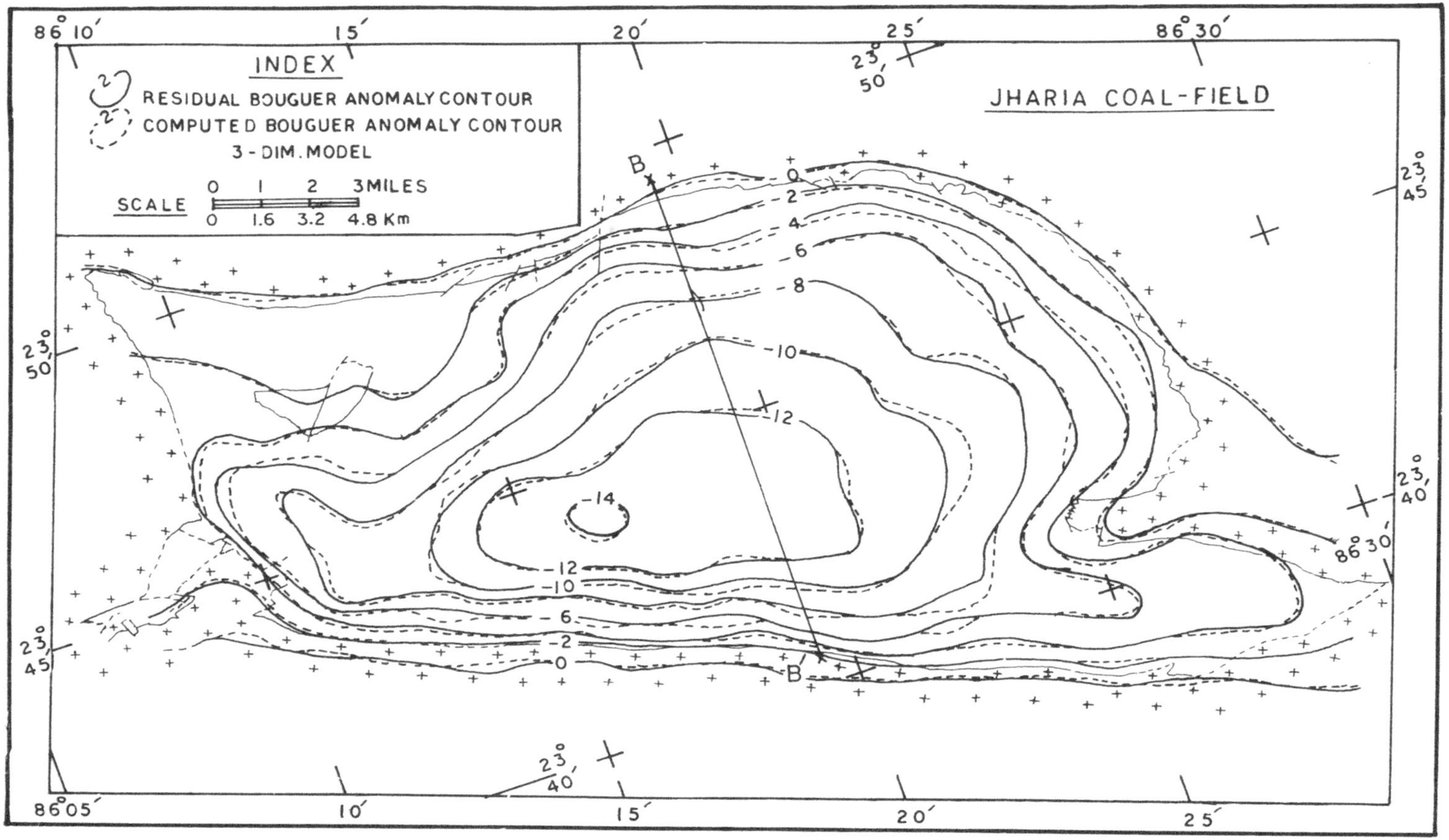

Fig. 4.10. Residual Bouguer anomaly map of the Jharia coalfield (after Verma *et al.*, 1979). The computed anomaly contours based on a three dimensional model are also shown in the figure. The gravity minimum of-14 mgal near Mohuda-Kumardih area suggests the location of the deepest part of the basin in the area.

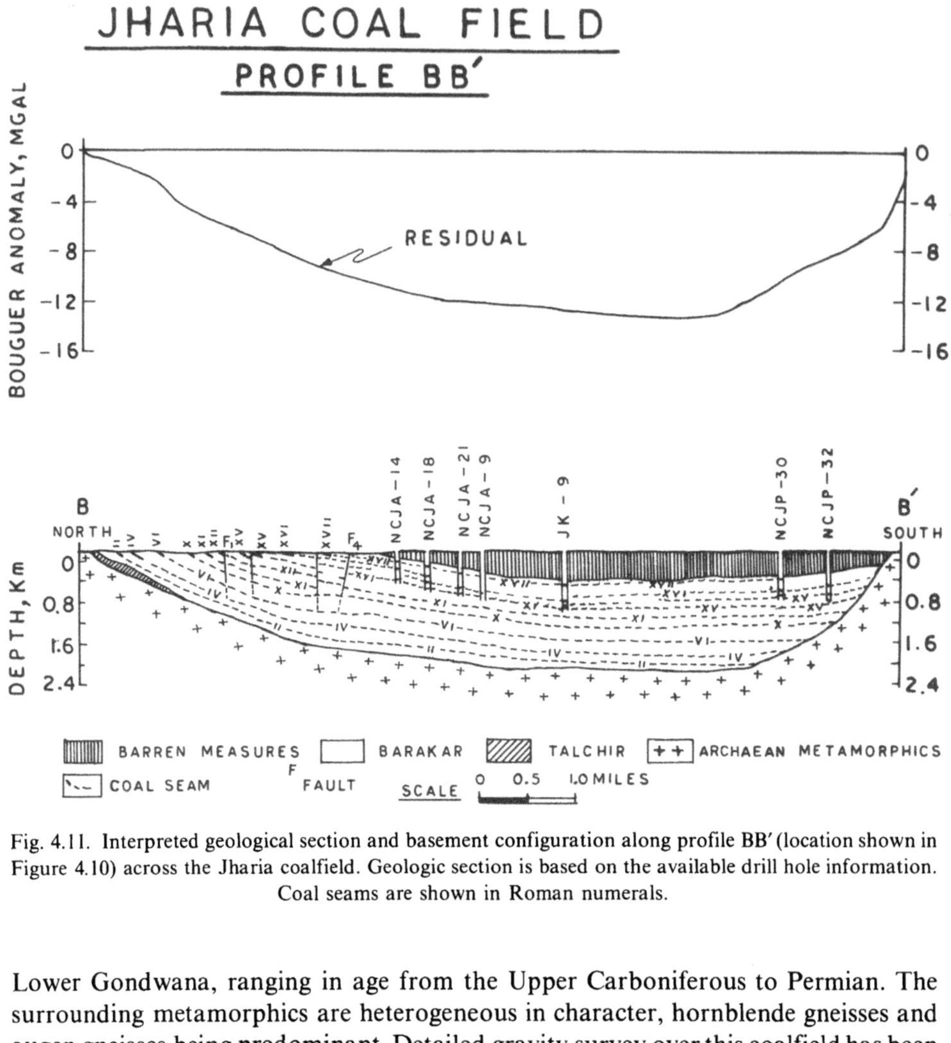

Fig. 4.11. Interpreted geological section and basement configuration along profile BB′ (location shown in Figure 4.10) across the Jharia coalfield. Geologic section is based on the available drill hole information. Coal seams are shown in Roman numerals.

Lower Gondwana, ranging in age from the Upper Carboniferous to Permian. The surrounding metamorphics are heterogeneous in character, hornblende gneisses and augen gneisses being predominant. Detailed gravity survey over this coalfield has been carried out by Verma and Ghosh (1974).

Figure 4.10 shows residual Bouguer anomaly map of the coalfield prepared by using graphical method (Verma *et al.*, 1979). The anomaly contours clearly outline the shape of the basin. Close contours on the south side suggest a faulted basement. A gravity minimum of −14 mgal near Mahuda-Kumardih area suggests the location of the deepest part of the basin in the area. The residual anomaly map has been interpreted in terms of a three dimensional model for the basin (Verma *et al.*, 1979). The model shows that the thickness of sediments increases from the margins towards the central part of the basin, reaching a maximum of about 2.2 km (below sea level) in the Mahuda-Kumardih area. Several geologic sections across the coalfield have been prepared based on the three dimensional model and the available drill hole information. Figure 4.11 shows an interpreted geologic section along profile BB′ (location shown in Figure 4.10) across the basin.

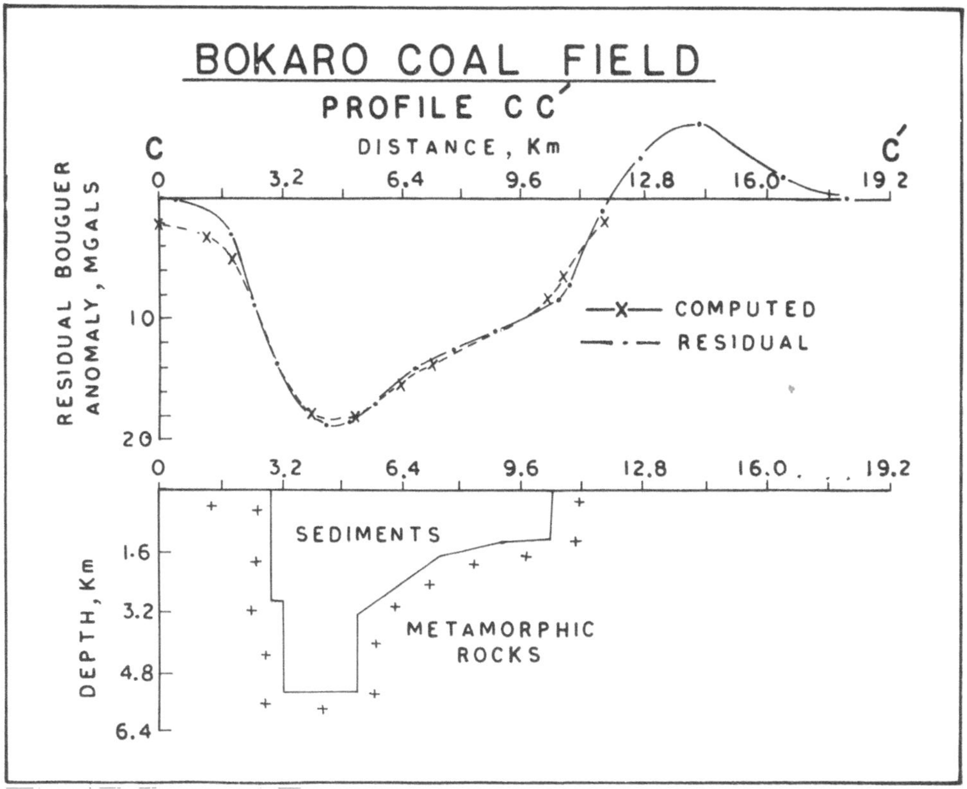

Fig. 4.12. Interpreted section along profile CC′ (location shown in Figure 4.8) across middle Bokaro low in Bokaro coalfield. The section shows that the coalfield is bounded by steeply dipping normal faults. It is a good example of a double walled graben.

The figure shows that the basement dips gently from north towards south and the basin is bounded by a steeply dipping fault at its southern margin. On the basis of gravity interpretation the Jharia coalfield appears to a be a single walled graben, faulting along the southern boundary having been mostly responsible for subsidence of the basement and the consequent deposition of sediments.

4.7.3. BOKARO COALFIELD

The coalfield (No. III in Figure 4.8) is situated between 23° 40′ to 23° 48′ N and 83° 15′ E to 84° 10′E and has an east-west strike. The Gondwana formations ranging in age from Upper Carboniferous to Upper Triassic are present in this coalfield. The surrounding metamorphics are mostly granites and gneisses.

Verma and Ghosh (1977) have discussed the results of gravity surveys carried out over this coalfield. Bouguer anomaly map over the area is shown in Figure 4.8. The coalfield is characterised by three gravity lows, of the order of −8, −19, and −10 mgal. Gravity anomaly along several different profiles has been interpreted on the basis of two-dimensional models. One of the interpreted profiles CC′ (location shown in Figure 4.8), across the Middle Bokaro low is shown in Figure 4.12. The figure shows that the

Bokaro coalfield is bounded by steeply dipping normal faults on its northern as well as southern sides. It appears to be a good example of a double walled graben.

The results of gravity studies over the Gondwana basins in India show that almost all the Gondwana basins are bounded by normal faults. Faulting appears to have played a major role in the deposition of sediments in the basins. Most of the Gondwana basins are double-walled graben structures, while some (such as Jharia coalfield) may be termed as single walled grabens. The Godavari Valley appears to have a rift valley structure.

CHAPTER 5

GRAVITY FIELD AND ITS RELATIONSHIP TO STRUCTURES IN DHARWAR AND BASTAR PROVINCES OF INDIA

5.1. Introduction

The Precambrian shield of South India over which crystalline rocks are found is a mosaic of three structural provinces over which a large variety of igneous and metamorphic rocks ranging in age from about 500 m.y. to 3000 m.y. are exposed. These provinces are: Dharwar, Bastar and the Eastern Ghats, whose regional distribution and extent is shown in Figure 5.1. Extensive geological investigations have shown that Karnataka and Bastar are essentially low-grade greenstone belts, dominated by granite intrusives, while the Eastern Ghats is a high grade metamorphic belt, distinct in its association with rock types of granulite facies regional metamorphism. Much earlier, Fermor (1936) classified these as non-charnockite and Charnockite regions respectively based on the absence of prevalence of charnockite type of rocks.

The appraisal of the individual gravity fields associated with the structural provinces under investigation in the southern Indian shield has been quite scanty till recently. Very few attempts have been made to interpret quantitatively the geological structures in these provinces. Each province exhibits its own distinct lithology, structural style, metamorphism and tectonics and as such, one would expect these characteristics to be reflected in their associated gravity fields. These investigations have been carried out for the Dharwar and Bastar provinces in the southern Indian shield by Subramanyam (1978a, b) and Subramanyam and Verma (1982). Some important results of these studies are outlined in this Chapter. The physiography, geology, gravity field and its interpretation for the two provinces under consideration, are presented in the forthcoming sections. Similar studies have been carried out for the Canadian shield (Tanner, 1969; Thomas and Tanner, 1975; Gibb and Thomas, 1976) as well as the African shield (Pretorius, 1974).

5.2. Physiography

Physiography of the southern shield areas (south of 16° N) is dominated by the Mysore Plateau, the Western Ghats, the Eastern Ghats and the southern hill masses. These two Ghats follow the strike of the West Coast and the East Coast of India respectively and merge in the southern tip, forming several hill masses, such as the Nilgiris, the Shevroys and Palni-Cardamom group of hills, the latter being separated from the former by the Palghat Gap. These hill masses and parts of the Eastern Ghats Belt are composed of charnockite suite of rocks. This physiographic aspect had formed a basis for Fermor (1936) to suggest that the charnockites region represents a relatively uplifted region in the southern Indian shield.

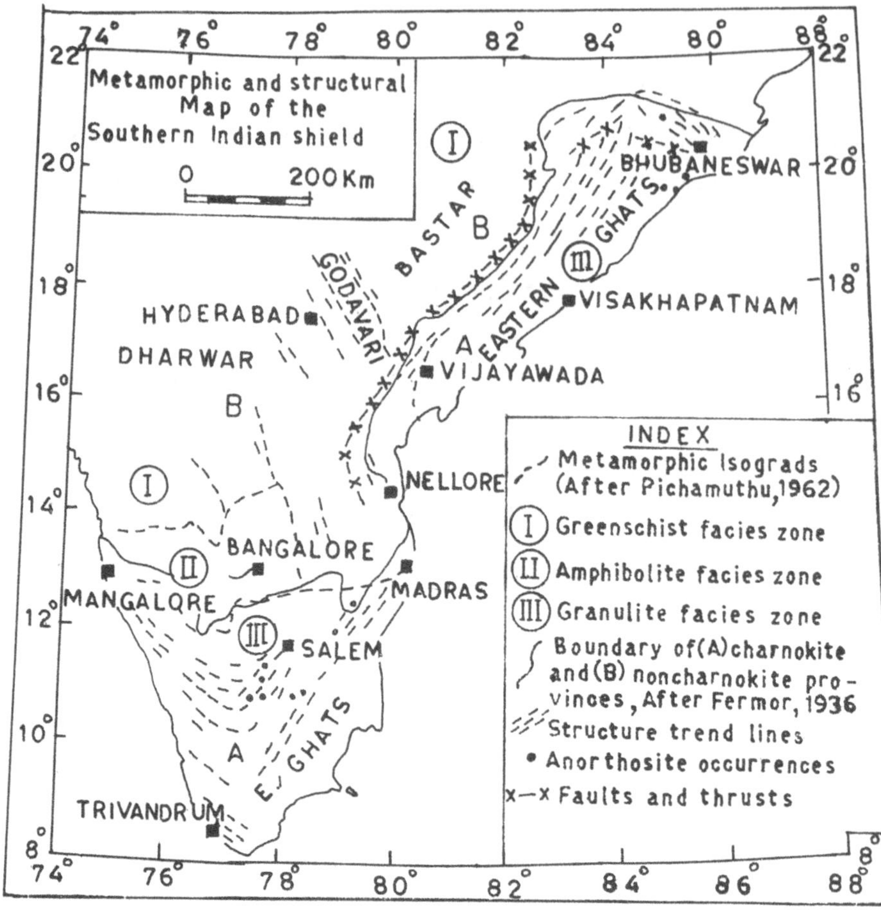

Fig. 5.1. Location of Karnataka, Bastar and Eastern Ghat provinces in southern Peninsular Shield of India. Approximate boundary between charnockite and non-charnockite provinces after Fermor (1936) is also shown.

5.3. Geology of Dharwar Province

This huge cratonic landmass is bounded on the north by the Deccan traps, on the north-east by the Godavari valley, to the east and south by the high grade Eastern Ghats belt, while it extends right up to the coast on the west. Geologically this is one of the best studied areas of the Indian shield. Figure 5.2 shows the geology of the area.

The craton has a wide distribution of greenstone belts (geologically known as the Dharwar schists) and intrusive granites which are engulfed in a sea of Peninsular Gneiss-Migmatite Complex. The greenstone belts are confined to three major groups of belts namely, the broad Dharwar-Shimoga-Bababudan on the west, the Gadag-Chitradurga in the center, and the scattered Sandur-Kolar-Ramgiri schist belts on the east. The three belts are believed to be separated by ridges or domal upwarps with granites forming core of the anticlinal structures. Pichamuthu (1962) opined that the Dharwars were originally deposited in a single basin and later granitisation processes and associated deformation have altered a substantial portion of the schists into the vast granite-gneiss complex, leaving behind large relics of Archaean Schists giving the impression of independent bands. Srinivasan and Sreenivas (1975) consider the possible evolution of the Dharwar formations in three separate geosynclines, viz., Dharwar-Shimoga, Gadag-Chitradurga and Kolar-Ramgiri-Hutti belts. Swaminath *et al.* (1976) divide the Karnataka craton into the western and eastern geotectonic blocks, with Closepet granite occupying the geosuture zone between the two blocks. According to Narayanaswami (1975) the Closepet granite represents a major geanticlinal ridge running along the central part of the Dharwar Craton.

On the basis of all available stratigraphic, geotectonic and radiometric ages, Sarkar (1980) has suggested that the Dharwar complex of Karnataka is older than 3200 m.y., the Peninsular Gneiss complex was formed between 2950-2600 m.y., while the Dharwar Supergroup consisting of orthoquartzites, various types of schists, conglomerates, banded magnetite quartzites, phyllites, limestones, amphibolites, various types of lavas unconformably overlies the peninsular gneisses and is about 2600-2900 m.y. old. Closing of Dharwar orogeny is marked by intrusions of Chitradurga granite (2450 m.y.) and Closepet granite (2380-2000 m.y.).

5.4. Gravity Field of Dharwar

Bouguer anomaly map of Dharwar and its surrounding areas (after Subramanyam, 1978a) superimposed on geology is shown in Figure 5.2. It may be noted that the gravity anomalies are predominantly negative in the area, varying from −60 mgal in the north to about −130 mgal close to Kudremukh. Prominent gravity lows are found southeast of Belgaum (−100 mgal), southeast of Kudremukh (−130 mgal) and south of Closepet (−110 mgal). These and other lows of smaller magnitude appear to be associated with outcrops of granitic bodies or the gneisses. Greenstone belts in general are characterised by gravity highs. When the axes of the gravity highs and lows are joined, they are found to have a general NNW–SSE strike (same as that of the Dharwar) and extend for several hundred kilometres as shown in Figure 5.3. The highs and lows alternate and extend up to the margin of the Cuddapah basin in the east. Prominent highs and lows are shown in Figure 5.2. Over the Shimoga belt the gravity contours of low L1 show a tendency to continue southward and merge with those of anomaly L3. This implies that almost the entire western half of the Shimoga belt may be underlain by granitoid rocks and schists in this part may be relatively thin. Only in the eastern part, where the gravity field becomes more positive, the schists may attain considerable thickness.

It is of interest to interpret the gravity anomalies quantitatively in terms of thickness of the greenstone belts and the associated granites. For this purpose gravity anomalies along four profiles AA′, BB′, CC′ and DD′ (shown in Figure 5.2) across the schist belts are considered.

The gravity anomaly in any area is invariably influenced by the regional gravity field, caused by deep-seated structures or mass inhomogeneities. This field, has got to be filtered out in order to obtain the residual anomaly, which is mostly of interest to us, and has to be interpreted in terms of the underlying geological features. The gravity field when averaged over $1° \times 1°$ areas was found to be an effective filter in eliminating the

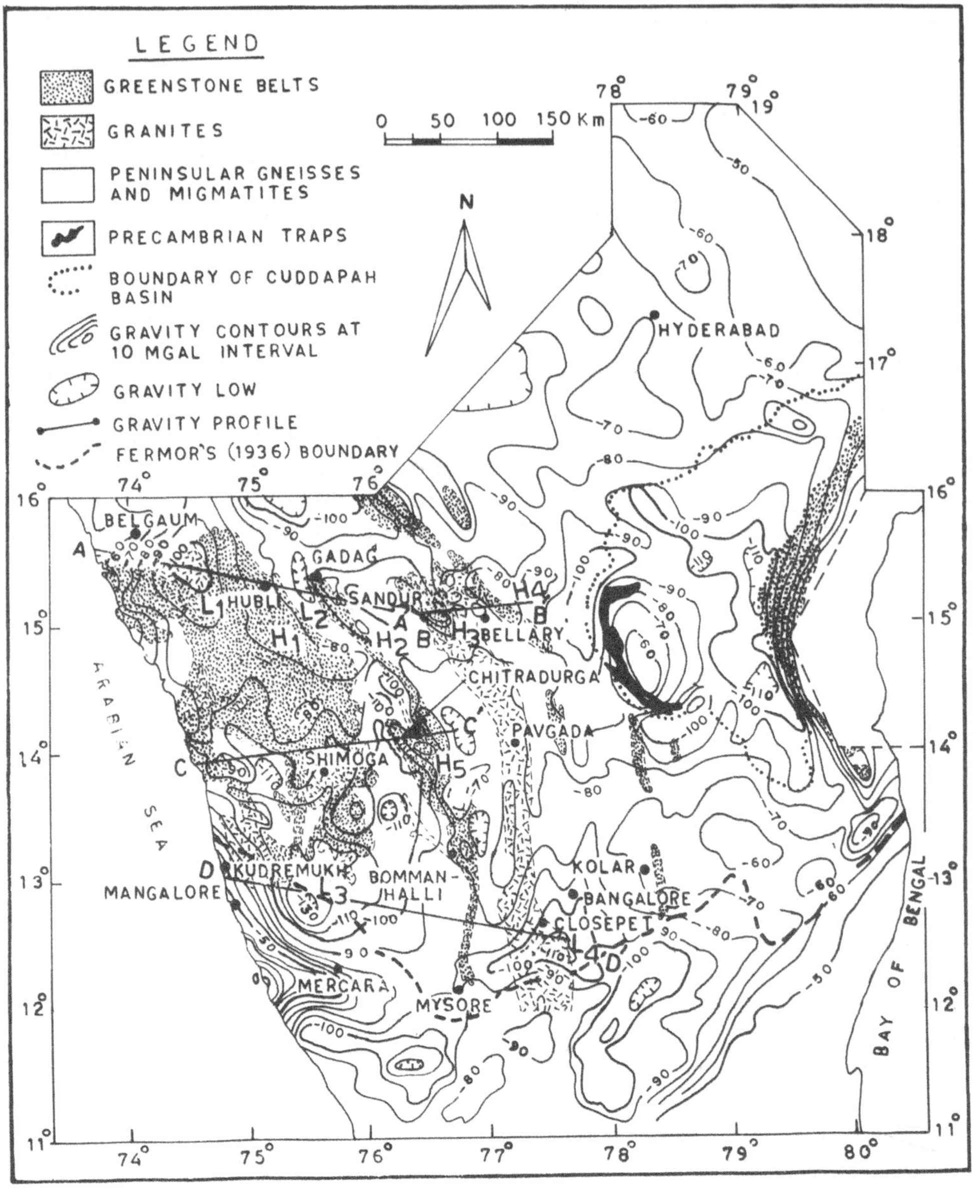

Fig. 5.2. Bouguer anomaly map of Karnataka province (after NGRI) superimposed on geology (after GSI). Prominent highs and lows are marked on the map. Locations of profiles AA′, BB′, CC′, and DD′ interpreted in terms of two-dimensional models are also shown.

local anomialies due to near surface geological features (Subrahmanyam and Verma, 1982). This field was taken as a guide for the regional. However, this was modified at few places on account of geological considerations. Figures 5.4 and 5.5 (after Subramanyam, 1978a) show the regional and residual anomalies along profiles AA′ to DD′ across the various schist belts and adjoining granite-gneiss terrain. The

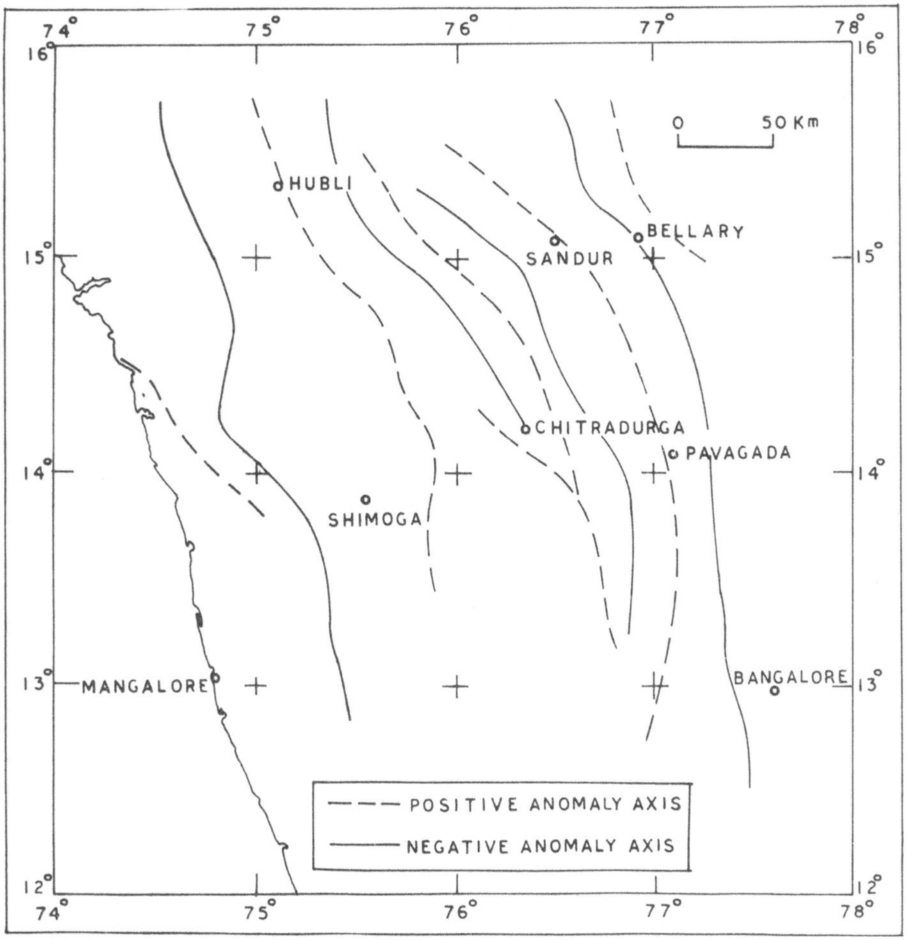

Fig. 5.3. Major trends of gravity highs (positive anomaly axes) lows (negative anomaly axes) observed over the Dharwar province.

residual anomaly so obtained has been intepreted in terms of two dimensional models, by the polygonal method (Talwani *et al.*, 1959) using a computer program. The density values for the greenstone belts, granites, gneisses and migmatites used in the model are given in Table 5.1 and also shown in Figures 5.3 and 5.4.

5.4.1. PROFILES ACROSS SHIMOGA AND CHITRADURGA BELTS

5.4.1.1. *Profile AA′*

The profile and its interpretation is shown in Figure 5.4a. It clearly brings out gravity low L1 few km southeast of Belgaum, a high H1 over the Shimoga greenschist belt close to Hubli, another low L2 southwest of Gadag and a high H2 over Chitradurga greenstone belt. The gravity low L1 can be explained in terms of an intrusive granite body which is about 2 km thick on the western side and about 6 m thick on the east with outward slope. The residual high H1 over Shimoga belt is 50 km wide and is about 8

TABLE 5.1

Densities of various rock types from Karnataka and Eastern Ghats

Rock type	No. of samples	Density values (gm cm^{-3})		
		Range	Mean	Std. Dev.
Schists and Gneisses				
Dharwar schist	82	2.60–3.35	2.84	0.20
Biotite gneiss	135	2.55–2.80	2.68	0.05
High grade gneiss	38	2.70–3.24	2.90	0.11
Migmatite	21	2.70–3.14	2.89	0.10
Khondalite	25	2.59–3.33	2.89	0.18
Basic Rocks				
Anorthosites	64	2.58–3.28	2.85	0.13
Dolerite	82	2.74–3.19	3.03	0.07
Acid Charnockite	140	2.57–2.89	2.71	0.07
Intermediate Charnockite	114	2.66–3.15	2.78	0.08
Basic Charnockite	124	2.67–3.41	2.99	0.15
Ultrabasic Charnockite	24	2.84–3.36	3.12	0.13
Acidic Rocks				
Granites	371	2.54–2.83	2.65	0.04
Combined				
Gneisses and granites (undivided)	544	–	2.67	–
Gneisses, granites and Dharwar Schists	735	–	2.75	–
Granulite Facies rocks including Charnockites	550	–	2.85	–

mgal in amplitude. The interpreted model suggests a broad basin-type structure, about 3 km deep for the Shimoga greenstone belt. The low L2 can be interpreted in terms of a concealed granite, about 2 km thick and 20 km wide underlying the gneiss. The gravity high, H2 over the Chitradurga belt has an amplitude of about 18 mgal, and is interpreted in terms of a deep basin-type structure for the greenstone belt. In contrast to the model for the Shimoga belt, which is broad and relatively shallow, the Chitradurga belt is narrow and of considerable thickness. The concealed granite between the two greenstone belts may represent the core of an anticlinal gneissic ridge that separates the Shimoga and Chitradurga belts.

5.4.1.2. *Profile CC′*

In order to examine the variations in the thickness of the Shimoga and Chitradurga belts from north to south, interpretation of gravity anomaly along profile CC′ in the southern part of Shimoga belt has been also studied. The residual anomaly along the profile is shown in Figure 5.4b. The profile shows a gravity high of small magnitude associated with the Shimoga greenstone belt and a prominent high H5 over the Chitradurga belt close to Chitradurga. The model constructed suggests a thin sheet-

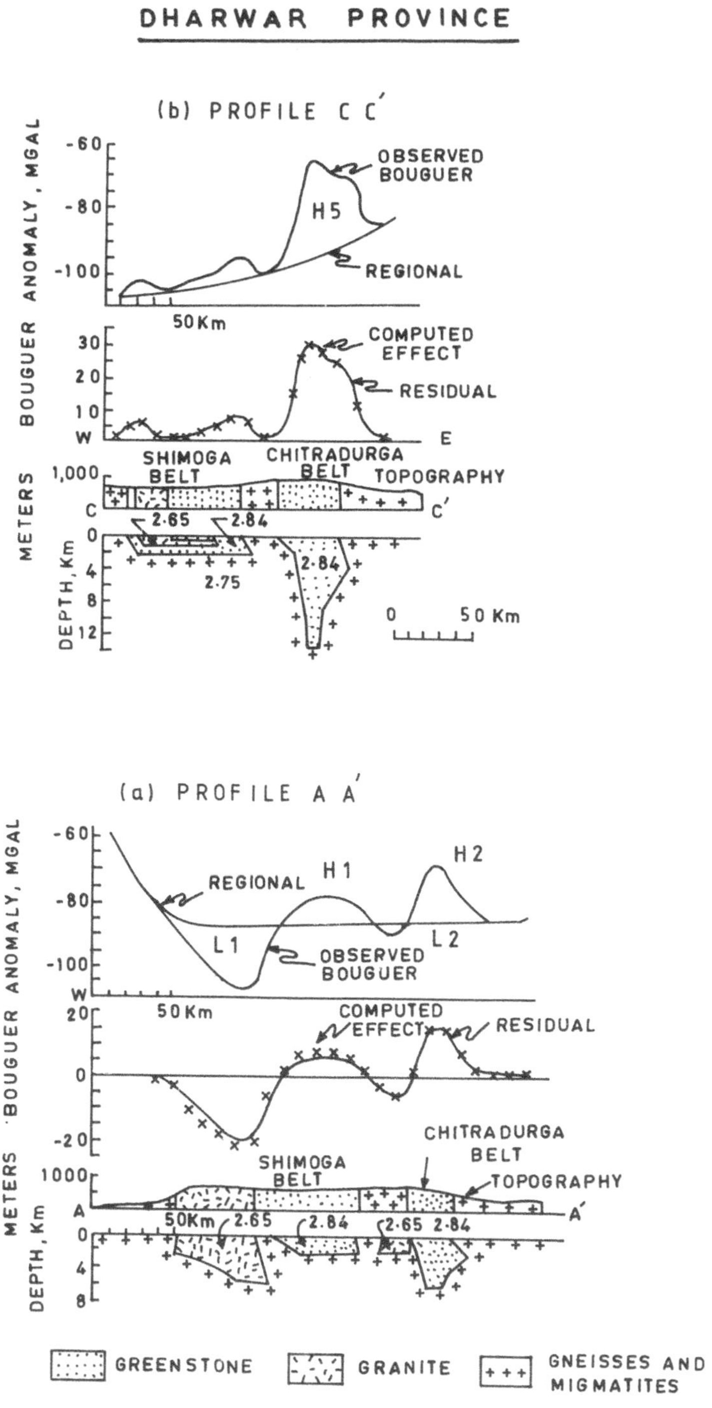

Fig. 5.4. Profiles AA′ and CC′ across Shimoga and Chitradurga schist belts and their interpretation in terms of geological configuration of the schist belts.

like granite body, less than 2 km in thickness emplaced within the Shimoga schist belt, which is only 3 km thick but 90 km wide. The contacts of the schist belt slope eastward with steep dips. The configuration of the Chitradurga belt is again in marked contrast to that of the Shimoga belt as observed along profile AA′. This is nearly 13 km thick in this area.

5.4.2. SANDUR BELT AND THE ADJOINING AREAS

Near Sandur, Bellary and further north, few isolated patches of greenstone schist belts belonging to the Sandur-Kolar group are noticeable. These are all associated with gravity highs. The Sandur anomaly is boat shaped, similar to the outcrop of the schist belt. To the east of the Sandur belt, a granite outcrop is associated with a mild negative anomaly. Interpretation of residual gravity anomaly along profile BB′ passing through the two schist belts east of Sandur and gravity highs H3 and H4 is shown in Figure 5.5a. The interpretation suggests deep trough-like structure for the two belts, the troughs being nearly 12 and 8 km thick east of Sandur and Bellary respectively. The granite body between them is nearly 2 km thick. The configuration of these belts is very

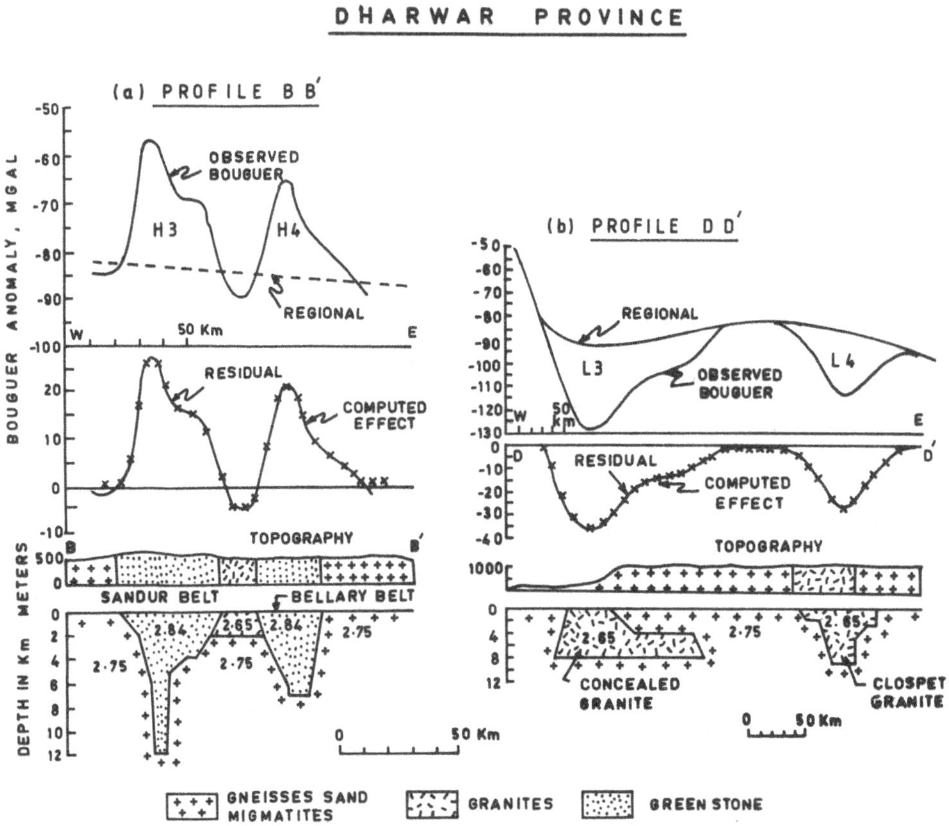

Fig. 5.5. Profiles BB′ and DD′ across Sandur Schist belt and Closepet granite respectively and their interpretation. The figure shows the configuration of the Sandur and Bellary schist belts and that of the Closepet granite as interpreted from gravity profiles.

similar to that of the Chitradurga belt discussed earlier, and is in marked contrast to that of the Shimoga belt.

5.4.3. PROFILE ACROSS CLOSEPET GRANITE

The meridional Closepet granite has a linear extent of over 300 km from south of Closepet, through Pavagada up to Sandur on the north. It has an average width of about 30km. From Figure 5.2 it may be seen that both positive and negative gravity anomalies characterise parts of this granite batholith. A prominent negative anomaly, having a magnitude of about −30 mgal and indicated by a gravity closure of −110 mgal (L4) has been observed in the southern part near closepet.

This anomaly is almost triangular in shape and appears to be the only well defined anomally associated with the Closepet granite. LANDSAT Imagery data interpreted by Sreenivasan and Sreenivas (1977) suggest that the Closepet granite is not a single batholith, but a composite of isolated patches of granite outcrops with intervening gneissic country. It may be noted that a positive anomaly represented by −70 mgal contour occurs along the western margin of the outcrop.

A quantitative interpretation of the major gravity anomaly associated with the Closepet granite (L4) is shown in Figure 5.5.b, along with profile DD′ which also cuts across the anomaly L3. Residual gravity shows two prominent gravity lows, L3 and L4 of magnitude −35 and −30 mgal respectively. Both the anomalies have been modelled in terms of low-density granite bodies. The anomaly associated with L3 is concealed under the gneissic cover, has a maximum thickness of 8 km and is dome shaped. The Closepet granite has a maximum thickness of 9 km and appears to be funnel-shaped.

5.5. Geological Implications of the Models

Interpretation of gravity anomaly along four different profiles across various Dharwar Schist belts has thrown some light on the geological structure of the Dharwar formations. It has been found that the Shimoga belt represents a broad and a shallow basin which is grossly different from the Chitradurga and Sandur group of schist belts, which are narrow and have considerable depth extends. It is possible that the different schist belts represent different depositional environments as suggested by Swaminath *et al.* (1976). Perhaps, some were deposited in shallow basins, while others represent fracture-based repositories developed in linear gashes. These ideas are supported by the inferred gravity models for Shimoga, Chitradurga and Sandur basins. The linear extent of the Chitradurga belt and the associated gravity field suggests that it may represent a prominent rift zone within the Dharwar Craton.

Granite emplacement in the areas separating the main greenstone synclinal structure, and in the southern parts near Kudremukh and Closepet is an important aspect of the gravity interpretation. In the latter case, inferred granite emplacement is essentially under the gneissic terrain, perhaps formed due to metamorphism under regional amphibolite facies environments. Gibb (1968) and Gibb and Halliday (1974), based on Miyashiro's (1961) arguments, have suggested that the amphibolite facies regional metamorphism largely accommodates considerable movement of volatiles and this results in granite intrusion on a larger scale. In the light of these arguments, inference of large granite bodies, about 8 to 10 km in thickness on the basis of gravity data seems to be justified.

5.6. Gravity Field over Bastar Province

The Bastar province extends from about 18° N to about 22° N and 79° E to 84° E. It is bounded by the Eastern Ghats on the east, the Mahanadi valley graben on the north, the Godavari valley graben on the south, and Deccan basalts on the west. The geology of the area is still not very well known, mostly on account of inaccessibility of the area. Before discussing the gravity field in Bastar province and its interpretation, the geology of the area is briefly outlined here.

5.6.1. GEOLOY OF THE AREA

The area is underlain by a widespread distribution of Archaean schistose rocks, granitic gneisses, granite, basic and acid effusives and areas of extensive sedimentary deposition and has been studied by several officers of the Geological Survey of India. Fermor (1940) gave a detailed stratigraphy of the Sausar and Sakoli Series which consist of various types of meta-sediments including quartzites, phyllites, gneisses, schists and metavolcanics. The Sausar Series consist of quartzites, various types of gneisses, schists, dolomites and marble. In the central part of the area a massive granite batholith, called the Dongargarh granite, is recognisable with occurrences of acidic and basic effusives along its northern and eastern margins. The rocks belonging to the Sakoli Series occur immediately to the west of the Dongargarh granite and the Sausar Series further towards its northwest.

Crookshank (1963) classified the Archaean schistose rocks of Bastar into Sukma, Bengpal and Bailadila Series, which are intruded by granitic and basic intrusives. Granitic gneisses are the most widespread formation in the area and appear to have formed the basement of all sedimentary deposition in the craton. Scattered occurrences of charnockites have been reported by Ghosh (1941) in the eastern part of the area. A large portion of the craton is covered by sediments of Precambrian age, considered to be the equivalent of Cuddapah. The Chattisgarh basin in the north and Bastar basin in the south mark the sites of these depositions.

A notable difference between the Dharwar and Bastar areas is in regard to the distribution of the greenstone belts. Although these belts are abundantly developed in Karnataka (Dharwar), they are limited in extent in Bastar.

5.6.2. BOUGUER ANOMALY MAP OF BASTAR AREA

Bouguer anomaly map of the area (after Subramanyam, 1978a) superimposed on geology is shown in Figure 5.6. The anomalies range from −15 mgal on the north-east to −85 mgal on the south-eastern part of the area. The gravity field is again high on the north-western part of the area. General level of Bouguer anomaly is −50 mgal as shown by the predominance of this contour in the entire area. The Bouguer anomaly map is characterised by several gravity highs as well as lows which bear a reasonably good correlation to the underlying geology.

Prominent highs (H1 and H2) are located over the outcrops of greenstone belts belonging to the Sausar and Sakoli Series. The high H3 underlies granites and rhyolites and is probably caused by basic igneous rocks underlying them. At few places highs (H5 and H6) are located over Chhatisgarh basin, which appear to be related to its underlying structure. A prominent gravity low (L1) is located over Dongargarh

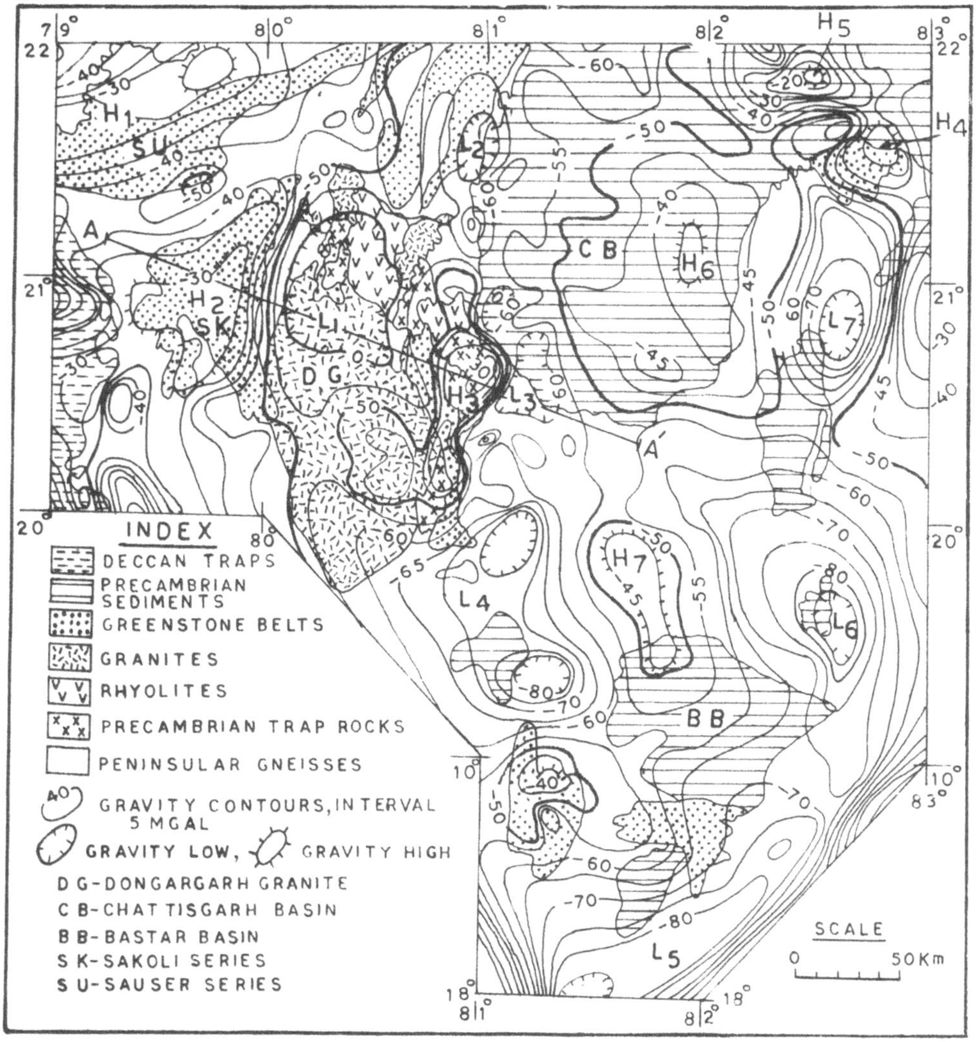

Fig. 5.6. Bouguer anomaly map of Bastar province (after Subramanyam 1978a) superimposed on underlying geology.

granite. Several other lows (L2, L3, L4) are possibly caused by granitic intrusions associated with greenstone belts or gneissic terrain.

The anomalies L5, L6, and L7 form a more or less continuous belt along the eastern part of the Bastar province. The gravity low reaches a minimum value of −85 mgal in the area. The lows are mostly underlain by the gneissic rocks and there seems to be no simple explanation for these based on surface geology. The close proximity of the anomalies to the boundary between the Bastar province and the Coastal Granulite Terrain (CGT) of the Eastern Ghats (Chapter 6) suggests that the source of these anomalies could be something quite different from a mere surface geological expression. Similar belt of negative anomalies is found to be associated with the

Dharwar Craton and Coastal Granulite Terrain (Eastern Ghats). This low anomaly belt is most likely related to the tectonics associated with the boundary between the Eastern Ghats belt and the Dharwar/Baster provinces. A detailed interpretation for this anomaly in terms of crustal composition will be considered in Chapter 6, where the gravity anomalies associated with boundaries of structural provinces in the southern Indian shield are discussed.

Profile AA′

Quantitative interpretation of anomalies along profile AA′ passing through the high H2, H3 and lows L1 and L3 in terms of two dimensional models has been attempted (Subramanyam, 1978a) and is shown in Figure 5.7. Graphical regional has been drawn on the basis of geological considerations. Density values used for the greenstone belts are the same as those for Dharwar province (See Table 5.1). Gravity high H2 is explained in terms of a greenstone belt, about 50 km wide and 5 km thick, developed in the form of a synclinal structure. The anomaly H3 on the east is explained by a high density basic intrusive body underlying the trap rocks. Gravity low L1 can be explained by the Dongargarh granite, which is about 5 km thick, below sea level. Further east, the low L3 can be explained by a concealed granite underlying the gneisses and migmatites. It may be recalled here that in Dharawar Craton also a number of concealed granitic bodies have been invoked in order to explain several gravity lows found over the gneissic rocks.

5.7. Nature of Upper Crust in the Archaean Terrain of South India

There are broadly two types of Archaean terrains exposed today (Windley and Bridgwater, 1971), those dominated by high grade gneisses and migmatites, and those by low grade greenstone belt and associated younger granites. For many years the popular opinion has been that the oldest continental rocks are the volcanics and sediments belonging to the greenstone belts. However, recent radiometric age determinations show that the oldest rocks in most of the continents are the high grade metamorphic gneisses and that most of the greenstone belts are comparatively young.

Windley and Bridgwater (1971) have interpreted the greenstone belt/granitic terrain as having formed at high crustal levels, whereas the granulite/gneiss terrain formed at relatively deeper levels. Archaean granulities have relatively low contents of H_2O, K, U, and Th compared to the amphibolite facies rocks. According to Windley (1973), if the model of Fyfe (1970) were applied to the developments of the Archaean crust, the granulites would be regarded as a lower crustal level residue from which granites were derived. These migrated upward to form borders of the overlying greenstone belts.

Windley (1973) has proposed an interesting model for the development of greenstone belts in Precambrican terrains. He envisages the following four stages in the development of these bodies.

(1) Formation of high grade gneissic terrain
(2) Formation of rift valleys, resulting in the deposition of sediments and volcanic eruptions.

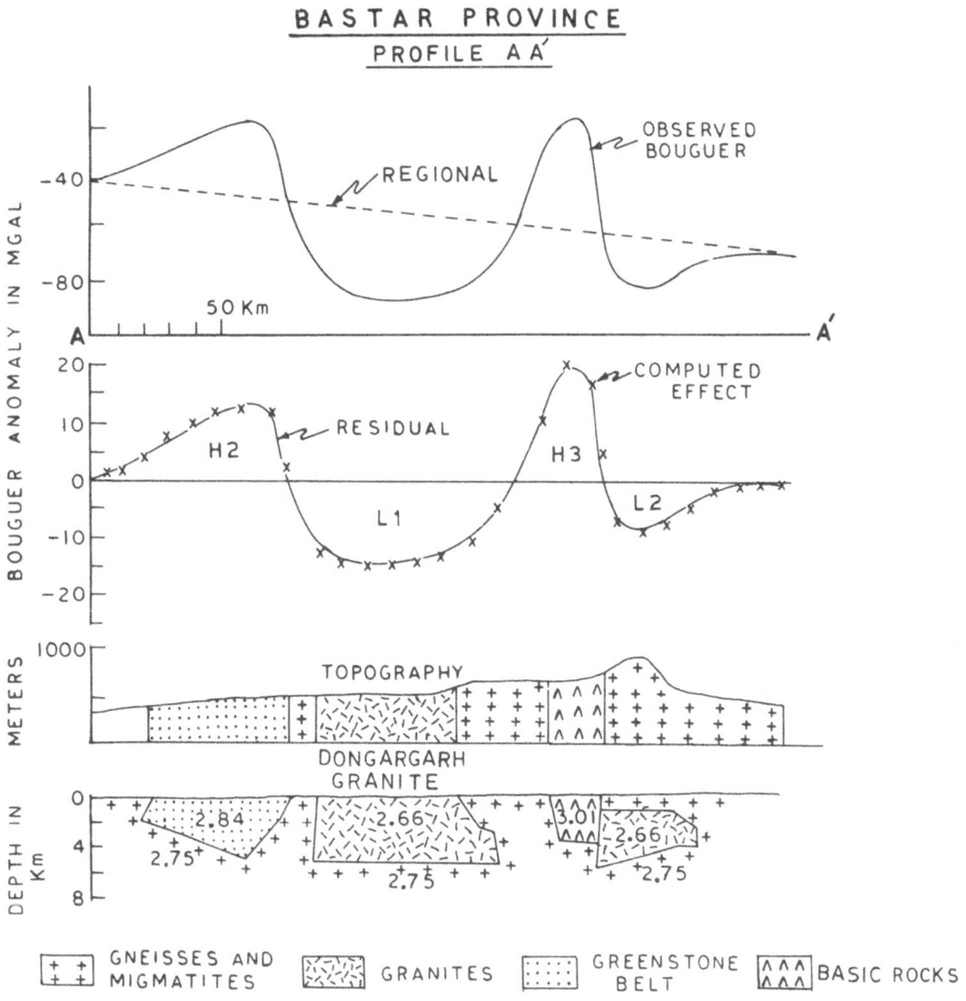

Fig. 5.7. Interpretation of profile AA′ (location shown in Figure 5.6) across Baster province, showing the possible configuration of the underlying geological masses including the Dongargarh granite.

(3) Increase in thickness of sediments and formation of granites at depth, folding and week metamorphism.

(4) Large scale folding, large scale granitic intrusions in the belt, and regional metamorphism.

The Dharwar greenstone-granite terrain is broadly comparable to similar terrains in shield areas of Australia, South Africa and Canada. The geophysical characteristics of these low grade terrains also bear some similarities (Mathur, 1974; Gibb and Halliday, 1974; Reeves and Hutchins, 1975). Usually the schist belts have synclinal structures, which give rise to well defined gravity highs and appear to be considerably deformed due to younger granite intrusives. The thickness of schist belts as estimated geophysically is of the order of 5 to 10 km.

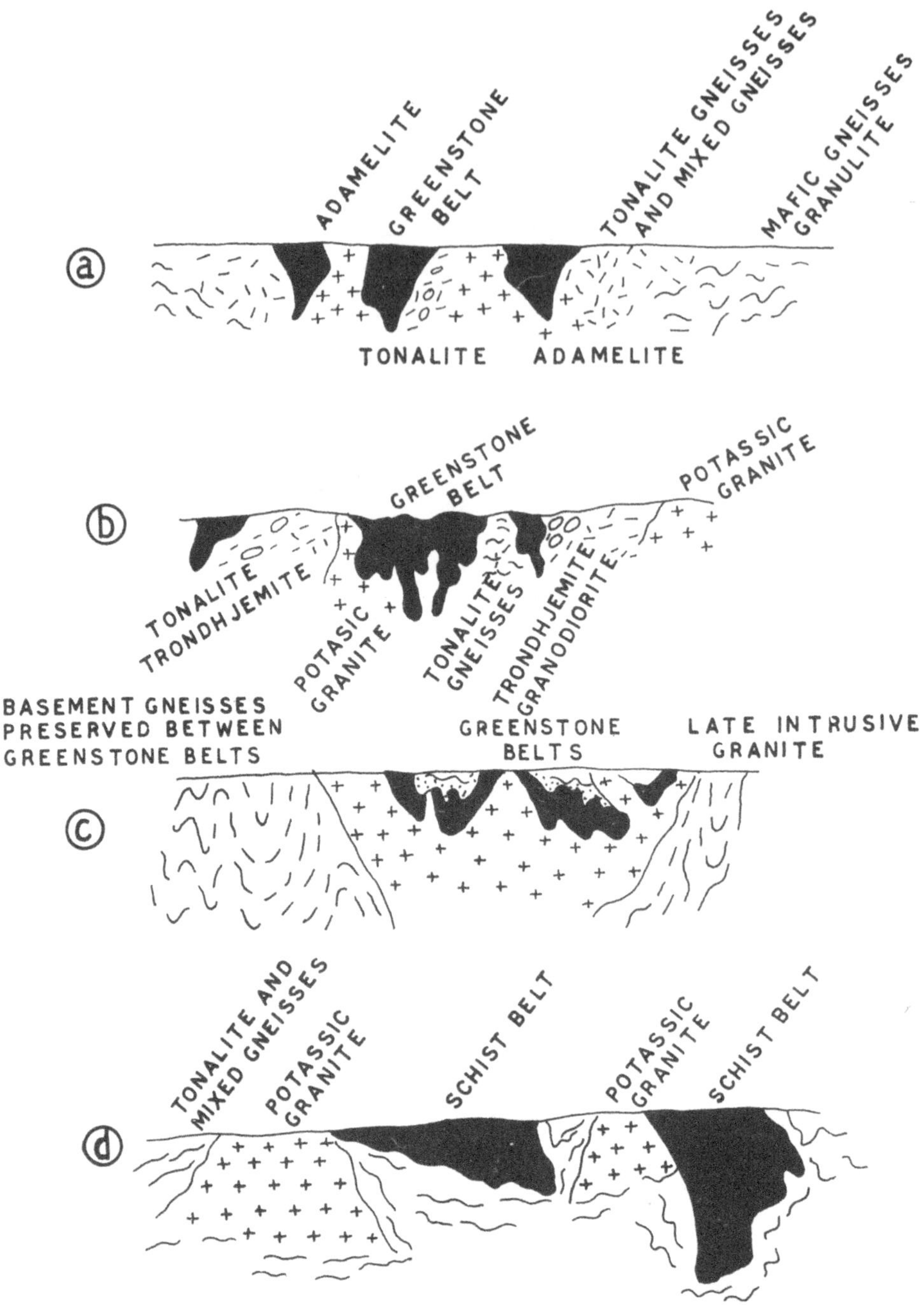

Fig. 5.8. Models for the evolution of the greenstone belts according to (a) Gibb and Halliday (1974) for Canada, (b) Anhaeusser (1973) for S. Africa, (c) Windley (1973) for Precambrian shields and (d) Subramanyam (1978a) for Dharwar Schist belts.

Different workers have presented models for the developments of greenstone belts in Canada (Gibb and Halliday, 1974), Africa (Anhaeusser, 1973) and Australia (Windley, 1973). These are shown in Figure 5.8 and have much similarity to each other. A common feature of all the models is the presence of large scale granitic intrusions in all the belts.

The Dharwar province is essentially a terrain of low grade regional metamorphism, ranging from greenschist to amphibolite facies, over which greenstone belts and granties occur in abundance. Radiometric ages for various rocks in the shield have shown that the gneissic rocks are older than the greenstone belts and the intrusive granites are younger than these. In view of the results obtained from the present gravity studies, a schematic model for the greenstone belts in the Dharwar Craton of India is shown in Figure 5.8d. The model can explain the gravity high obtained over the greenstone belts and lows over the granitic bodies. It is somewhat similar to the models prepared by other workers for the greenstone belts of Canada, Africa, and Australia.

CHAPTER 6

GRAVITY STUDIES IN THE EASTERN GHATS BELT

6.1. Geology of the Area

This belt has a great regional extent, nearly 1800 km long, and runs mostly along the East Coast of India, from Bhubaneswar on the northeast, through the southern parts of the peninsula, following the coastal regions of Andhra Pradesh and extending right up to the west coast, passing through Madras city, Nilgiri Hills, Mangalore and Cape Comorin. The charnockites together with the khondalite suite of rocks constitute the most important rock types in this belt. Figure 5.1 shows the location of this belt and its relationship to adjoining Dharwar and Bastar provinces.

The charnockite rocks consist mostly of hupersthene-diopside-hornblende gneisses and migmatites. They show a combination of characters of igneous and metamorphic rocks. As early as 1900, Holland described the charnockite series of rocks from the Eastern Ghats terrain and classified the charnockites into acid, intermediate, basic and ultrabasic types, depending upon their silica content. These were ascribed an igneous plutonic origin by a process of magmatic differentiation. Half a century later, Pichamuthu (1953) put forward his concept of charnockitisation of Peninsular Gneisses by metasomatic processes.

The charnockite suite of rocks are believed to be overlain by khodalites which consist of garnet–sillimanite–graphite–quartz schists and gneisses (Narayanaswami, 1975). Fermor (1936-40) has shown that khodalites were metamorphosed under deep seated katamorphic conditions but owe their present position to regional uplift. He has contended that the western margin of the Eastern Ghats must be a faulted one.

The above association to charnockites and khondalites seem to indicate a distinct eugeosynclinal sequence of meta-volcanics (charnockite suite) and meta-sediments (khondalites) which have been highly folded and migmatised. The assemblage of charnockites and khondalites is distinctly different from that of the schistose rocks of Dharwar region.

The charnockite-khondalite suite of rocks are engulfed in a sea of granulite facies gneisses and migmatites, which form the next most important lithologic unit in the Eastern Ghats Belt. Narayanaswami (1970) has suggested their origin to be a result of extensive granitisation and migmatisation. Although these high grade gneisses are clubbed with gneisses of the Dharwar region and commonly called Peninsular gneisses, the high grade metamorphism of the former distinguishes them from the latter.

In addition to the charnockites, khondalites and the gneisses, limited occurrences of several other rock types such as mafic granulites, granites, eclogites, kodurites, nepheline syenites, carbonatites and anorthosites have been reported from the Eastern Ghats Belt. The most significant of these are probably the massive and layered anorthose–gabbro complexes, distributed along the entire length of the Eastern Ghats bearing similarity with other high grade terrains of the world. Some of these occurrences are shown in Figures 6.1. and 6.4.

The Eastern Ghats belt is a region of high grade metamorphism as evidenced by occurrences of charnockite, garnet and sillimanite. Reviewing the geologic history of the Eastern Ghats belt, Murthy *et al.* (1971) have arrived at the following conclusions regarding the evolution of this belt. The extensive occurrence of the sedimentary facies rocks viz khondalites (paragneisses) represent relics of sediments deposited in major geosycline during the Archaean. The sediments were then intruded by basic charnockites as sills and both were later folded and metamorphosed under granulite facies conditions during the Eastern Ghats orogeny, closing around 2600 m.y. (Sarkar, 1980). Subsequently the acid–intermediate charnockites were emplaced. These rocks represent magma formed at depth and crystallised under granulite facies condition. According to Sarkar (1980) the Eastern Ghats orogeny accompanied by second phase of metamorphism and granitic intrusions ended around 1600 m.y. It is apparent that within the Eastern Ghats belt, sedimentation, folding, metamorphism and igneous activity have taken place (as is typical of orogenic belts), during the various phases of the orogeny.

There is a structural discontinuity for at least 240 km in the Eastern Ghats belt between Madras and Nellore on the east coast, where rock types belonging to the Karnataka province intervene. This has led to a subdivision of the belt into southern and coastal sectors (Narayanaswami, 1970, 1975). On the basis of radiometric age determinations (Aswathanarayana, 1964, 1968; Crawford, 1969; Ramamurthy and Sadashivaiah, 1967) it has been found that the southern belt is much older (3100–2500 m.y.) in comparison to the coastal belt (1650–1300 m.y.). A subdivision of the Eastern Ghats into Southern Granulite Terrain (SGT) and Coastal Granulite Terrain (CGT) has been proposed (Subrahmanyam, 1978) and is retained for present purposes.

6.2. Gravity Field over Southern Granulite Terrain

Figure 6.1 shows Bouguer anomaly map of the Southern Granulite Terrain (after Subrahmanyam, 1978) superimposed on geology. It may be seen from the figure that the gravity field maintains a general north-easterly trend along the east coast and a northwesterly trend along the west coast. In the central part, the gravity field is appreciably negative, of the order of −60 to −120 mgal and is characterised by several gravity highs and lows. Some of the prominent highs and lows are discussed below. Density values for various rock types found in the Eastern Ghats belt are given in Table 5.1 (Chapter 5).

6.2.1. PROMINENT HIGHS AND LOWS

A prominent high H–1, in the northwestern part of the area lies between Mangalore and Cannanore. The anomaly is almost triangular in shape and bears a close association with the outcrop of charnockites. Detailed geological investigations of the area by Vidyadharan *et al.* (1977) indicate occurrences of high density rocks like gabbroic anorthosites, gabbros and pyroxene granulites (basic charnockites) with a synformal structure. The anomaly is probably caused by high density pyroxene

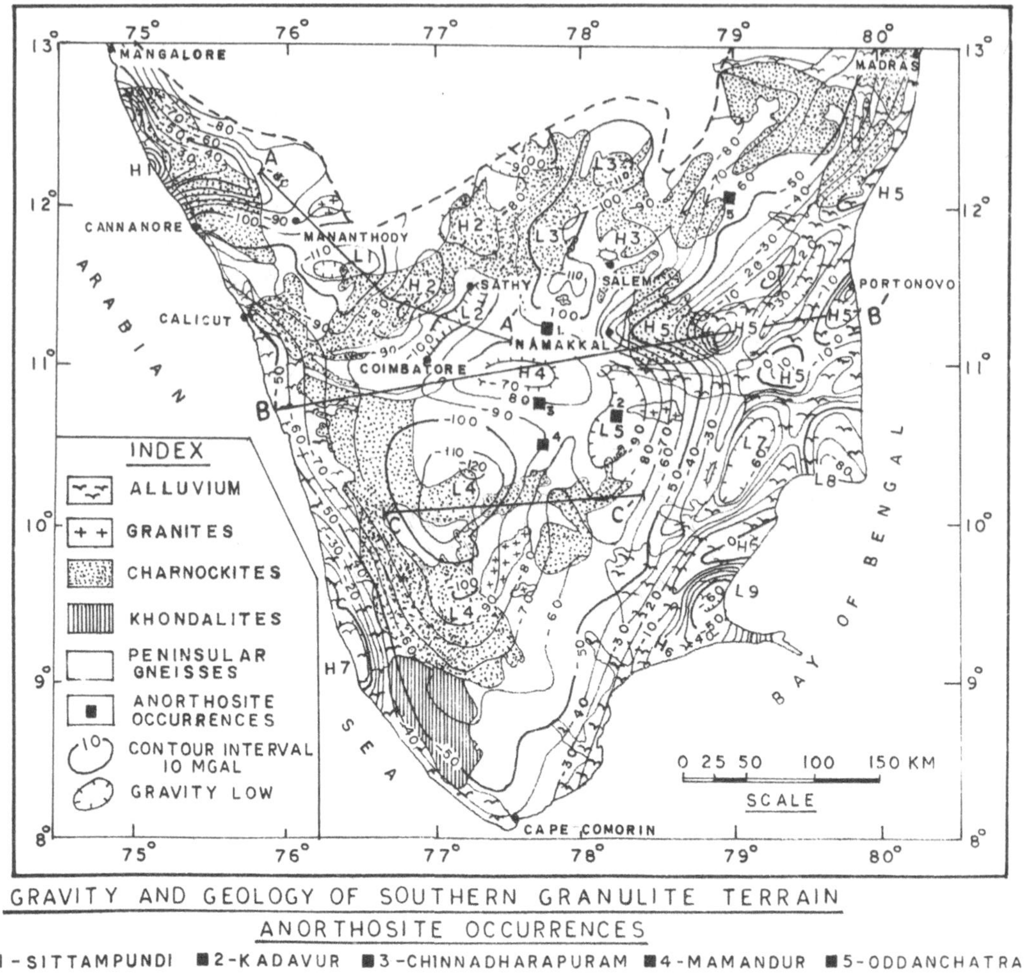

Fig. 6.1. Bouguer anomaly map of the Southern Granulite Terrain superimposed on geology. Locations of prominent highs and lows as well as profiles AA′, BB′ and CC′ is also shown in the figure.

granulite and gabbroic anorthosites filling up a syncline type structure in the area.

6.2.1.1. *Nilgiri Hills and Surroundings*

Anomaly H-2 is a prominent gravity high, marked by −70 to −80 mgal closures along the length of the Nilgiri charnockite massif in the north central part of the area. The anomaly characterizes the highly elevated topography of the Nilgiri hills, where elevations exceed 2.5 km. The behaviour of the anomaly is quite contrary to the general observation, i.e. Bouguer anomaly becomes more positive with increasing elevation. The maximum anomaly recorded is −70 mgal NW of Coimbatore.

Viswanathan and Tareen (1970) have carried out detailed sampling of Nilgiri charnockites on a profile passing through Ooty and reported extensive occurrences of pyroxene granulites. Narayanaswami (1970) has suggested that the Nilgiris represent a folded belt with its axis trending in a NE–SW direction, same as that of the positive anomaly associated with it. It is therefore reasonable to infer that the positive anomaly is a result of upthrusting of high density rocks like pyroxene granulites in a folded belt trending northeast.

Profile AA′ (Figure 6.2) across gravity lows L–1 and L–2 and high H–2 shows clearly the nature of the gravity field in this area. The Low L–1 has been interpreted by Qureshy *et al.* (1969) and Krishnabrahmam and Kanungo (1976) in terms of a huge concealed granite batholith which outcrops in a small area near Mananthody. The granite body appears to extend right up to the west coast, where the anomalies are strongly negative, reaching a peak value of −100 mgal.

Anomaly L–2 flanks the eastern margin of the Nilgiris and extends in a NE–SW direction between Coimbatore and Sattey. It is a linear belt, having a width of about 20 to 30 km and an extent of about 100 km. The anomaly is spread over a vast gneissic country rock and it is known that a suite of igneous rocks called the Sivamalai Syenite series outcrop in a cluster of small hillocks around Coimbatore (Pascoe, 1950). Density data reported for the syenites in this area suggests that their density is of the order of 2.66 gm cm^{-3} which is appreciably lower than the density of the gneisses surrounding it. In view of above consideration a quantitative interpretation of the residual gravity anomaly along profile AA′ in terms of two-dimensional model (Subrahmanyam, 1978) is shown in Figure 6.2. The model suggests that the Niligiri hill mass is underlain by charnockites (with pyroxene granulite composition) extending to a depth of about 5 km (below sea level). The figure also shows that the Niligiri high is about 40 km wide and has a peak amplitude of about 25 mgal. The granitic body exposed on the northwest, is about 4 km thick and is concealed under gneisses towards southeast. The anomaly L–2 is explained in terms of a concealed syenite body with its western contact being nearly vertical and the eastern contact having a steplike structure.

The gravity field in continuation of the high H–2 changes its character from high to low in northeasterly direction, shown in Figure 6.1. as L–3. The lowest anomaly in the area is −110 mgal and is found partly over charnockites and partly over high grade gneisses and migmatites, since the anomaly is associated with charnockite outcrops and there are no granite exposures in the area, the most probably explanation for the anomaly is in terms of acidic charnockites present in the area. These charnockites are quite similar in composition to granites and have an average density of about 2.71 gm cm^{-3}, which is appreciably lower than the average value of 2.85 gm cm^{-3} for the high grade gneisses.

6.2.1.2. *Palni-Cardamom Hills*

These hills form an important part of the southern granulite terrain. They are characterized by a prominent gravity low L–4 on the gravity map (Figure 6.1). The anomaly is roughly oval shaped and extends from about 9° N to about 10° 45′ N, covering mostly the charnockite outcrops. The minimum anomaly is −120 mgal in the area. Granitic bodies are exposed towards the southeastern part of the area. However, these are not likely to be the main cause of the anomaly. The hilly terrain is essentially composed of acid and intermediate charnockites. It appears that the most probable cause of the anomaly is the presence of acid charnockites present in the area, as in the case of L–3. A quantitative interpretation of the gravity field over the southern hill

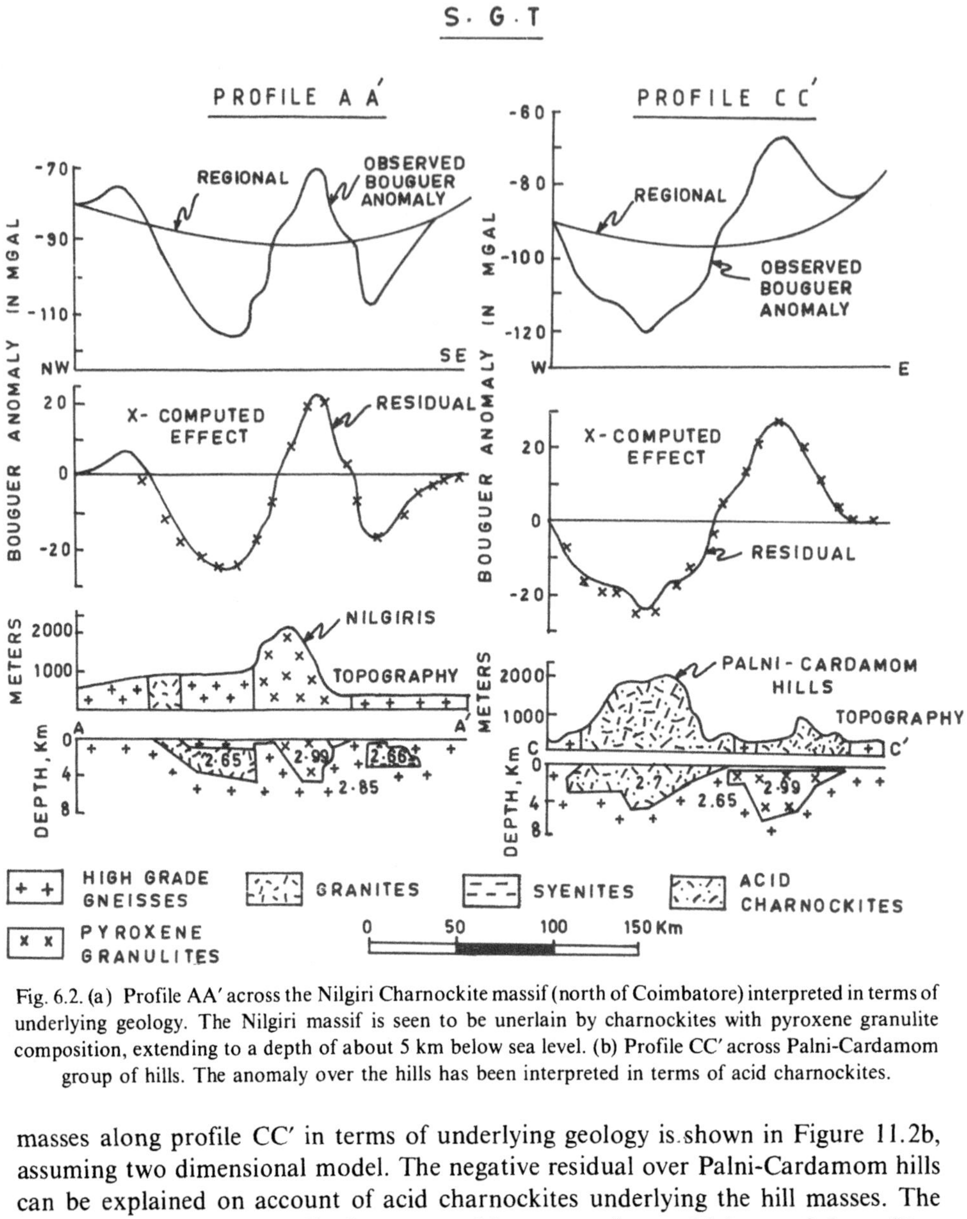

Fig. 6.2. (a) Profile AA′ across the Nilgiri Charnockite massif (north of Coimbatore) interpreted in terms of underlying geology. The Nilgiri massif is seen to be unerlain by charnockites with pyroxene granulite composition, extending to a depth of about 5 km below sea level. (b) Profile CC′ across Palni-Cardamom group of hills. The anomaly over the hills has been interpreted in terms of acid charnockites.

masses along profile CC′ in terms of underlying geology is shown in Figure 11.2b, assuming two dimensional model. The negative residual over Palni-Cardamom hills can be explained on account of acid charnockites underlying the hill masses. The charnockites appear to be dipping west and have a maximum thickness of about 5 km underneath the hills. The positive residual east of the hills can be accounted for in terms of a pyroxene granulite body developed in the form of a synformal structure, somewhat similar to that obtained for the Nilgiri granulite body.

6.2.1.3. *Other Gravity Highs and Lows*

Besides the gravity highs and lows discussed above, several others appear prominently over the SGT. A gravity high H-4 is located in the central part of the area between

Coimbatore and Namakkal. The gravity high H-3 lies partly over charnockites and partly over the Cauvery basin. Gravity lows L-5 and L-6 have northeasterly strike (same as that of Eastern Ghats). They appear prominently over the Cauvary basin. Figure 6.3 shows the nature of gravity anomalies over H-4, H-5 and over part of the Cauvary basin up to Portonovo along profile BB′. There is no immediate explanation for a gravity high H-4 in terms of the surface geology as it is spread over a vast gneissic-migmatite country. However, a number of anorthostie-gabbro complexes (sometimes associated with ultramafic rocks) known to occur in the SGT, outcrop close to and within the area occupied by this anomaly. It appears that the gravity high could be explained in terms of a gabbro-anorthosite mass underlying the gneissic rocks at shallow depths, whose surface expressions are the basic complexes exposed in the area.

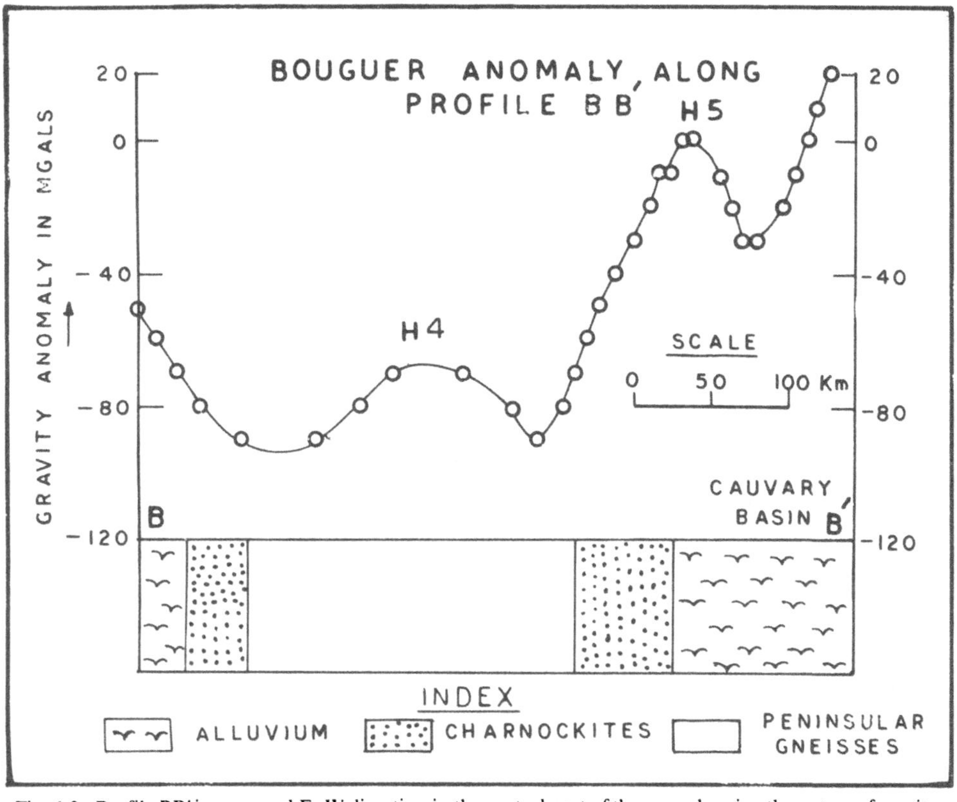

Fig. 6.3. Profile BB′ in a general E–W direction in the central part of the area, showing the nature of gravity anomaly. The high H4 is probably due to an anorthosite–gabbro body, while H5 near Namakkal is probably due to presence of basic charnockites in the area.

Gravity highs over the Cauvery basin can be explained in terms of basement ridges underlying the Tertiary sediments, while the lows could be explained in terms of basin type structures.

Interpretation of gravitry field over SGT shows that the anomalies exhibit considerable variations over the charnockites themselves. Gravity highs are associated with pyroxene granulites (basic charnockites) and lows are observed over acidic charnockites. It is quite possible that the pyroxene granulites consititute the basement into which the acid charnockites are intrusive, as suggested by Subramaniam (1967). This also implies that the pyroxene granulites are older than the acidic charnockites.

Granulite occurrences over the SGT are associated with positive gravity field. This is in good agreement with the results obtained over other shield areas of the world (Tanner, 1969; Gibb and Halliday, 1974; Mathur, 1974; Thomas, 1974; Reeves and Hutchins, 1975), where gravity highs have been observed over high density granulite facies rocks or thick and dense crustal blocks.

Several anorthosite–gabbro complexes associated with ultramafic rocks are known to occur over the SGT. It has been observed that a prominent gravity high is associated with some of these outcrops, which may be underlain by a huge gabbroic mass, underneath the gneisses.

There appears to be a clear contrast in the nature of the gravity field observed over the crystalline rocks belonging to the SGT and the Cauvery basin. The former is characterised by appreciably negative values compared to the latter. The line of demarcation between the two areas roughly passes through Madras-Namakkal-Cape Comorin line. The abrupt change in the nature of the gravity field suggests changes in crustal parameters such as thickness or mean density or both.

6.3. Gravity Field over the Coastal Granulite Terrain

The Coastal Granulite Terrain (CGT) extends for a considerable distance along the east coast from about 14° N to nearly 21° N. It is rather narrow, about 50 to 60 km in southern part but broadens to about 200 km in the northern part. On the western side of the CGT lies the Cuddapah basin in the south and Bastar province in the north.

Bouguer anomaly map of the CGT superimposed on the underlying geology is shown in Figure 6.4. The map is characterised by general northeast-southwest trending contours, the anomaly ranges from −50 mgal on the western margin to +30 mgal near the east coast. The trend of the contours is interrupted at two places, close to lat. 17° N by the NW trending Godavari valley graben and near 21° N by the Mahanadi valley with the same general trend. Near the southern margin of the Mahanadi valley, the trend of gravity contours changes almost to east-west. A number of gravity highs and lows are found along the entire length of the CGT.

The Gravity field over the CGT is appreciably positive −50 to +30 mgal) as compared to that over the SGT (−50 to −120 mgal). This suggests basic differences in the nature of the crust and mantle relationships in the two regions. The main rock types in this area are the unclassified gneisses, khondalites, charnockites, and Tertiary sediments. Several gravity highs are found over khondalites and charnockites as well as alluvial covered areas. Qualitative and quantative interpretation of some of these is discussed below. Density values for various rock types have been listed in Table 5.1 (Chapter 5).

6.3.1. GRAVITY HIGHS AND LOWS

Gravity high H-1 is located in the southern part of the area and trends in NE–SW direction. Two −20 mgal closures delimit this anomaly, which is spread over outcrops

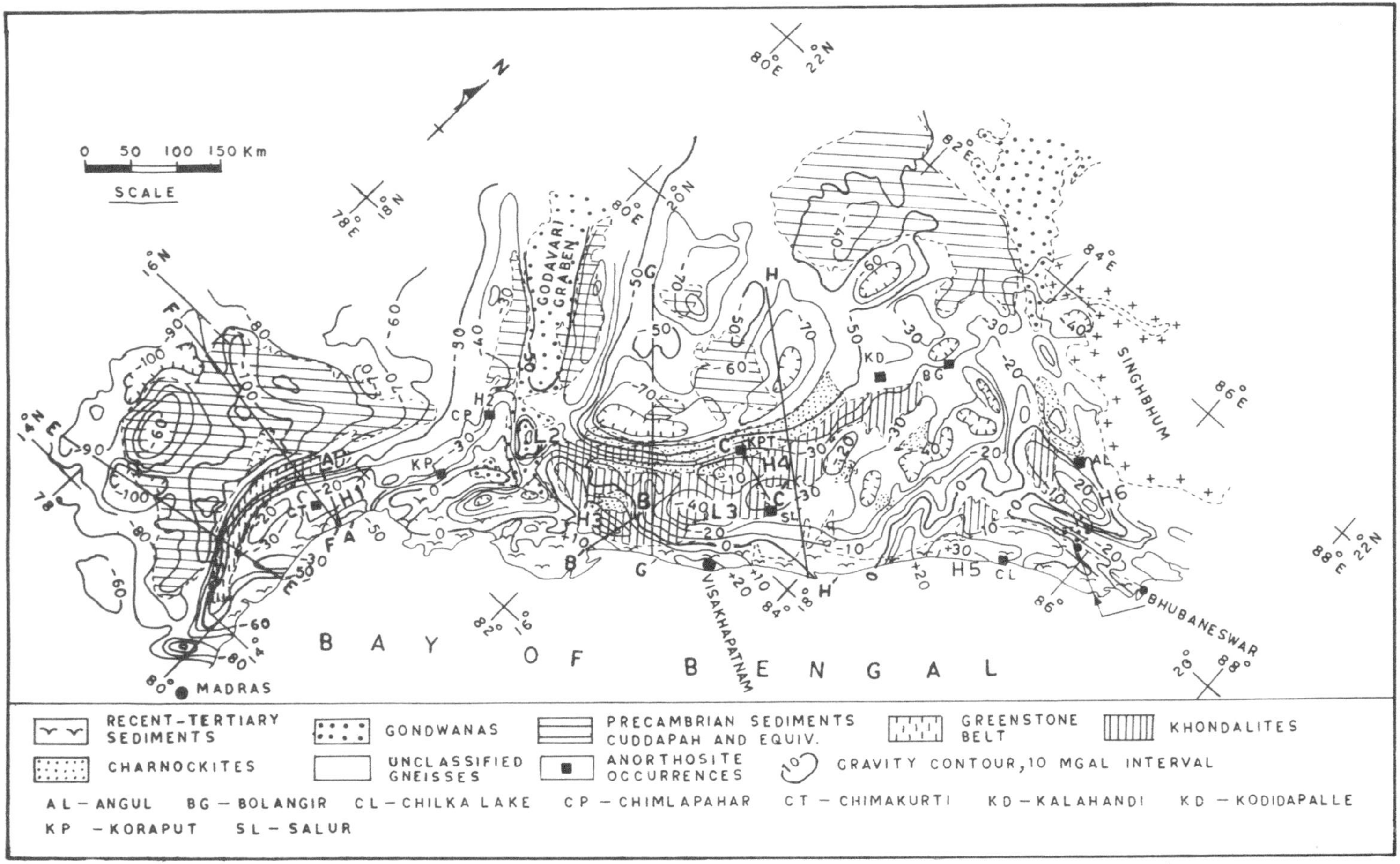

Fig. 6.4. Bouguer anomaly map of the Coastal Granulite Terrain (CGT) superimposed on the underlying geology. The figure shows the locations of several highs and lows as well as profiles AA′, BB′ and CC′ interpreted using two dimensional models. Anorthosite occurrences after Dey (1969).

of schists, gneisses as well as alluvial covered areas. In the northern part of the area several isolated outcrops of gabbroic rocks have been reported near Chimakurti, which are probably the rocks causing the anomaly. It is probable that the entire positive anomaly is caused by the underlying gabbroic mass. A quantitative interpretation of the anomaly along profile AA′ across H–1 is shown in Figure 6.5a. The causative body is inferred to be an underlying gabbroic mass of about 4.0 km thickness.

Similar explanations can be given for gravity highs H2, H4, H5, and H6 which are located close to anorthosite complexes of Chimlapahar, Koraput-Salur, Chilka-Lake and Angul respectively. Two dimensional models for highs H3 and H4 are shown in Figure 6.5b and c. The high H3 appears to be due to a concealed gabbro complex underlying Khondalite suite of rocks.

A number of gravity lows are found in the area of the CGT. Some of these such as L–1 are found over the alluvial covered areas and L–2 over areas underlain by the Gondwana sediments. These lows can be explained in terms of the low density of sedimentary formations as compared to the gneisses or khondalites and charnockites surrounding them. There is however a very prominent gravity low L–3 found over the outcrops of khondalites, gneisses and charnockites. The available density data (see Chapter 5) indicate that the charnockites have the least density of all these formations. It is possible that this prominent low is caused by a series of charnockitic intrusives along its length as suggested by Subrahmanyam (1978a).

6.3.2. SUMMARY OF RESULTS

The gravity anomaly map of the Coastal Granulite Terrain does not register any large scale variations as in the case of the Southern Granulite Terrain. However, the map reveals several narrow anomaly belts running for hundreds of kilometers along the length of the Eastern Ghats terrain. The parallelism between the gravity contours and the structural grain of the CGT is quite striking. As we have seen a number of gravity highs could possibly be explained in terms of underlying gabbro/anorthosite masses. The emplacement of these masses along the Coastal Granulite Terrain could probably be related to the existence of fracture zones of great linear extent along the terrain of Eastern Ghats.

The idea of attributing positive gravity anomalies to basic or ultrabasic intrusives in orogenic belts is certainly not a new one. Brooks (1966, 1970) has put forward such a concept much earlier. The CGT is a part of the Eastern Ghats belt which has been recognized as a true mobile belt dating back to Archaean times (Murthy *et al.*, 1971). The Grenville province in the Eastern Canadian shield is very similar to the Eastern Ghats belt. There also several anorthosite outcrops have been found to be associated with positive gravity anomalies, which have been interpreted in terms of gabbroic intrusions underlying them (Tanner, 1969; Thomas 1974). In the light of the results obtained from other similar shield areas of the world, the interpretation of gravity highs discussed above seems to hold good. However, detailed gravity studies over these areas need to be undertaken in order to confirm these interpretations.

Some of the prominent negative anomalies along the belt have been interpreted in terms of acidic charnockite intrusives. Narayanaswami (1975) has suggested that the Eastern Ghats belt (particularly the area of the CGT), has experienced several syn- and

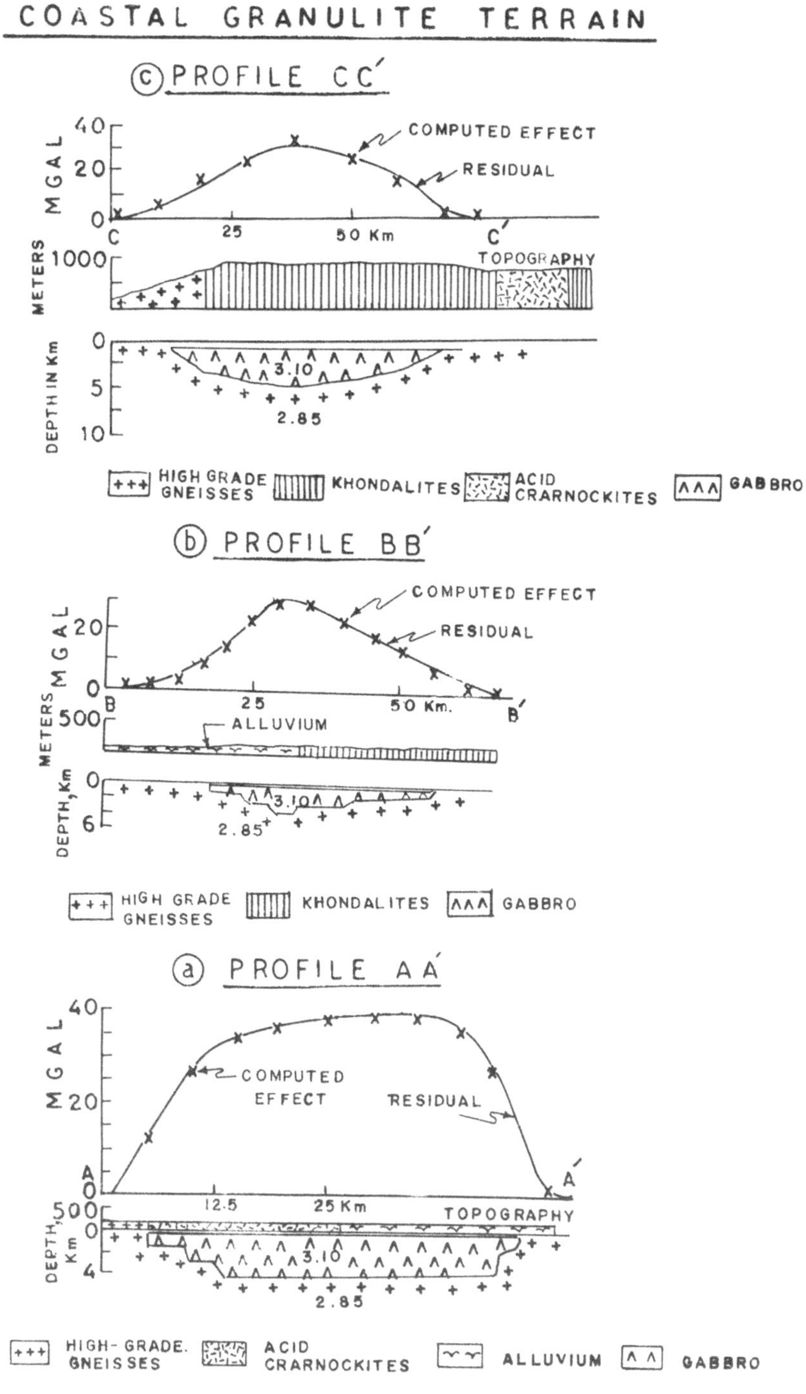

Fig. 6.5. Interpretation of gravity highs along profile AA′, BB′ and CC′ (locations shown in Figure 6.4) in terms of underlying masses assuming two-dimensional models.

post-tectonic granite intrusives (hypersthene granites and gneisses). Most probably some of these could be the acid charnockites inferred from the negative anomalies.

On the whole, the study of gravity field in the Eastern Ghats belt throws some light on the processes that have taken place in this belt during the period of its evolution.

6.4. Nature of the Boundary Between Eastern Ghats and Dharwar/Bastar Provinces

The nature of the boundary between the Eastern Ghats and the Dharwar greenschist province in the north has been a subject of speculation and investigation for a considerable time. Fermor (1936) was probably one of the first to realise the geological significance of the problem, as he rightly observed the total contrast in physiography, lithology, metamorphism and structural style of the charnockite and non-charnockite provinces. He suggested that the entire charnockite terrain as a whole has been block-uplifted and thrust over the adjoining parts of non-charnockite terrain along a deep seated fault. He contended that this terrain represents a deep level crust, now uplifted and exposed due to extensive erosion. However, Crookshank (1938) from detailed geological investigation of a portion of this boundary in Bastar province, disagreed with Fermor's concept of a faulted contact, as he could not find any evidence to support such a view. Although, this problem has not been rigorously studied in detail, some of the later workers have postulated a faulted contact between the Eastern Ghats belt and the Dharwar and Bastar provinces (Nautiyal, 1966; Subramaniam, 1967; Iyengar, 1971; Murthy *et al.*, 1971). On the other hand, observing the gradational increase in metamorphism in the Kolar and Sargur schist belts, which lie in the proximity to the boundary between the low grade Karnataka/Dharwar province and the SGT, Shackelton (1976) considered the possibility of the granulite terrains continuing underneath the Archaean Dharwar craton towards north.

The problem of close spatial relationship between the high grade granulite terrains and the low grade greenstone-granite terrains is one of the most significant ones concerning the Precambrian sheilds. Gravity studies have contributed significantly to a better understanding of the nature of the boundaries between the structural provinces in Canada (Gibb and Thomas, 1976), South Africa (Pretorius, 1974) and W. Australia (Glikson and Lambert, 1976; Wellman, 1978). The work of Gibb and Thomas (1976) in studying the relationships between different structural provinces in Canada is of particular significance. These authors have studied the nature of the gravity field in Grenville, Churchill, Slave and Superior provinces and noted that the gravity field over the Grenville and Churchill provinces (which are younger in age, about 1000 m.y and 1800 m.y. respectively) is much more positive as compared to that over the Superior province (which is much older in age, about 2560 m.y.). They have studied the nature of the field transition across the Grenville/Churchill provinces to the Superior province and Slave to Churchill province and noted that the gravity profiles across these provinces are remarkably similar in character. The gravity anomaly decreases by about 40 to 50 mgal along a distance of about 210 km as we come from the older province towards the younger one and then rises sharply with a steep gradient by about 80 to 100 mgal near the contact of the two provinces within a distance of about 80 km. The gravity field over the younger province is higher than that of the regional background over the older province by about 24 mgal. In order to explain this anomaly, they have envisaged a downflexuring of the crust of normal density underlying the older province towards the boundary of the younger province which accounts for the negative anomaly and a thicker as well as denser crust underlying the younger province to account for the positive field. Thus, the steep gradient between the positive and negative pair of anomalies reflects a discontinuity separating the two crustal blocks with different densities–underlying the older and younger provinces. Typical gravity profiles across the structural provinces in Canada and their geological intepretation is shown in Figure 6.6. The younger province in Canada such as Grenville represents a high grade metamorphic terrain. Similar terrains in other countries are the Limpopo in South Africa, Eastern Ghats in India and Albany–Fraser belt in Western Australia. Since the Eastern Ghats is a high grade metamorphic terrain (corresponding to granulite facies), the nature of the gravity field from this terrain to the adjoining Dharwar and Bastar provinces can be analysed in view of the relationships obtained by Gibb and Thomas (1976) for type anomalies between similar provinces in Canada.

6.4.1. EVIDENCE FROM GRAVITY ANOMALIES

The Coastal Granulite Terrain of Eastern Ghats differs from the Southern Granulite Terrain in its geological and geophysical characteristics. It is characterised by a positive gravity field, both of regional and local extent. The regional gravity field over the CGT ranges between −50 to −10 mgal compared to −50 to −80 mgal over Bastar and −80 to −100 mgal over Dharwar provinces. This change in the gravity field from the CGT to Bastar and Dharwar provinces is marked by a sharp gradient as seen in Figure 6.4. This variation in the regional gravity fields of CGT and the adjoining low-grade cratons probably suggests a variation in crustal parameters, such as density/composition or thickness.

It has generally been recognised that the reworked, polymetamorphic granulite facies terrains of Archaean shields are usually underlain by a thicker and denser crust giving rise to positive regional anomalies over them (Tanner, 1969; Berry and Fuchs, 1973; Mathur, 1974; Thomas, 1974; Glikson and Lambert, 1976; Gibb and Thomas, 1976). By analogy, it may be suggested that the CGT/Eastern Ghats may also have a denser and thicker crust. The gravity field over the CGT is practically continuous from the southern tip of the Cuddapah basin to the Mahanadi Valley, except for a small interruption in the Godavari Valley. The Godavari Gondwana lineament developed much later (during Paleozoic-Mesozoic-period), cutting across the Archaean Eastern Ghats lineament.

An interesting feature of the gravity field over the CGT and its surrounding Karnataka/Dharwar and Bastar areas is that the gravity anomalies are strongly negative near the western margin of the CGT but rise sharply near the boundary of the two provinces, and then become increasingly positive towards the eastern boundary of the coastal terrain. This can be very well seen from profiles EE′ and FF′ from the Cuddapah basin (Dharwar craton), to the Eastern Ghats and GG′ and HH′ from Bastar to the Eastern Ghats (see Figures 6.4 and 6.7). It may be noted that the gravity field decreases on an average by about 25 mgal from the Dharwar/Bastar provinces towards the Eastern Ghats and then rises sharply by 50 to 60 mgal close to the boundary between the two provinces. The sharp gradient of anomalies near the

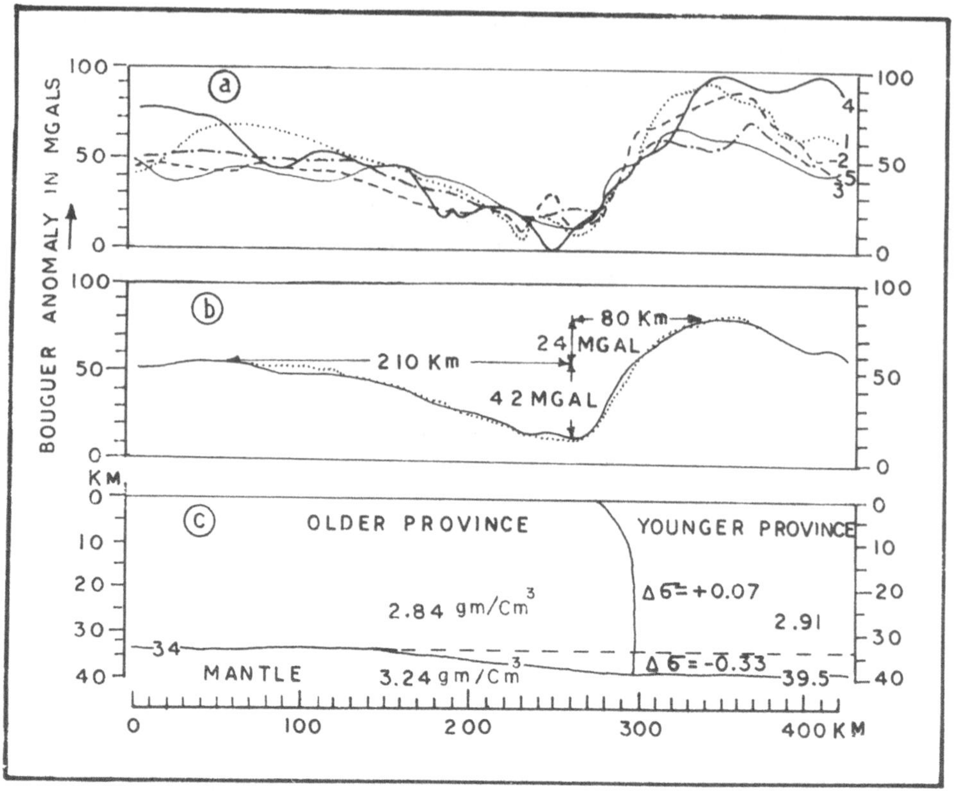

Fig. 6.6. Nature of gravity field across Superior to Chirchill/Greenville province or Slave to Churchill province in Canada after Gibb and Thomas (1976) and its interpretation in terms of crustal blocks with different composition. No. 1-5 in figure (a) refer to profiles across 1 and 2. Superior to Churchill boundary, 3. Superior to Greenville boundary, 4. Churcill to Greenville boundary and 5. Slave to Churchill boundary. Reproduced from Nature, V. 262, No. 124-131, p. 136 with permission of R. A. Gibb, M. D. Thomas and MacMillan Journal Ltd., Copyright, 1976.

western margin of CGT suggests that its boundary with Bastar and Dharwar cratons is a faulted one.

These profiles are very similar to those obtained by Gibb and Thomas (1976) across structural provinces between Superior and Churchill/Grenville or Churchill and Slave provinces (as seen in Figure 6.6). The amplitude of anomalies observed across the provinces is about 60 to 80 mgal, with low near the contact of two provinces and high over the younger province. They have inferred from the models constructed that the younger province is densier by about 0.07 gm cm^{-3} compared to the older province and that the boundary between the two dips towards the younger province. As seen from figure 6.6 isostatic equilibrium between the two provinces is maintained by increase in crustal thickness under the younger province by about 4 km. Crustal thickness also increases to some extent under the older province, suggesting that the compensation is regional in nature.

Similar positive and negative (called dipole) anomalies have been encountered across several structural provinces in Australia, such as Arunta block to Lander

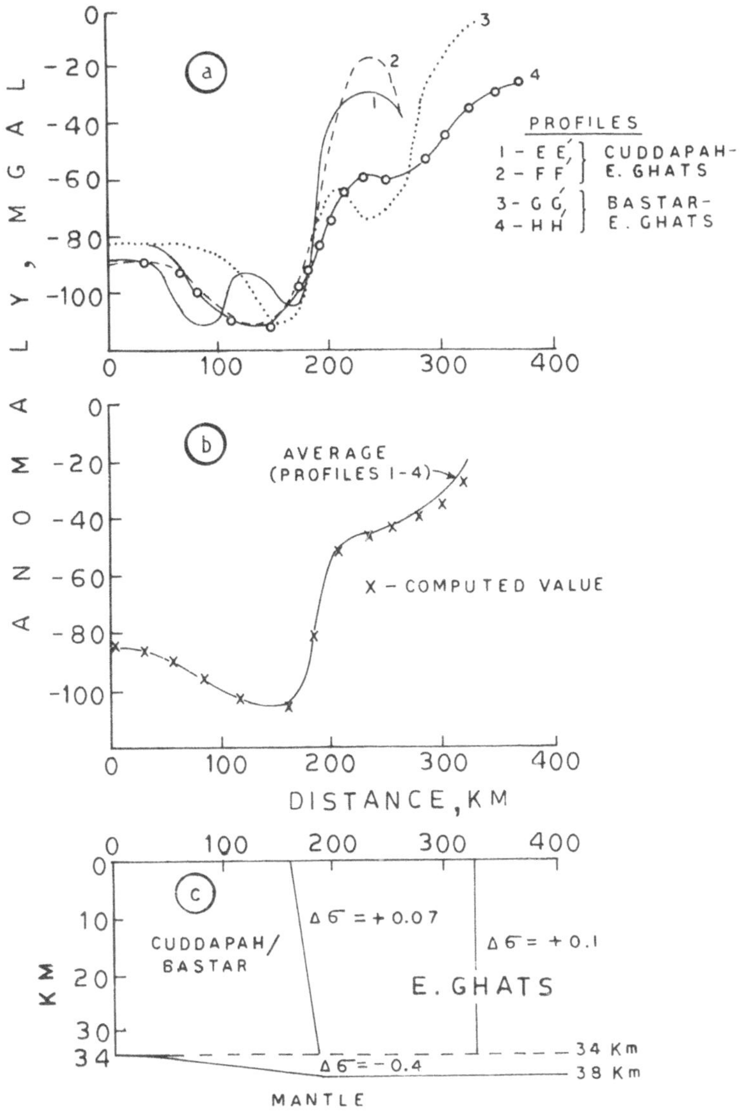

Fig. 6.7. Nature of the gravity anomaly from Bastar to Eastern Ghats along profiles GG′ and HH′, and from Cuddapah basin to Eastern Ghats along profiles EE′ and FF′. Locations of profiles are shown in Figure 6.4.. The transition in gravity field across these provinces is interpreted in terms of change in composition in the crustal blocks from west to east.

Trough and Amadeus basin, Pilbara block to Albany Fraser Domain and Paterson province to Pilbara block and few other places (Wellman, 1978). The amplitude of anomalies across these areas is found to be of the order of 60 to 120 mgal. Wellman (1978) has interpreted all major dipole anomalies in terms of crustal blocks with different densities and associated isostatic compensation, resulting in higher crustal thickness underneath the higher density block. The contrast in mean density between the different blocks is inferred to be about 0.1 gm cm^{-3}.

Taking the mean of the anomalies obtained along the profiles EE′, FF′, GG′, and HH′ (see Figure 6.4) as representative of the anomalies obtained from Cuddapah/Bastar areas to the coastal areas of Eatern Ghats, a crustal section has been interpreted as shown in Figure 6.7. The density of the crust over greenstone areas has been taken as 2.88 and that over CGT 2.95 gm cm^{-3}, increasing further to 2.98 gm cm^{-3} near the east coast, thus having a density difference of 0.07 and 0.1 gm cm^{-3} with respect to the areas on the west. The thickness of the crust over the older Bastar/Dharwar provinces has been taken to be nearly 34 km in the western part, based upon the results obtained by Kaila *et al.* (1979) from D.S.S. studies in the western part of the Cuddapah basin. The model constructed shows that the crustal thickness increases towards the eastern side and reaches a maximum value of about 38 km underneath the Eastern Ghats in the coastal areas.

The model also suggest that the crustal blocks are in isostatic equilibrium by a process of regional compensation. These results are consistent with Woollard's (1959) observations that as the crust thickens, the density also increases to maintain isostatic equilibrium. The density discontinuity separates crustal blocks of diverse thickness, density, age as well as metamorphism. These results are in agreement with the seismic results of Berry and Fuchs (1973) in Canada and Mathur (1974) in Australia for similar high grade granulite terrain, like Grenville in Canada and Wheat Belt in Australia respectively, where in general the high grade terrains are underlain by denser as well as thicker crusts.

CHAPTER 7

ANALYSIS OF GRAVITY FIELD OVER SINGHBHUM AND THE ADJOINING AREAS

7.1. Introduction

Singhbhum is an area of complex geology, located in eastern part of the Indian peninsula, approximately between latitude 21° to 23° 15′ N and longitude 84° to 87° 30′ E. The area has much economic importance on account of sulphide and uranium mineralisation along the Singhbhum Thrust Zone and iron ores in the area. The gravity field in the area has been considerably influenced by the history of the Iron Ore Geosyncline starting from about 3200 m.y. including sedimentation, volcanism, basic and ultrabasic intrusions, granitic intrusions of batholithic dimensions, differentiated granites as well as thrust movements of major dimensions. According to Sarkar (1980), the area has experienced at least three orogenic cycles during its history. The various events that have taken place have left their imprint in the form of gravity 'highs' and 'lows'. A study of the gravity field in the area throws a considerable light on the complex subsurface geology and igneous activity that affected the area.

Regional gravity surveys in the area were first conducted by Qureshy *et al.* (1972), which brought out information on some major structures present in the area. Detailed gravity surveys were conducted in the area by the Department of Applied Geophysics, Indian School of Mines, Dhanbad during 1976-1979. The results have been partly published (Verma *et al.*, 1978, 1984) and presented in the form of a Ph.D. thesis (Sarma, 1980). Before discussing the results and their significance, it is desireable to outline briefly the geology of the area.

7.2. Geology of the Area

The geoloy of the area was studied systematically by Jones (1934), Dunn (1929, 1937, and 1940) and Dunn and Dey (1942). On account of the mineral deposits associated with different formations of the area, several geologists have studied the area in detail: Sarkar and Saha (1966, 1977), Sarkar *et al.* (1969), Saha (1964, 1972, 1975), Iyengar *et al.* (1964, 1965), Prasad Rao *et al.* (1964), Naha (1965), Acharya (1964), Kanungo and Mallick (1967), Ray (1976) and others. Jadavpur University Publication (Deb, 1966) deals with extensive discussions on the geology of the Singhbhum region. And outline of the geology of the area is shown in Fig. 7.1. Stratigraphic sequence in the area is given in Table 7.1. Various rock units in the area in stratigraphic sequence are the following:

7.2.1. OLDER METAMORPHIC GROUP (OMG)

The oldest rock formation in the area is Older Metamorphic Group which consists of

TABLE 7.1

Stratigraphic succession of rocks in Singhbhum and the surrounding areas (after Sarkar *et al.*, 1977)

Alluvium		
Tertiary Gravel		
Laterite		
Baripada Beds		
---- End of Singhbhum Orogenic cycle (850 m.y.) ----		
Newer Dolerites (1600–950 m.y.)		
Mayurbhanj Granite (1200 m.y.)		
Biotite granite gneiss, granophyres, Soda granite	1500–1100 m.y.	
Chakradharpur granite-gneiss, Kuilpal granite, Nilgiri granite gneiss.		
Anorthosite-Gabbro (1470 m.y.) with pyroxinite and picrite-peridotite		
Ultramafic intrusions		
Kohaṉ Group (1600–1500 m.y.)		
Quartzite, conglomerate, sandstone, limestone and shales etc.		
Dhanjori Group (1700–1600 m.y.)		
Dalma-Dhanjori-Simlipal-Jagannathpur lavas along with quartzites, conglomerates, sandstones, phyllites etc.		
---- Unconformity ----		
Dhalbhum stage	Mica schist, orthoquartzite-schist, shale hornblende schist, calc-chlorite schist	Gangpur Group
Chaibasa stage	Biotite-muscovite schist etc.	Conglomerates, quartzites, shale, phyllite, sandstone dolomite, marble, limestone, slates etc.
---- Unconformity ----		
Singhbhum Granite (2950 m.y.)	Mainly biotite, granodiorite, adamellite, hornblende granodiorite, diorites etc.	
---- Iron Ore Orogenic Cycle ----		
Iron Ore Group	Upper shale, volcanics, shales, BHJ/BHQ, Lower shales, mafic lavas, orthoquartzites, conglomerates, sandstones, alternating with phyllites	
---- Unconformity ----		
---- OLDER METAMORPHIC CYCLE ----		
Older Metamorphic Gneiss (3200 m.y.)	Banded amphibolites such as in the Singhbhum granite	
Older Metamorphic Group	Banded catagneiss, mica-schist, quartz schist, tonalite granite gneiss.	
Basement 3300 m.y.		

metasediments associated with basic intrusives. Although the type area lies near Champua (22° 04′ N, 86° 40′ E), rocks belonging to this group are found in patches practically all over the area. According to Basu *et al.* (1981) the OMG tonalite was formed 3800 m.y. ago and is the oldest rock so far found in India.

7.2.2. IRON ORE GROUP

The Iron Ore Group (IOG) are the oldest sedimentary formations exposed in this region. These rocks are exposed all around the Singhbhum granite batholith and extend towards the west up to 84° E longitude, abutting against the younger Gondwana formations of Mahanadi Valley. The rocks cover a large area and are associated with workable iron ore deposits which form a part of the Banded Hematite Jaspers (BHJ) and Banded Hematite Quartzite (BHQ) formations. The type area for this group of rocks is the Koira basin (near 21° 20′ N, 85° 17′) exposed in a horseshoe shaped syncline. The IOG is mostly made up of shales, schists, phyllites, BHJ / BHQ, lying in sequence with volcanics, overlying the basal conglomerate, sandstone and quartzite layers. The volcanic rocks surround the sediments of IOG, particularly on the south side forming a U-shaped exposure. On the basis of radiometric studies, age given for this group is about 3000 m.y. (Sarkar and Saha, 1977). Although at present a major part of the Iron Ore basin lies to the west of the Singhbhum granite batholith, the original extent of the basin must have been considerable, as seen from the presence of Iron ores to the north, east, as well as south of the Singhbhum granite.

7.2.3. SINGHBHUM GRANITE BATHOLITH

This large granite body extends for more than 150 km in N–S direction and more than 70 km in E–W direction between latitude 21° to 22° 45′ N and longitude 85° 30′ to 86° 30′E and forms a major part of the area. Extensive studies of this batholith by Saha (1966, 1972 and 1975) have revealed that this granite batholith is a late tectonic, epizonal-mesozonal batholith, consisting mainly of biotite granodiorite, adamallite, together with biotite-muscovite and leucogranite and consists of numerous granitised rocks. On the basis of petrological studies, the granite body has been divided into 11 different units (Saha, 1966).These are: Rajanagar-Kengali and Dalima belonging to the first phase, Hatgamaria-Rairangpur-Onlajori belonging to the middle phase and Haludpukur-Chapra, Gorumahisani, Gamaria-Maluka near Karanjia belonging to the last phase. The entire granite body is traversed by Newer Dolerite dykes and other basic intrusives. Beyond the northern limit of the Singhbhum granite exposure, Singhbhum group of rocks are present. Between these two lies the Singhbhum thrust zone. The radiometric age determined for the Singhbhum granite is nearly 2950 m.y. (Sarkar, 1980).

7.2.4. SINGHBHUM GROUP OF ROCKS

After the deposition of Iron Ore Group of rocks and the intrusives of Singhbhum granite, stable conditions were established in the area for quite sometime.

The next major period of sedimentation was witnessed in the area towards north and west of Singhbhum granite, approximately between latitude 22° 45′ to 23° 15′ N and longitude 85° to 87° E, where the Singhbhum Group of rocks consisting of the Chaibasa and Dhalbhum formations were deposited between 2000 and 1700 m.y.

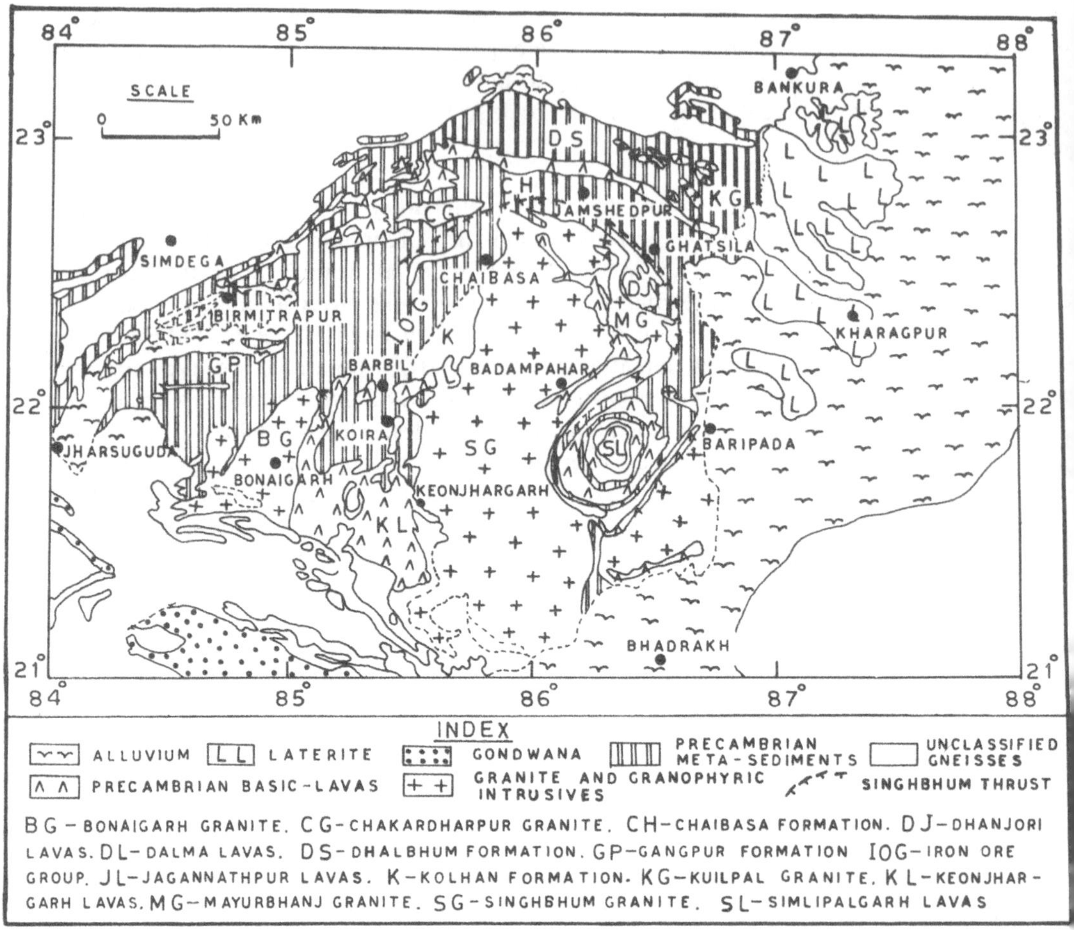

Fig. 7.1. An outline of geology of Singhbhum and the surrounding areas.

(Sarkar, 1980). During the period of sedimentation, the Dalma lavas were erupted. The rocks were metamorphosed during the Singhbhum orogenic cycle, and now consist of various types of metamorphics such as mica schist, hornblende schist, tuff, chert, orthometamorphic schist along with sericite phyllite, chlorite phyllite, conglomerate, quartzite etc. The rocks have been folded into the Singhbhum anticlinorium and the Dalma synclinorium.

7.2.5. KOLHAN, DHANJORI AND SIMLIPAL BASINS

The next period of sedimentation was witnessed in the areas west as well as east of the granite batholith. The rocks deposited in this area belong to the Kolhan Group, deposited along the western margin of the Singhbhum granite and those deposited in Dhanjori and Simlipal basins to the east of the Singhbhum granite. Sedimentation in this area is believed to be 2100 m.y. in age (Sarkar and Saha, 1983). The rocks belonging to the Kolhan Group are mainly sandstones, quartzites, conglomerates, grits, limestones and shales. They are essentially flat lying and shallow in depth. The contact between the Iron Ore Group and Singhbum Group on the west is covered by these rocks. The sedimentation in the Dhanjori and Simlipal basins was affected considerably by volcanism during the Singhbhum orogenic cycle, between 2100-2200 m.y. according to Sarkar and Saha (1983).

The Dhanjori lavas form a part of the Dhanjori basin, exposed between latitude 22° 30′ to 22° 45′ N and longitude 86° 15′ to 86° 30′ E in the northeastern part of the area, overlie the quartzites, conglomerates and consist of tuffaceous basalts, mica schists, metagabbros close to the thrust zone.

The Simlipal lavas forming a major part of the Simlipal basin are exposed between latitude 21°30′ to 22° N and longitude 86° to 86°30′ E. These have been studied in detail by Iyengar and Banerjee (1964). The lavas overlie the phyllites of IOG with basal conglomerate and quartzite beds. The area has been affected by post Dhanjori orogenic cycle resulting in intrusion of gabbro-anorthosite and granophyre mass in Dhanjori as well as Simlipal areas.

7.2.6. JAGANNATHPUR LAVAS

The volcanic activity withnessed by Dalma lavas in the north, Dhanjori and Simlipal lavas in the east (between 1700 to 1600 m.y.) is represented by Jagannathpur lavas in the west. These lavas are exposed between 22° to 22° 15′ N and 85° 30′ to 85° 45′ E in a rectangular shape adjoining the Singhbhum granite and are tholeiitic in composition. These leaves overlie the older lavas belonging to the IOG.

7.2.7. GANGPUR GROUP

This group of rocks has been studied by Krishnan (1937), Kanungo and Mallick (1967), Kanungo (1972), Banerjee (1968) and others. The group is exposed to the north-west of the IOG and consists of phyllites, carbonaceous shales, quartzites, mica-schists, marbles, limestones and is intruded by basic sills and dykes. Sarkar *et al.* (1969) have assigned an age of 900–940 m.y. for this group.

7.2.8. ULTRAMAFIC INTRUSIONS

Apart from basic rocks discussed earlier, mafic and ultramafic rocks are found in the Singhbhum region and its surroundings. These are represented by serpentinite, dunite, gabbro-anorthosite rocks in the form of sills and dykes, associated with IOG. Prominent exposures of these rocks are near Gorumahisani (22°20′ N, 86° 17′ E) and along the periphery of Simlipalgarh basin, as well as east of Anandpur (21° 13′ N, 88°07′ E). On the basis of radiometric ages, these rocks are assigned an age of 1470 m.y. which makes them younger than the Dalma-Dhanjori lavas.

7.2.9. MAYURBHANJ AND OTHER GRANITES

Apart from the main Singhbhum granite batholith, several other granites, much younger in age are found in the area. These include the Mayurbhanj granite located between 22° to 22°30′ N and 86° 15′ to 86° 45 E′ Kuilpal granite (near 22° 50′ N, 86° 38′ E) Chakradharpur granite (near 22° 40′ N, 85° 30′ E), and soda granites near the Singhbhum thrust zone. On the basis of radiometry, the ages of these are found to vary from 1500 to 1100 m.y. (Sarkar, *et al.*, 1969).

7.2.10. NEWER DOLERITES

The entire Singhbhum region and its surroundings are traversed by dolerite dykes ranging in age from 1600 to 900 m.y. (Sarkar *et al.*, 1977). These are seen conspicuously as intrusive into the Singhbhum granite batholith and the Bonaigarh granite.

7.2.11. SINGHBHUM THRUST ZONE

A major event in the geological history of the Singhbhum area is thrusting of the Singhbhum Group of younger and high grade metamorphic rocks over the older low grade sediments and metavolcanics belonging to the IOG. The zone along which thrusting has taken place is known as the Singhbhum Thrust Zone. It is arcuate in shape, convex towards north and is nearly 200 km in length, starting from 22° 15′ N, 86° 30′ E and ending somewhere near 22° 35′ N, 85° 30′ E (Sarkar *et al.*, 1969). This has been a major zone of weakness along which base metal mineralisation took place practically all along the belt. Soda granites (Table 7.2) are mostly associated with the thrust zone.

7.3. Observed Bouguer Anomaly Map

The observed Bouguer anomaly map of the entire area, based on about 1900 gravity stations established by the Department of Applied Geophysics of Indian School of Mines, and Hawaii Institute of Geophysics is shown in Figure 7.2 (Sarma, 1980). It may be seen that here the Bouguer anomaly ranges from +10 mgal in the eastern part to about −60 mgal over the Singhbhum granite batholith. The anomaly map bears a good correlation to the surface geology. Two prominent contour trends are noticed, an east-west trend in the northern part prevailing over the Singhbhum Group of rocks and the Dalma volcanics, and a north-south trend prevailing over the IOG to the west as well as to the east of the Singhbhum granite, over the Dhanjori volcanics, the Simlipal basin and further south.

The map is characterised by the conspicuous presence of several gravity highs and lows. The lows are prominent in the central part, mostly over the Singhbhum granite batholite, while the highs prevail over the western, northern and the eastern sides of the batholith. The prominent highs and lows are discussed briefly.

7.3.1. GRAVITY HIGHS

7.3.1.1. *Koira-Jamda High, H_1*

This is a prominent gravity high, prevailing over the sediments and volcanics belonging to the IOG, west of the Singhbhum granite. The high has a magnitude of about 14 mgal and trends in a NNE–SSW direction, extending for about 40 km from near Koira in the south to beyond Jamda in the north. The strike of the anomaly follows that of the IOG formations.

7.3.1.2. *Dalma High, H_2*

This high is located between latitude 22° 45′ N to 23° N and longitude 85° to 86° 30′ E, and prevails over the Singhbhum Group of rocks including the Dalma volcanics. Magnitude of the high is nearly +20 mgal above surroundings. The trend of the high is

TABLE 7.2

Density values for various rock formations in Singhbum and the surrounding areas.

Sl. No.	Formation	Rock Type	No. of samples	Range ($gm\ cm^{-3}$)	Mean value ($gm\ cm^{-3}$)	Standard Deviation
1.	Older Metamorphics	Cal-Tremolite schists, hornblende schist, garnet tonalite gneiss	5	2.64–2.67	2.66	0.013
2.	Iron Ore Group	a) Shales, schists, phyllites, tuffs	18	2.5 –3.1	2.75	0.0132
		b) Banded Hematite Jasper, Banded Hematite Quartzite	25	2.66–4.67	3.2	0.507
		Weighted Average	43		2.82	
3.	Singbhum Group	Chlorite, mica schists and shales	26	2.61–2.97	2.78	0.083
4.	Basic Rocks	a) Dalma Metavolcanics	21	2.9 –3.2	3.05	0.074
		b) Dhanjori lavas	10	3.0 –3.07	3.05	0.022
		c) Simlipal lavas	23	2.7 –3.03	2.95	0.091
		d) Gabbro-anorthosites	13	2.88–3.09	3.01	0.057
		e) Amphibslites	18	2.7 –3.04	3.01	0.149
		f) Unclassified	96	2.9 –3.05	2.98	0.155
		All combined	181		3.01	
5.	Singhbhum granite	Biotite granite, granophyre and granite gneisses (17 locations)	213	2.59–2.72	2.68	

mostly east-west. However, it takes a sharp turn near longitude 85° 15′ E in accordance with the strike of the Chaibasa formations in the area. The highest anomaly is observed over the Dalma volcanics. The high extends partly over the alluvial covered areas to the east, suggesting the continuation of the Singhbhum Group of rocks in the area.

7.3.1.3. *Gorumahisani High* H_3

This is a very prominent gravity high, having a magnitude of about +30 mgal and located to the east of the granite batholith, centred near Gorumahisani (22°20′ N, 86°17′ E). The area covered by this high is underlain by the Dhanjori formations consisting of IOG, lavas, ultramafics and granitic rocks.

7.3.2.4. *Simlipalgarh High,* H_4

This high is located to the south of the Gorumahisani high and is centered (near 21° 52′

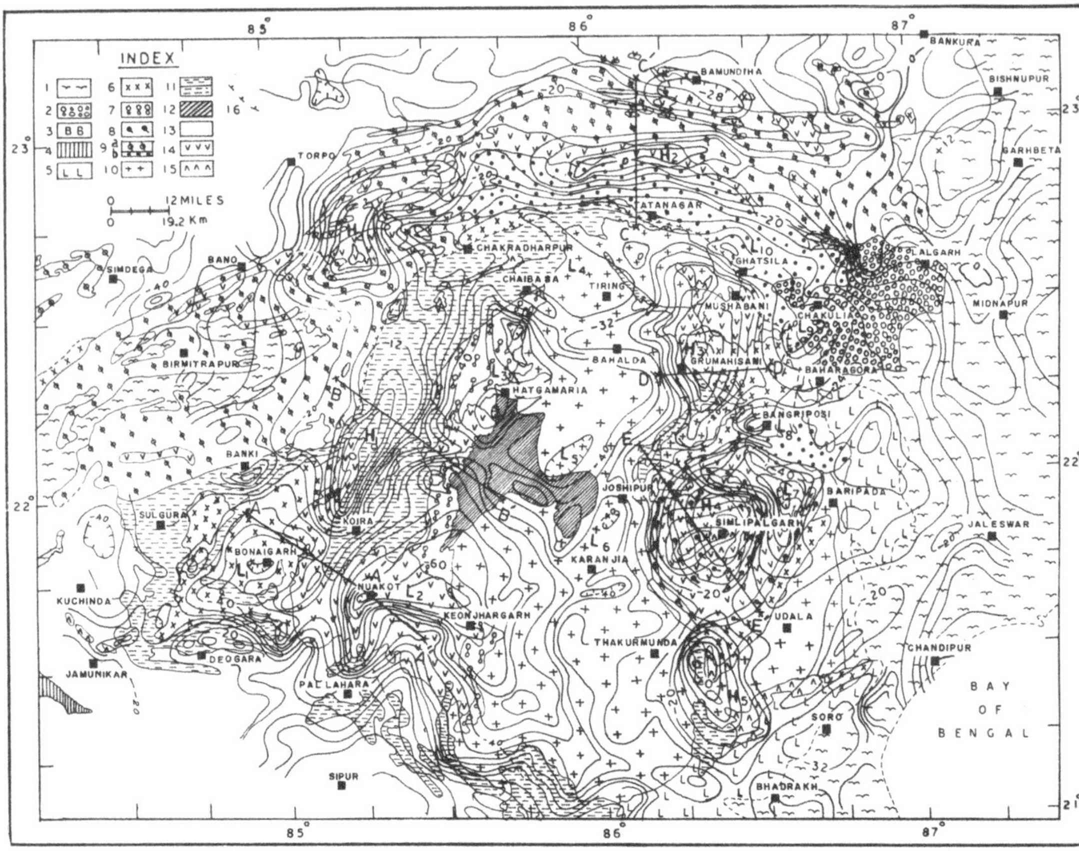

Fig. 7.2. Bouguer anomaly map of Singhbhum region (Sarma, 1980) superimposed on surface geology (Sarkar *et al.*, 1969). Geological index; (1) Alluvium, (2) Gravel, (3) Baripada beds, (4) Gondwana sediments, (5) Laterite, (6) Granite, (7) Sandstone, shale, quartzite (Kolhan Group), (8) Dhanjori quartzite, (9) Singhbhum Group (a) Dhalbhum (b) Chaibasa formation, (10) Singhbhum granite, (11) IOG, (12) Older Metamorphic Group, (13) Unclassified gneisses, (14) Volcanics, (15) Ultrabasics, (16) Singhbhum thrust. Contour interval, 4 mgal.

N, 86° 22′ E) over the Simlipal basin. The high has a magnitude of about +24 mgal. The area is underlain by the Simlipal volcanics, quartzites and conglomerates in an alternating sequence in the form of a ring complex. The high is connected to the Gorumahisani high towards north, and with the south Simlipal high towards its south.

7.3.1.5. *South Simlipal High, H_5*

An interesting and unexpected high towards south of the Simlipal is H_5, centered few km east of Thakurmunda. This high is mostly underlain by granitic rocks on the surface. However, a tongue of anorthosite-gabbro suite of rocks exposed in the form of U-shape passes through the central part of the anomaly. The high is most probably related to these rocks.

7.3.2. GRAVITY LOWS

A number of gravity lows are observed on the Bouguer anomaly map. Few of the prominent ones are discussed below.

7.3.2.1. *Bonaigarh Low, L_1*

This is a prominent gravity low to the west of the Singhbhum granite and is centered near Bonaigarh (21°48′ N, 84°47′ E). It is nearly rectangular shaped with an observed value of −48 mgal at the centre. The area is underlain by the Bonaigarh granite and gneissic rocks and partly by IOG formations. The anomaly is apparently related to the Bonaigarh granite.

7.3.2.2. *Nuakot Low, L_2*

This is one of the most interesting gravity lows encountered in the Singhbhum area. Here the lowest Bouguer anomaly observed is nearly −70 mgal (centered near 21° 42′ N, 85° 15′ E) near Naukot village. The anomaly has a general E–W trend. However, near Naukot the anomaly turns sharply to N–S direction. The area covered by the anomaly lies partly over the Singhbhum granite on the east side and partly over the volcanics belonging to the IOG. The shape as well as magnitude of the anomaly suggests the extension of Singhbhum granite underneath the volcanics.

7.3.2.3. *Hatgamaria Low, L_3*

This is a prominent gravity low located in the north-western part of the Singhbhum batholith. The low is split into two parts, north as well as south of Hatgamaria, characterised by Bouguer anomaly values of −52 mgal and −56 mgal respectively. The area covered by the low is underlain partly by the Singhbhum granite and partly by the rocks belonging to the Kolhan series. The nature of the anomaly suggests the extension of the Sighbhum granite towards west underneath the sediments of IOG.

7.3.2.4. *Central Singhbhum Lows: L_4 L_5, L_6*

The central part of the Singhbhum granite batholith is characterised by a number of gravity lows such as L_4, L_5, and L_6 aligned in N–S direction. The magnitude of these lows is of the order of −10 to −15 mgal. These lows suggest compositional variations within the Singhbhum granite batholith (Verma *et al.*, 1978).

7.3.2.5. *Baripada Low, L_7*

This is an interesting gravity low, lying to the east of Simlipalgarh basin. The low has an amplitude of about −20 mgal. There is no apparent explanation for this low, since the area is underlain by the Simlipal basic volcanics. However, the exposures of granitic rocks to the east and NW suggest that the volcanics are underlain by the Singhbhum or associated granites.

7.3.2.6. *Bangriposi Low, L_8*

This low is located to the north of Baripada low, and is associated with the exposure of Mayurbhanj granite to its noth as well as south.

7.3.2.7. *Chakulia-Ghatsila Lows: L_9, L_{10}*

These lows are located close to the Singhbhum thrust zone. Here the surface rocks belong to the Singhbhum Group. However, exposure of soda granites close to the thrust zone suggest that these lows could be caused by the soda granite bodies (Table 7.2) underlying this area.

7.3.3. DENSITY VALUES OF VARIOUS ROCK FORMATIONS IN THE AREA

The highs and lows in general are caused by the variation of density of the near surface rock formations. This is apparent from steep gradients associated with localized contours. The IOG consisting mostly of iron bearing sediments as well as Singhbhum Group of rocks, which are mostly high grade gneisses, schists, and volcanics, are of higher density than the basement complex and hence cause the gravity highs. The Singhbhum granite batholith may be assumed to form the basement complex practically for the entire area. It consists of a number of granitic bodies of varying composition and density, the average for 213 samples being 2.68 gm cm^{-3} (Table 7.2). This value has been taken to be representative basement density for the entire area. The belt of gravity highs covers the Singhbhum batholith practically from all the sides. The density values of various rock formations from the area have been determined (Verma *et al.*, 1978; and Sarma, 1980) and are given in Table 7.2. Nearly all the gravity lows are located over the Singhbhum granite batholith or its margins. A few of these are also seen over the alluvial covered areas to the east. The lows over granitic areas can be explained in terms of the low density of the granites. Density values for granites from 17 different locations over the Singhbhum batholith (Sarma, 1980) shows that their average density varies from 2.59 gm cm^{-3} to 2.69 gm cm^{-3}. Such large variation in density can cause significant gravity anomalies, of the order of 10 to 20 mgal or more, since the thickness of granitic bodies is quite appreciable, being of the order of 5 to 10 km (Verma *et al.*, 1978, 1984; and Sarma, 1980).

7.4. Interpretation of Anomalies in Terms of Two/Three Dimensional Models

One of the main objectives of gravity studies is to interpret the anomalies in terms of geological models of the underlying rock formations. However, for this purpose the regional anomalies, due to deep seated bodies, have got to be filtered out. For such a large area as the Singhbhum, the regional anomalies can be effectively filtered out by fitting a third order or a 4th order polynomial surface. Regional anomaly maps of the entire Singhbhum area using 1st to 4th order polynomial surfaces have been prepared by Sarma (1980). These maps show that the 4th order polynomial surface gives a reasonable picture of the regional gravity field. It filters out anomalies of long wave length, of the order of 170 to 200 km, which are deep seated without removing any near surface anomalies which are of immediate interest from geological point of view. Subtraction of the regional from the observed provides the residual anomalies, which have been interpreted in terms of two dimensional models using Talwani *et al.*'s (1959) polygonal method as well as three dimensional models, using Cordell and Henderson's (1968) iterative computer program.

Since the geological formations in the present area are in general elongated along the strike direction and have appreciable length compared to their depths, most of the anomalies can be treated as two dimensional in character. Geological models interpreting some of the important anomalies across the major rock formations of the area have been prepared and are discussed below.

7.4.1. PROFILE AA'A"

This profile (location shown in Figure 7.2) is located in the southwestern part of the area and passes over the Bonaigarh granite on the west, IOG in the central part and ends over the Keonjhargarh volcanic in the east. The residual anomaly corresponding to a 4th degree surface was modified slightly on the basis of geoligical considerations. The residual anomaly obtained on the basis of assumed regional has been interpreted. The profile shows a residual low (of the order of −10 mgal) over the Bonaigarh granite, a high over IOG rocks and a low (of the order of −20 mgal) over the volcanics near Nuakot. The interpreted model (Figure 12.3) suggests that the gravity high can be explained by the IOG and the underlying Keonjhargarh volcanics, which have a thickness of the order of 4 km. It is possible that the deposition of IOG rocks was preceded by large basic volcanic eruptions, which are widespread in the area.

The gravity low near Nuakot cannot be explained in terms of the Keonjhargarh volcanics which are high density rocks. However, the granitic rocks belonging to Singhbhum granites, near Keonjhargarh are found to have a lower density (2.63 gm cm^{-3}) as compared to the basement complex of density 2.68 gm cm^{-3} (Sarma, 1980). The model suggests an appreciable thickness for the low density granitic body present near Nuakot underneath the volcanics and possibly connected with the Bonaigarh granite to the west. The granite is found to be intrusive in nature as seen in Figure 7.3. It is conical in shape with outward dipping sides.

7.4.2. PRIFILE BB'

This profile is located in the western part of the area and runs in a NW–SE direction for about 70 km. The profile crosses some of the major rock formations in the area, including the Singhbhum Group of rocks, the IOG (including the Kolhan series) and Older Metamorphics, ending over the Singhbhum granite towards the southeast. A characteristic feature of profile is a gravity high of about 32 mgal amplitude on the western side, and a low of about 18 mgal towards the east. The interpreted model (Figure 7.4) shows that the grvity high is caused by the Singhbhum/Gangpur and IOG rocks, forming a synclinal structure, the latter being underlain by volcanics, which are exposed to the east of the IOG. It may be noted that the volcanics are an inseparable

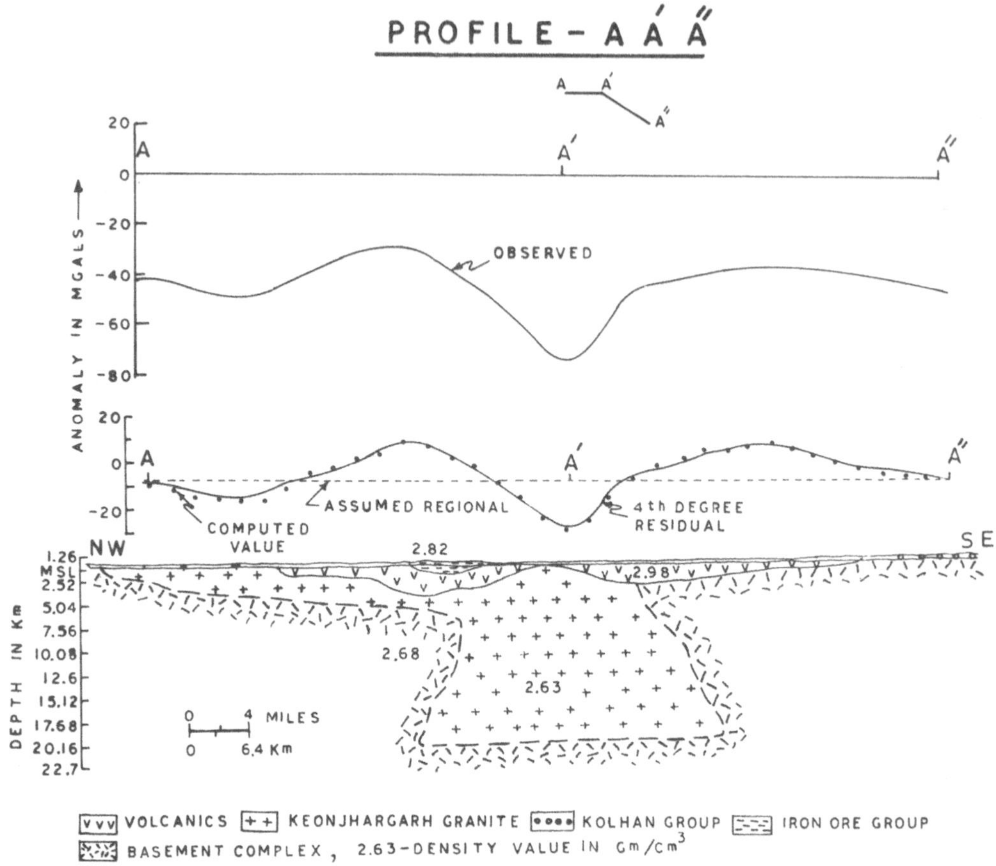

Fig. 7.3. Profile AA′A″ acros Bonaigarh granite, IOG rocks and Keonjhargarh volcanics in southwestern part of the area. The two dimensional model explains the gravity low in terms of a massive granite underlying volcanics.

part of the IOG and are associated with the sediments everywhere. The thickness of sediments and volcanics is of the order of 8–9 km in the area and may further increase somewhat towards southwest, as the magnitude of anomaly increases in this direction. On the basis of geological considerations the thickness of IOG rocks is estimated to be 12 to 13 km.

The gravity low in the eastern part few km southwest of Champua can be satisfactorily explained in terms of a low density grantic body (having a density of about 2.63 gm cm^{-3} as found locally) with a thicknes of about 10 km. The granite body is narrow at the top but broad-based at the bottom. It may have been formed by differentiation from the main mass.

7.4.3. PROFILE CC′

This profile is located in the northern part of the area and cuts across the Singhbhum basin including the Dalma volcanics in a general N–S direction. It ends in the southern part near the margin of the Singhbhum granite batholith. A characteristic feature of

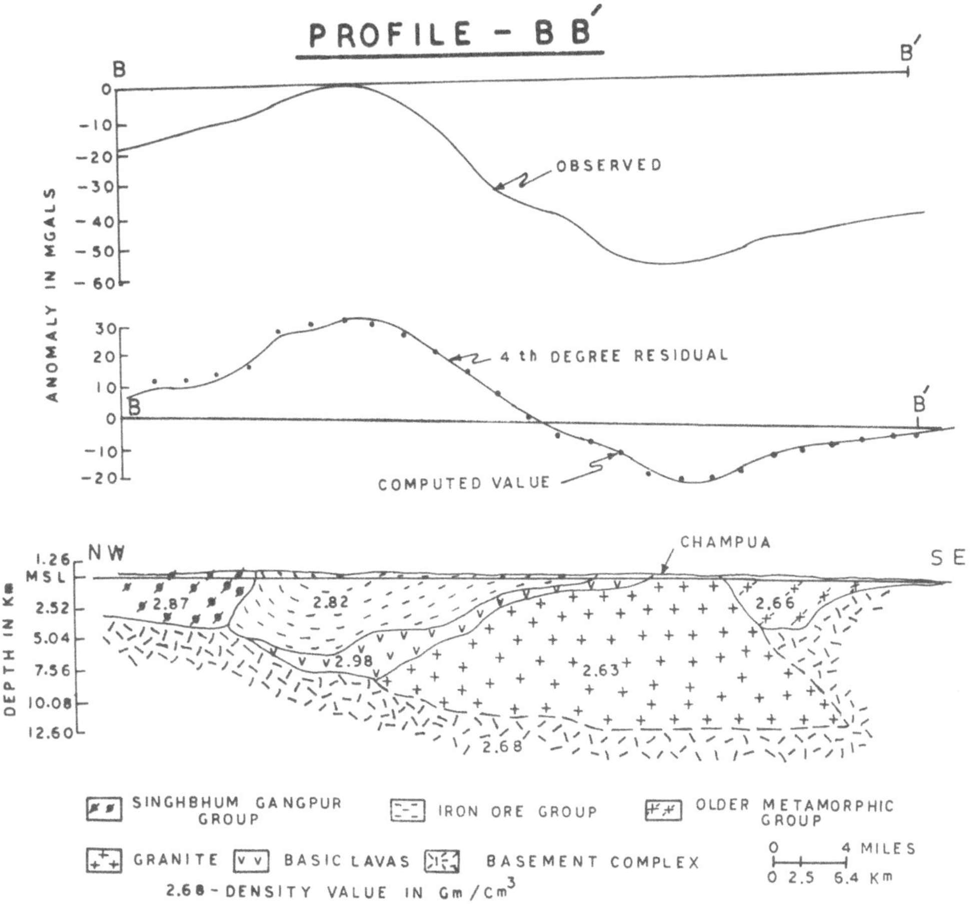

Fig. 7.4. Profile BB′ across the IOG rocks, Kolhan Series and Older Metamorphics in the western part of the area. The model for the profile shows the nature of Iron Ore basin and granitic body underlying to the east.

the profile is a gravity high of about 30 mgal amplitude (with respect to the assumed regional), located mostly over the Singhbhum Group of rocks. The two dimensional model for the anomaly is shown in Figure 7.5.

The gravity high is apparently caused by the metasediments belonging to the Singhbhum Group of rocks (which have a higher density as compared to the basement complex) developed in the form of a synclinal structure with Dalma volcanics near the central part.

The model suggests that the thickness of metasediments and interbedded volcanics is quite appreciable, (about 4.0 km below sea level) in the central part of the Singhbhum basin. The metasediments are thrust against the granitic body with a dip of about 45° towards north. The Dalma lavas have appreciable thickness in the central part of the basin and are interbedded with sediments of Singhbhum Group. The granitic body in the southern part, is locally known as the Saraikela granite, which is probably an extension of the Chakradharpur granite further towards the west.

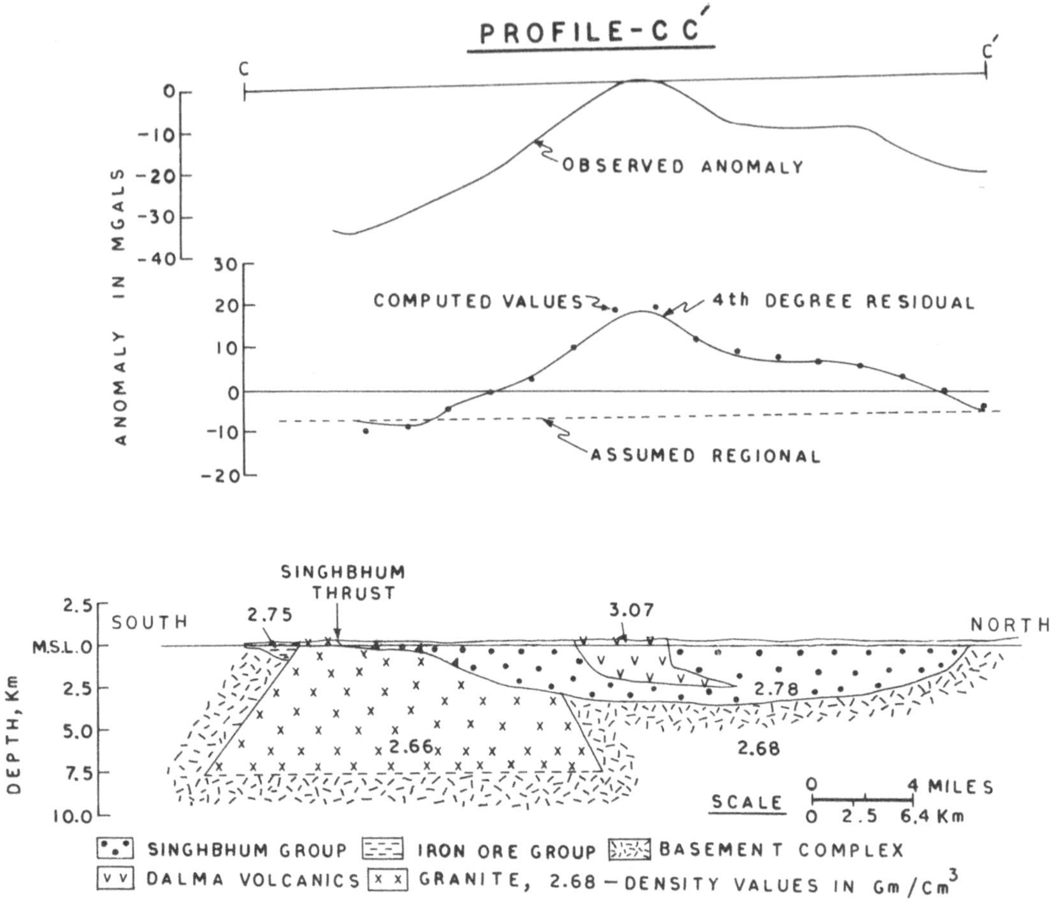

Fig. 7.5. Profile CC′ in a general N–S direction across the Singhbhum basin, along 86° E longitude, showing the nature of the basin and Dalma volcanics in the central part.

7.4.4. PROFILE DD′

This profile is located in the northeastern part of the area (Figure 7.2) and cuts across the Dhanjori basin in a general E–W direction. The main feature of the anomaly is a gravity high of about +32 mgal amplitude in the central part of the basin.

The Dhanjori as well as Simlipal basins have had their origin as a part of the geological history of Dhanjori Eugeosyncline and iron ore deposits of Mayurbhanj. In these basins sediments were deposited after the culmination of the Singhbhum orogeny (Iyenger and Banerjee, 1964). The Dhanjori basin was filled with phyllites, quartzites and lavas and was later intruded by an anorthosite-gabbro mass. Pyroxene granite, riebeckite granite and granophyres found in the area are closely related to this mass and represent an acid counterpart of this intrusion. These occur at the top of the intrusive body according to Iyenger and Banerjee (1964).

It is reasonable to expect that the gravity high is caused by the main gabbro-anorthosite mass which is underlying the granophyres, lavas and quartzites. The interpreted anomaly in terms of a two dimensional model is shown in Figure 7.6. The thickness of gabbro-anorthosite mass and associated volcanics is estimated to be 2.75 km.

The gravity high over Dhanjori basin has also been interpreted in terms of a three dimensional model (Sarma, 1980). The model does not differ much from the two dimensional model except in the central part, where the estimated thickness of the gabbroic mass is nearly 3.5 km.

7.4.5. PROFILE EE′

This profile is located in the eastern part of the area and cuts across the nearly circular shaped anomaly over the Simlipal basin in a NW–SE direction. The residual anomaly along the profile has a magnitude of about +32 mgal, which can be well explained in terms of Simlipal volcanics and interbedded quartzites. The volcanics are believed to be of submarine origin (Iyenger and Banerjee, 1964). The model constructed to explain the anomaly is shown in Figure 7.7.

The thickness of volcanics and quartzites obtained on the basis of two dimensional model is 3.25 km in the central part of the Simlipal basin. However, on the basis of

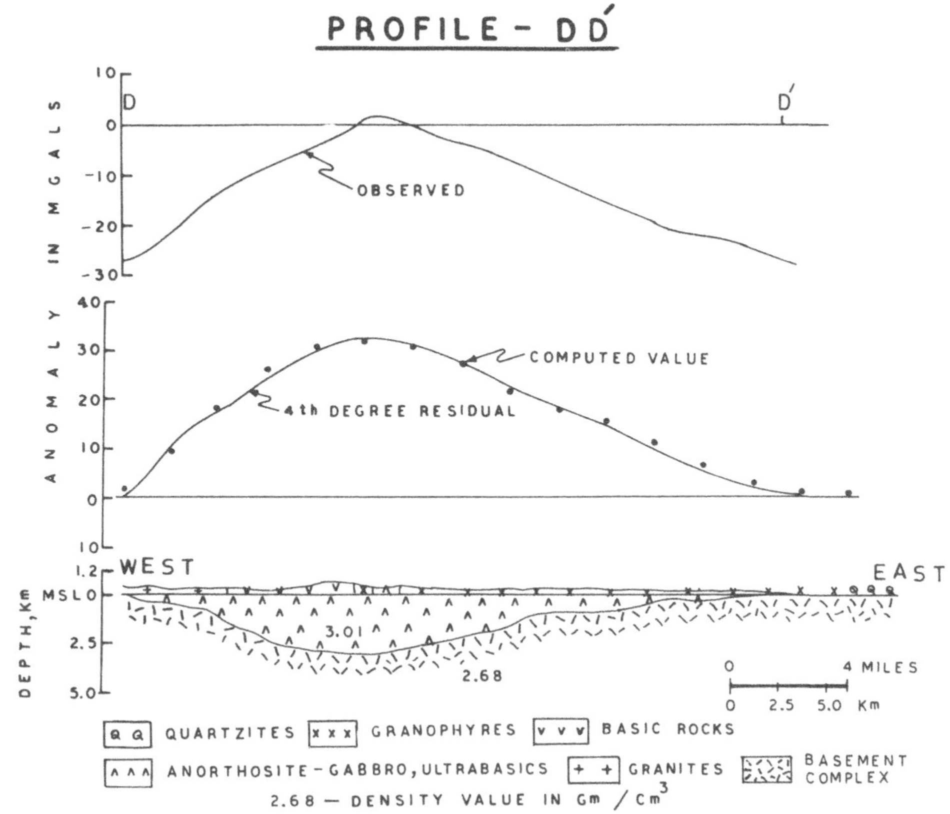

Fig. 7.6. Profile DD′ across the Dhanjori basin in E–W direction. The gravity high is explained in terms of an anorthosite - gabbro body underlying the area.

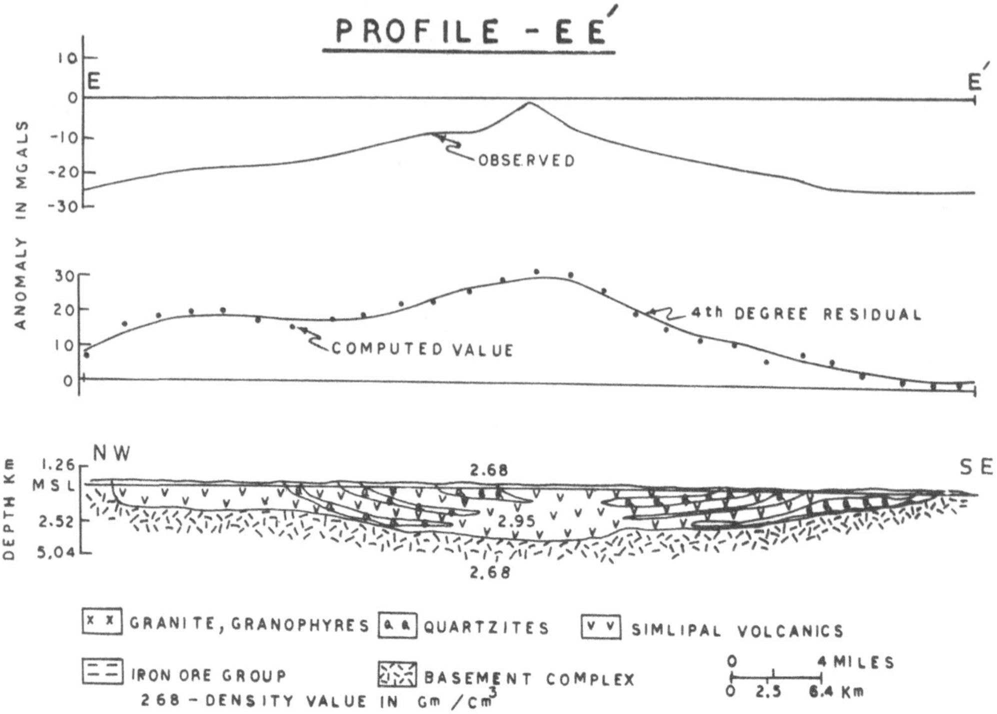

Fig. 7.7. Profile EE′ across Simlipal basin in E–W direction. The anomaly can be explained in terms of basic intrusives interbedded with quartzites belonging to IOG.

three dimensional model the maximum thickness is nearly 5.5 km. The various episodes of volcanic activity have been assumed on the basis ot geological evidence (Iyenger and Alwar, 1965).

The gravity high close to Champajhar (south of Simlipalgarh) is very similar to the high over Simlipal basin. Although most of the anomalous area is underlain by granitic rocks, the nearby exposure of basic intrusives suggests that the causative body could possibly be a gabbro-anorthosite mass underlying the area.

7.5. Significance of Results in Terms of Geologic History of Singhbhum

It is interesting to note that all the gravity highs in the area are associated with synclinal structures filled with sedimentary or metasedimentary formations interbedded with basic intrusives in the form of lavas or gabbro-anorthosite masses. This is supported by the gravity highs observed over the IOG, the Singhbhum Group, Dhanjori and Simlipal basins. The basic intrusives must have taken place during different orogenic cycles.

It is probable that the Iron Ore Group of rocks were desposited over volcanics, which could have been a few km thick in the central region of the area (now lying west of the Singhbhum granite batholith), folded and uplifted. In the anticlinal part, the Singhbhum batholith has intruded. During geological times, the uplifted part has been eroded, exposing the grantic batholith over a large area leaving IOG rocks practically all around it. Large thickness of sediments belonging to the IOG and the Singhbhum Group suggest that geosynclinal conditions prevailed in the area intermittently during the period 3200 m.y. to about 1600 m.y. Since the volcanic rocks either underlie the sediments or are interbedded, the basic igneous activity could have taken place before or during the period of sedimentation.

Presence of gravity highs over Dhanjori, Simlipal and S. Simlipal areas in N–S direction suggests, that an impressive amount of volcanic activity took place on the eastern side of the Singhbhum batholith over the areas occupied by the Dhanjori eugeosycline (Iyenger and Banerjee, 1964). During the later stages of volcanic activity, mafic and ultramafic intrusions took place in this area. These are also responsible for causing gravity highs in the area. Nature of gravity field suggests that there were at least two separate basins having similar history but somewhat isolated from each other on the eastern side of the Singhbhum batholith.

All the gravity lows in the area (as over the Singhbhum granite) are due to the granitic masses which are sometimes associated with anticlinal structures, suggesting their intrusive nature. The granitic bodies have different density in different areas and cause gravity lows of varying amplitude, as in Keonjhargarh, Karanjia, Rairangpur and Haludpukur. The low density granites could have formed during different metamorphic cycles or through magmatic differentiation of a preexisting sialic crust. Gravity low observed over Keonjhargarh suggests that granitic rocks underlie at least part of the Iron Ore Group west of exposed limit of Singhbhum granite and probably extend as far as Bonaigarh or even beyond.

The gravity work done so far in the area only brings out broad features of the various geological masses. Much more detailed gravity surveys are needed to be undertaken in order to throw more light on the underlying geology and relationship of various masses to each other.

CHAPTER 8

ISOSTASY IN INDIAN PENINSULA

8.1. Historical Development of Isostasy in India

Prior to the study of isostasy in India, the question of deep structure of the earth's crust was first raised as a result of astronomic and geodetic observations carried out by Pierre Bouguer from France during his memorable expedition to the Andes during 1735 to 1744. These observations showed that there was a marked deficiency of mass underneath the Andes.

The study of isostasy in India was initiated as a result of the precise measurement of arc by triangulation in India by Col. George Everest, Surveyor General of Survey of India (SOI). During 1830 to 1847 Col. Everest published the results obtained in the measurement of "Great Arc" which had been established in India along the meridian 77° 30′. He computed differences between geodetic and astronomic latitudes of three stations, Kaliana, Kalianpur and Damaridga. Astronomically, the distance between Kaliana and Kalianpur was found to be 5° 23′37.058″, while from triangulation or geodetic mesurements, the distance was found to be 5° 23′42.294″. Thus the difference between the two measurements was found to be 5.236″ which comes to nearly 500 ft. (or 151 m). Figure 8.1. shows the location of these stations.

Col. J.T. Walker of SOI had suggested that the difference was due to mass deficiency underneath the Himalaya and not due to error in measurements. The results of these measurements indicated that the gravitational effects of mountains on plum lines (which affects the astronomicl observations but not the geodetic measurements) could not be neglected. This suggested that the problem of attraction of the mountains had to be investigated. In 1852, Sir Andrew Wagh, the Surveyor General of India asked Archdeacon of Calcutta, John Henry Pratt to study the effect of the attraction of plum line in India.

Pratt proceeded to determine the deflecting influence of the high Himalaya and Tibetan areas on the plum line at Kaliana and Kalianpur. He found the deflection to be 27.853″ at Kaliana near foothills, and 11.968″ at Kalianpur in Central India, the difference being 15.885″. This difference was nearly three times the observed difference of 5.236″, between the astronomical and geodetic latitudes. This showed that nearly 2/3rd of horizontal attraction of the great Asian land-mass was compensated by mass deficiency of some sort.

Pratt presented the results of his studies before the Royal Society of London on December 7, 1854. He concluded that, "the difference of 5.236″ must be attributed to some other cause rather than error in geodetic measurements. A very probable cause is the attraction of the superficial mountains which be in such abundance to the north of the India arc".

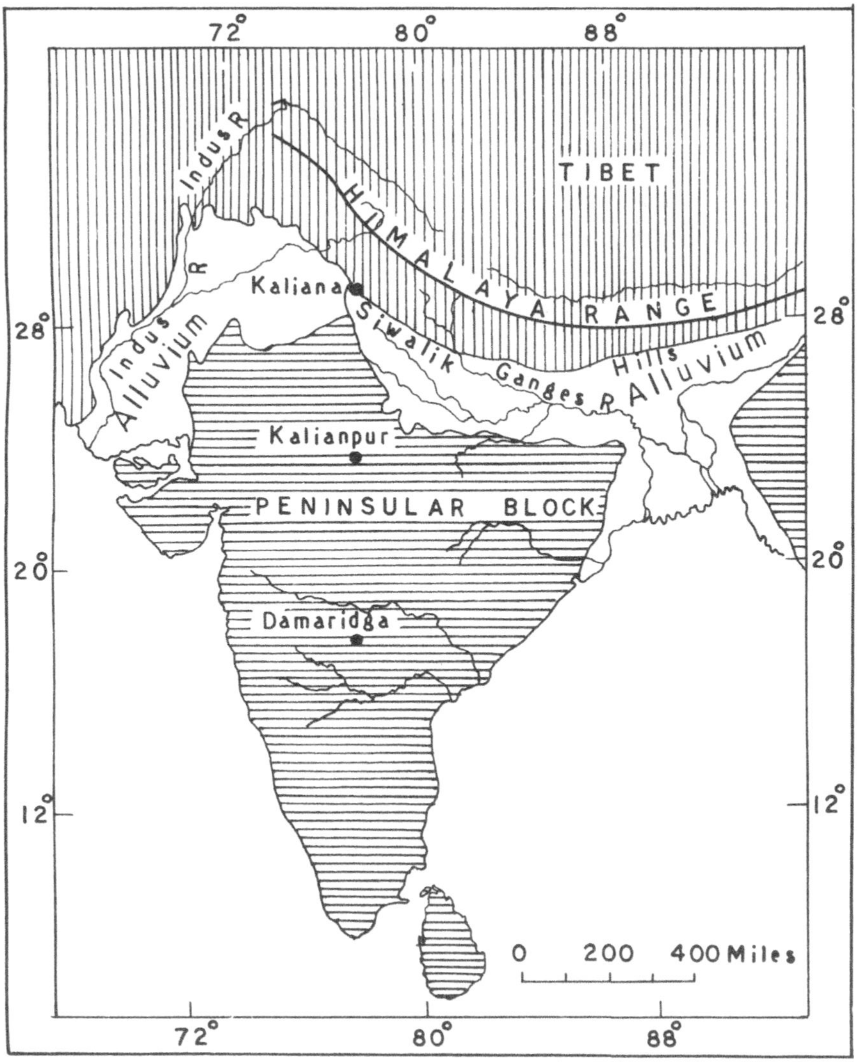

Fig. 8.1. Locations of geodetic stations Kaliana, Kalianpur and Damaridga, used for triangulation purposes by Survey of India. Figure reproduced from Daly (1969), with permission of M/s. Prentice-Hall Inc. U.S.A., Copyright 1940.

He further summarised,

"It appears to me to be unquestionable, that the geodetic operations are in no way sensibly affected by mountain attraction and therefore need no correction. It is the astronomical operation of observing the difference of latitude which requires the correction. It is only when the plumb is brought into use to determine the vertical angles of stars that the effect of attraction becomes sensible, and never in geodetic calculation, where only the horizontal angles or extremely minute vertical angles are observed", (Daly, 1969, p. 40, reproduced with permission of M/s. Prentice Hall Inc., U.S.A., Copyright 1940).

Giving his explanation of the phenomenon observed, Pratt suggested that during the formation of the crust of the earth concentration and expansion must have taken place, resulting in elevation and depression of some of its parts. He suggested that as a result of these processes, the mountains have been formed and the masses thus raised above the level have produced corresponding attenuation of matter below. Thus he gave a basic concept, that the mountains are characterised by mass deficiency underneath.

In 1855, Airy wrote a classic paper on the Indian puzzle and sent it to the Royal Society. He conceived that the interior of the earth was molten and the earth has a thin crust, about 10 miles thick. He imagined a mountain 2 miles high above a table land and predicted that the mountain would break the crust and sink either the whole, or in part into the molten lava. His ideas are basic to the development of the concept of isostasy and may be quoted briefly. He wrote that,

"I conceive that there can be no other support than that arising from the downward projection of a portion of the earth's light crust into the dense lava; the horizontal extent of that projection corresponding rudely with the horizontal extent of the table land, and the depth of its projection downward being such that the increased power of its floatation thus gained is roughly equal to the increase of weight above its table land. It appears to me that the state of earth's crust lying upon the lava may be compared with perfect correctness to state of raft of a timber floating upon water; in which, if we mark one log as floating much higher above the surface of the earth, we are certain that its lower surface also lies deeper in water than the lower surface of the others", (Daily, 1969, p. 45, reproduced with permission of M/s. Prentice Hall Inc., U.S.A., Copyright 1940).

Thus by his root theory, Airy offered an explanation for the stable support of the mountain chains, wide table lands and continents, each tending to be floated by a corresponding root or downward projection of rocks. Airy and Pratt's schemes of compensation are schematically illustrated in Figures 8.2 and 8.3 respectively. The desity values chosen for different blocks of the earth's crust are arbitrary, only to illustrate the scheme of compensation.

In 1889, C.E. Dutton introduced the word 'Isostasy' in a classic paper. He wrote,

"If the earth were composed of homogeneous matter, its normal figure of equilibrium without strain would be a true spheroid of revolution; but if heterogeneous, some parts were denser or lighter than others, its normal figure would no longer be spherical. Where the lighter material was accumulated, there would be a tendency to bulge, and where denser material existed, there would be a tendency for matter to flatten or depress the surface. For this condition of equilibrium of surface, to which gravitation tends to reduce a planetary body, irrespective of whether it be homogeneous or not, I propose the name Isostasy, which in Greek language means 'in equipoise with'. If the earth be not homogeneous, if some portions of the earth are lighter than others, then the isostatic figure is no longer a spheroid of revolution but a deformed figure, bulged where the matter is light and depressed where it is heavy", (Daly, 1969, p. 51, reproduced with permission of M/s. Prentice Hall Inc., U.S.A., Copyright 1940).

The ideas of Maj. C.E. Dutton were further elaborated by J.F. Hayford and W. Bowie, the successive heads of U.S. Coast and Geodetic Survey.

In India, systematic gravity observations using pendulums were initiated in 1865, at the suggestion of Sir Edward Sabine, then president of the Royal Society of London. Capt. Basevi of SOI was put incharge of the operations. He took 30 observations between 1865 and 1871, mostly along the Great Arc. Basevi crossed the mountain passes into Ladakh with the aim of measuring gravity at high altitudes in the Himalayas. He reached More Plains and established a station at latitude 34° 20′ N, longitude 79° 22′ E, at an altitude of 5200 m. The results of Capt. Basevi's observations strengthened the basis of Pratt's discovery of the phenomenon of isostasy, and were published in London in 1868 in Pratt's publication of "Figure of the Earth".

A new program of gravimetric observations was undertaken by the Survey of India in 1903 using Sternech pendulums. The aim was to study the problem of compensation in the Himalayan chain and to confirm the existence of a 'Hidden Range' as postulated by Burrard, south of the Ganges depression.

8.2. Various Schemes of Isostatic Compensation

According to the phenomenon of isostasy as visualised by Airy, the mountainous areas and plateaus are characterized by root formation, while the oceanic areas are characterized by the presence of antiroots. These roots and antiroots are bound to influence the gravity field observed on the surface, or reduced to the level of the geoid by making suitable corrections. The phenomenon of isostasy can best be studied by applying correction to the observed gravity field for the root or antiroot (as the case may be) according to the Airy or Pratt concepts of isostasy. This correction is known as the isostatic correction. Application of this correction according to the two concepts is discussed below.

8.2.1. AIRY SCHEME OF COMPENSATION

Airy's scheme of compensation can be illustrated by picturing the earth's crust as floating in a dense magma, just as icebergs float in the oceans. Consider three areas, one a mountainous area with elevation 'h' above sea level, another at sea level and the third one under the oceans as shown in Figure 8.2.

Let ρ_0 be the density of the materials forming the crust and ρ the density of heavier

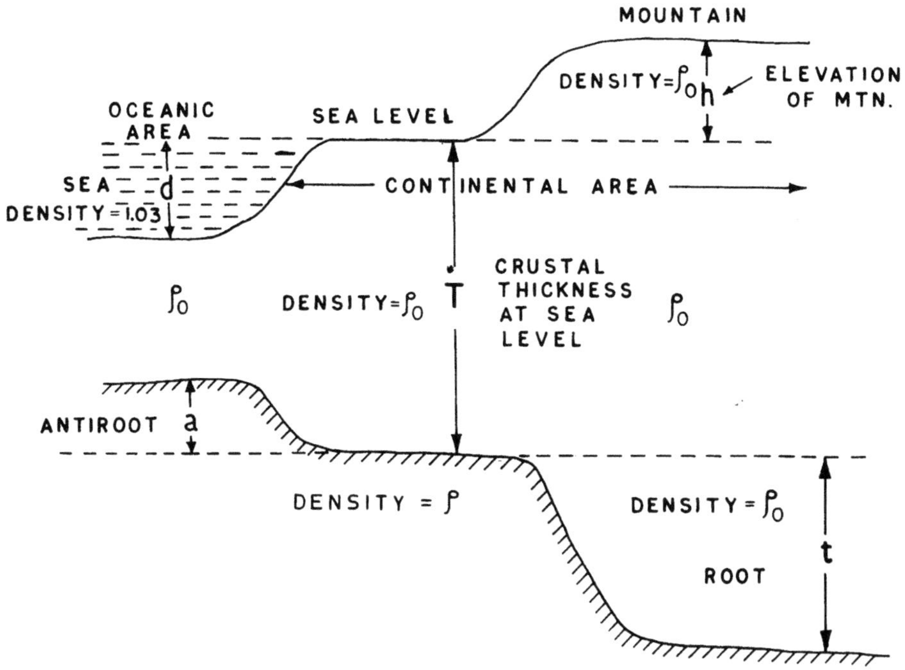

Fig. 8.2. Nature of earth's crust under continents forming roots under elevated areas and antiroots under oceans, according to Airy's concept of isostasy.

materials underlying it, the upper mantle. Let T be the thickness of the crust at sea level, d the depth of water column under the sea, h, the height of the moutain above the sea level. According to the Airy system, the mass of the mountain should be compensated by the presence of a root of lighter material (thickness t) protruding into the mantle of higher density. Similarly, the lighter mass of the water column under the oceanic areas will be compensated by the presence of an antiroot, of thickness 'a'. Now, if we consider masses/unit area under the three areas, they should be the same.

Equating the mass of the mountain per cm^2 above the sea level with the deficiency of mass present in the root, we get,

$$\rho_0 \cdot h = t\ (\rho - \rho_0) \tag{8.1}$$

or

$$t = \frac{\rho_0 \cdot h}{(\rho - \rho_0)} \tag{8.2}$$

Equation (8.2) shows that the thickness of the root (t) is directly proportional to the elevation of the mountain and inversely proportional to the density contrast between the crust and the mantle. If the mean density of the crust is taken as $\rho_0 = 2.88$ gm cm^{-3} and that of the mantle, $\rho = 3.31$ gm cm^{-3} (Woollard, 1962), we get a density contrast, $\Delta\rho = 0.43$ gm cm^{-3}. Taking the height of the mountain $h = 1$ km and using Equation (8.2), we see that the root of the mountain will be approximately 6.7 km. For high mountains like the Himalaya, where the average height over large areas may be as much as 5 km, the corresponding root may be of the order of 35 km.

The above mentioned values of densities for the crust and the mantle are by no means representative of all areas. The values are found to differ from place to place and are taken here only to illustrate the main point that the expected root formation under the mountains can be appreciable, of the order of 6 to 7 km for one km elevation.

Equating the masses cm^{-2} under the oceanic areas (taking density of 1.03 gm cm^{-3} for sea water) and the standard crust at sea level we get,

$$d(1.03) + a\ \rho + (T - d - a)\ \rho_0 = T.\ \rho_0 \tag{8.3}$$

or

antiroot,

$$a = \frac{d(\rho_0 - 1.03)}{(\rho - \rho_0)} \tag{8.4}$$

Taking the above mentioned densities for the crust and the mantle materials, we see that for one km depth of the water column, the antiroot formed will be about 4.3 km. In deep oceans, the depth of water is often of the order of 4 or 5 km. The corresponding antiroot expected for such areas is of the order of 17 to 21 km. If the standard thickness of the continental crust at sea level is taken as 30 km, for 4 km depth of water column the thickness of the crust under oceanic areas as predicted by Airy's theory is of the order of 9 to 10 km, which is appreicably lesser than that of the continental crust at sea level.

During the years 1924, 1931, and 1938 Heiskanen made the Airy assumption more precise and computed tables for computing isostatic correction. The assumptions of Airy-Heiskanen system are as follows.

(1) Isostatic compensation is complete.
(2) The compensation is local and directly under the topography.
(3) The density of the earth's crust is everywhere constant and is equal to 2.67 gm cm^{-3}.
(4) The density of the underlying material is 3.27 gm cm^{-3} or greater than that of the crust by 0.6 gm cm^{-3}.

The calculation of isostatic anomalies under these assumptions will be discussed later.

8.2.2. PRATT-HAYFORD SCHEME OF COMPENSATION

Under Pratt's scheme of compensation (modified by Hayford), we consider again three different areas, discussed earlier and shown in Figure 8.3.

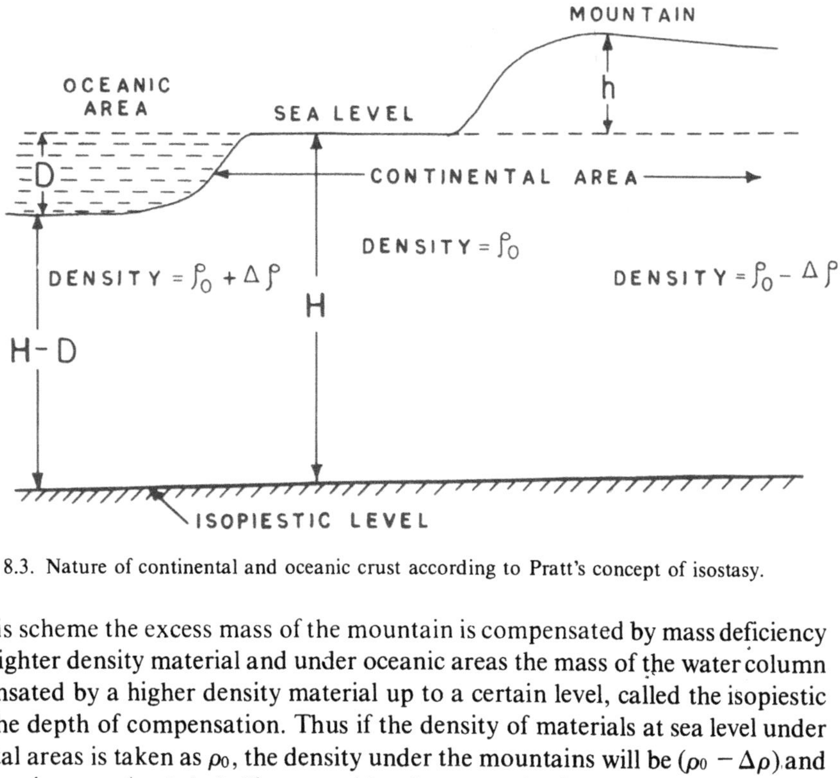

Fig. 8.3. Nature of continental and oceanic crust according to Pratt's concept of isostasy.

Under this scheme the excess mass of the mountain is compensated by mass deficiency due to a lighter density material and under oceanic areas the mass of the water column is compensated by a higher density material up to a certain level, called the isopiestic level or the depth of compensation. Thus if the density of materials at sea level under continental areas is taken as ρ_0, the density under the mountains will be $(\rho_0 - \Delta\rho)$ and under oceanic areas $(\rho_0 + \Delta\rho)$. If we consider the masses/unit area up to the depth of compensation, under the mountainous areas we get,

$$H \cdot \rho_0 = h\,(\rho_0 - \Delta\rho) + H\,(\rho_0 - \Delta\rho) \tag{8.5}$$

or

$$\Delta\rho = \frac{h \cdot \rho_0}{(H + h)} \, . \tag{8.6}$$

Equation (8.6) shows that the value of $\Delta\rho$ is directly proportional to the height of the mountain.

Considering the masses/unit area under the oceanic areas, we get,

$$H\rho_0 = 1.03\ D + (\rho_0 + \Delta\rho)\ (H - D) \tag{8.7}$$

or

$$\Delta\rho = \frac{D(\rho_0 - 1.03)}{(H - D)}\ . \tag{8.8}$$

This suggests that the value of $\Delta\rho$ under oceanic areas is directly proportional to the depth of the water column above.

8.2.3. VENING-MEINESZ REGIONAL ISOSTATIC SYSTEM

Isostatic compensation discussed earlier cannot be local since small topographic masses are not expected to be isostatically compensated. According to Bowie (1924), features which have an areal extent of about 10 000 km^2 (nearly 1° × 1°) appear to be compensated, while several smaller features do not appear to be in complete isostatic equilibrium.

Taking these factors into account Vening Meinesz (1931) modified the Airy theory and established his own system of regional compensation. The difference between regional and local compensation for a topographic mass such as a mountain is shown in Figure 8.4. It is quite apparent that in the case of regional compensation, a much wider area of the crust is affected by the mechanism of compensation.

In 1941, Vening Meinesz published tables for computing isostatic anomalies according to his system of regional compensation. The exact mechanism of compensation for a particular mountain range/system still falls under the purview of research. However, it is now generally accepted that regional compensation prevails in most of the cases.

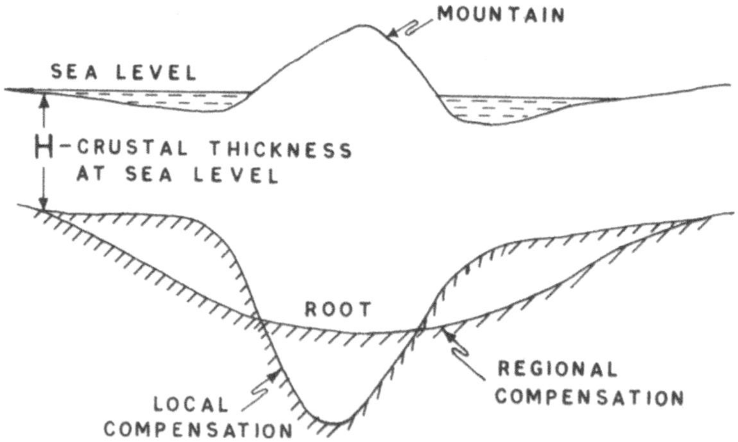

Fig. 8.4. Concept of local and regional compensation for mountain masses. It is quite apparent that in the case of regional compensation, a much wider area of the crust is affected, or roots under mountains may extend beyond the areas underlain by the topographic features.

8.3. Method of Computing Isostatic Anomaly

Computation of isostatic anomaly involves making corrections for the gravity effect of root (or antiroot) at a station, according to Airy scheme of compensation. According to the Hayford scheme, the gravity effect of the mass underlying the topography (negative in the case of elevated areas), up to the depth of compensation has to be computed for the station under consideration. For purposes of computations, Hayford introduced a system of circular zones with the radii given in Table 8.1. As seen in this table, each zone is divided into a number of compartments. Zones A to O extending up to a distance of 1°29′58″, are known as Lettered Zones and zones 18 to 1 are known as numbered zone. Thus for computing the isostatic anomaly, the entire topography around the globe is taken into consideration.

TABLE 8.1

Hayford zones and compartments

Zone	Outer Radius (m)	Number of compartments	Zone	Outer Radius (°) (′) (″)	Number of compartments
A	2	1	18	1 41 13	1
B	68	4	17	1 54 52	1
C	230	4	16	2 11 53	1
D	590	6	15	2 33 46	1
E	1 280	8	14	3 03 05	1
F	2 290	10	13	4 19 13	16
G	3 520	12	12	5 46 34	10
H	5 240	16	11	7 51 30	8
I	8 440	20	10	10 44	6
J	12 400	16	9	14 09	4
K	18 800	20	8	20 41	4
L	28 800	24	7	26 41	2
M	58 800	14	6	35 58	18
N	99 000	16	5	51 04	16
O	166 700	28	4	72 13	12
	(1° 29′58″)		3	105 48	10
			2	150 56	6
			1	180 00	1

Topographic mass within a zone (forming part of a cylinder) can be represented as shown in Figure 8.5. The gravity effect Δg_t of the mass enclosed within the zone at the station 0 is given by,

$$\Delta g_t = \phi \cdot G \cdot \left[\sqrt{r_1^2 + h^2} - \sqrt{r_2^2 + h^2} - r_1 + r_2\right] \qquad (8.9)$$

where ϕ is the angle subtended by the cylindrical segment at the station, r_1 and r_2 are the inner and the outer radii, h is the difference in elevation between the zone and the station, G is gravitational constant. For applying isostatic correction according to the Airy scheme of compensation, the gravity effect of the root corresponding to the topography has to be calculated at the point O. This is given by the relationship,

$$\Delta g_A = \phi \cdot G \cdot \Delta\rho\left[\sqrt{r_1^2 + (T+t)^2} - \sqrt{(r_1^2 + T^2)} + \sqrt{r_2^2 + T^2} - \sqrt{r_2^2 + (T+t)^2}\right] \tag{8.10}$$

where T is thickness of the crust at sea level, $\Delta\rho$ is the density contrast between the root and the surrounding material (Heiskanen and Vening Meinesz, 1958).

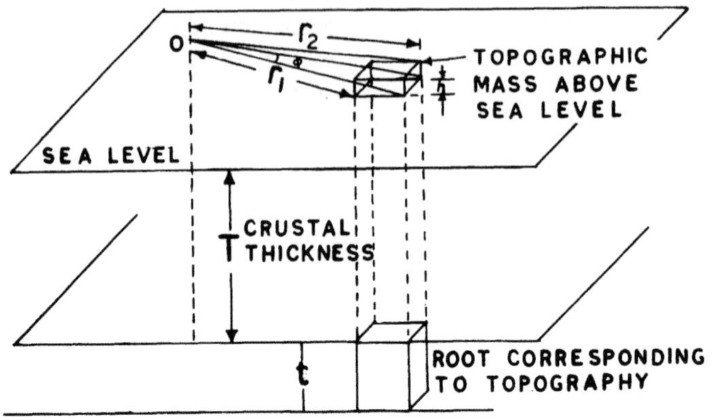

Fig. 8.5. Segment of a cylinderical mass above sea level and its corresponding root, for purposes of calculating isostatic correction.

Heiskanen (1938) has computed reduction tables for calculating the gravity effect of various Hayford zones (A to O) for different values of elevation and $T = 20, 30, 40$, and 60 km. These tables have been used extensively for the calculation of Airy isostatic anomalies. In preparing these tables, the zones A to G have been combined into one for purposes of simplifying the calculations. For applying topographic–isostatic correction accroding to Airy's scheme for $T = 30$ for zones 18 to 1, the maps prepared by Larki *et al.* (1961) for the whole world can be used. The maps instantly give the value of the correction for any particular latitutde/longitude of the world.

For calculating the isostatic anomaly over continental areas according to Hayford scheme, a slightly modified form of Equation (8.10) (Heiskanen and Vening Meinesz, 1958) has to be used i.e.

$$\Delta g_H = \phi \cdot G \cdot \left[\frac{h\rho_0}{H+h}\right]\left\{\sqrt{(r_1^2 + H^2)} - \sqrt{(r_2^2 + H^2)} - r_1 + r_2\right\} \tag{8.11}$$

where, H is the depth of compensation. Other parameters are the same as per Equation (8.8). It may be noted that in this formula, the density of the cylinderical column is replaced by the density deficiency under the topographic zone as calculated by using Equation (8.8).

For practical purposes, the topography around the station is divided into various zones (given in Table 8.1) using a suitable template on the same scale as the topographic maps. The mean elevation of each zone is obtained and isostatic correction for Airy or Hayford systems is computed using suitable tables, such as the ones published by Isostatic Institute of Helsinki (Heiskanen, 1958) or Netherlands Geodetic Commission (Vening Meinesz, 1941).

An example of isostatic reduction for a station (X) using Airy Heiskanen System is shown in Table 8.2. The isostatic correction has to be added along with other corrections to obtain isostatic anomaly, which is defined, as,

$$\Delta g_{iA} = \Delta g_B + \text{Airy isostatic correction for zones A to O} + \text{Topographic plus Airy isostatic correction for zones 18 to 1}$$

where, Δg_B is the Bouguer anomaly including topographic correction up to Hayford zone 0 and the curvature correction. In Hayford topographic-isostatic reduction, the effects of topographic mass anomalies and the compensating mass anomalies are considered from zone to zone. This reduction already contains both the Bouguer and terrain corrections. Hayford isostatic anomaly may be defined as,

$$\Delta g_{iH} = \Delta g_B + \text{Hayford isostatic correction for zones A to O} + \text{Topographic plus Hayford isostatic correction for zones 18 to 1}$$

TABLE 8.2

Typical example of Calculation of Airy Isostatic Anomalies ($T = 30$ km) for a Station X

Station No X, Latitude 16°5 N, Longitude 83° E, Elevation 46 m, Bouguer Anomaly, −20.0 mgal

Zones	Mean elevation (m)		Isostatic correction (mgal)
A–G	305		+0.2
H	396		+0.4
I	366		+0.8
J	427		+1.7
K	335		+2.7
L	411		+5.6
M	457		+13.7
N	442		+8.5
O	610		+8.5
A–O	–		+42.1
18–1	–		−10.0
A–1	–	(total)	+32.1

Bouguer anomaly (Δg_B) for Stn. $X = -20.0$ mgal.
Airy Isostatic anomaly (Δg_{iA}) for Stn. $X = +12.1$ mgal.

8.4 Isostatic Anomaly Map of Peninsular India

According to Airy's scheme of compensation, isostatic correction for elevated areas essentially is a correction for the gravity effect of the roots for masses above sea level. This is added to the Bouguer anomaly which does not take this factor into account. In areas characterised by isostatic compensation, the isostatic anomaly should approach near zero values. The Airy scheme of compensation predicts that a correlation between elevation and crustal thickness should exist i.e. areas with large elevations should have greater crustal thickness and vice versa. Studies of crustal thickness using Deep Seismic Sounding methods and seismological data have enabled such a correlation to be studied on a world wide basis (Garland, 1965; Heiskanen and Vening Meisenz, 1958; Woollard, 1962).

It has been found that a number of mountainous areas of the world have large crustal thicknesses, negative Bouguer anomalies but very low isostatic anomalies, indicating that crustal thickness variations are very close to those required for isostatic compensation. There are other areas however, where isostatic over or under compensation prevails, though on a World-wide scale the phenomenon of isostasy tends to be valid (Garland, 1965).

Isostatic anomaly map of India prepared by N.G.R.I. on the basis of Airy-Heiskanen scheme of compensation for $T=30$ km is shown in Figure 8.6. Hayford isostatic anomaly map for H=113.7 km prepared by Survey of India is shown in Figure. 8.7. The figure shows that dominant positive anomalies (highest, +80 mgal over Shillong Plateau) are found in the northern part of Peninsular India, while the southern part is predominantly negative (lowest negative anomaly being nearly −100 mgal near Madras).

Some of the major trends seen on the Bouguer anomaly map can also be seen on the isostatic anomaly map as well. These include the north-easterly trend of the Aravalli mountains, the easterly trend of the Satpura mountains, the north-easterly trend of the Eastern Ghats, the north-westerly trend of the Godavari Valley etc. (see Figure 3.2 for major trends in Peninsular India). The Aravalli mountains are characterized by appreciably positive isostatic anomalies (maximum, +50 mgal), compared to near zero anomalies on its margins. This feature together with sharp gradients of Bouguer anomalies observed on its margins (see Figure 3.5) supports the contention that the Aravallis is a horst-type strucutre. Similar appears to be the case for the Satpura mountains which maintain a positive trend (near 21° N) from 76° E to 84° E, isostatic anomalies reaching a maximum value of about +40 mgal.

In northern peninsula of India three mountain ranges, the Aravallis, the Vindhyans and the Satpuras have a distinct topographic expression. Some idea of the nature of isostatic compensation prevailing under these mountains can be gained by considering the nature of gravity anomalies prevailing over the Aravalli mountains, the Vindhyan and Satpura mountains as well as the Eastern Ghats along profile AA′ (see Figure 3.5). It may be noted from the figure that.

(a) the free-air anomaly is strongly positive over the Aravallis, Satpuras, the Eastern Ghats and bears a good correlation to topography. Isostatic anomalies over these areas are also positive reaching up to +50 mgal maximum over Aravallis and +40 mgal over the Satpuras. Over the Eastern Ghats the isostatic anomalies are low, of the order of +10 to −10 mgal.

(b) the inverse relationship expected between elevation and Bouguer anomally over these mountains, on the basis of isostasy is nearly absent. On the other hand Bouguer anomalies are positive over the mountains with respect to the surroundings,

(c) there is a considerable difference between the Bouguer anomalies and isostatic anomaly along the profile. The former is appreciably negative along the profile while

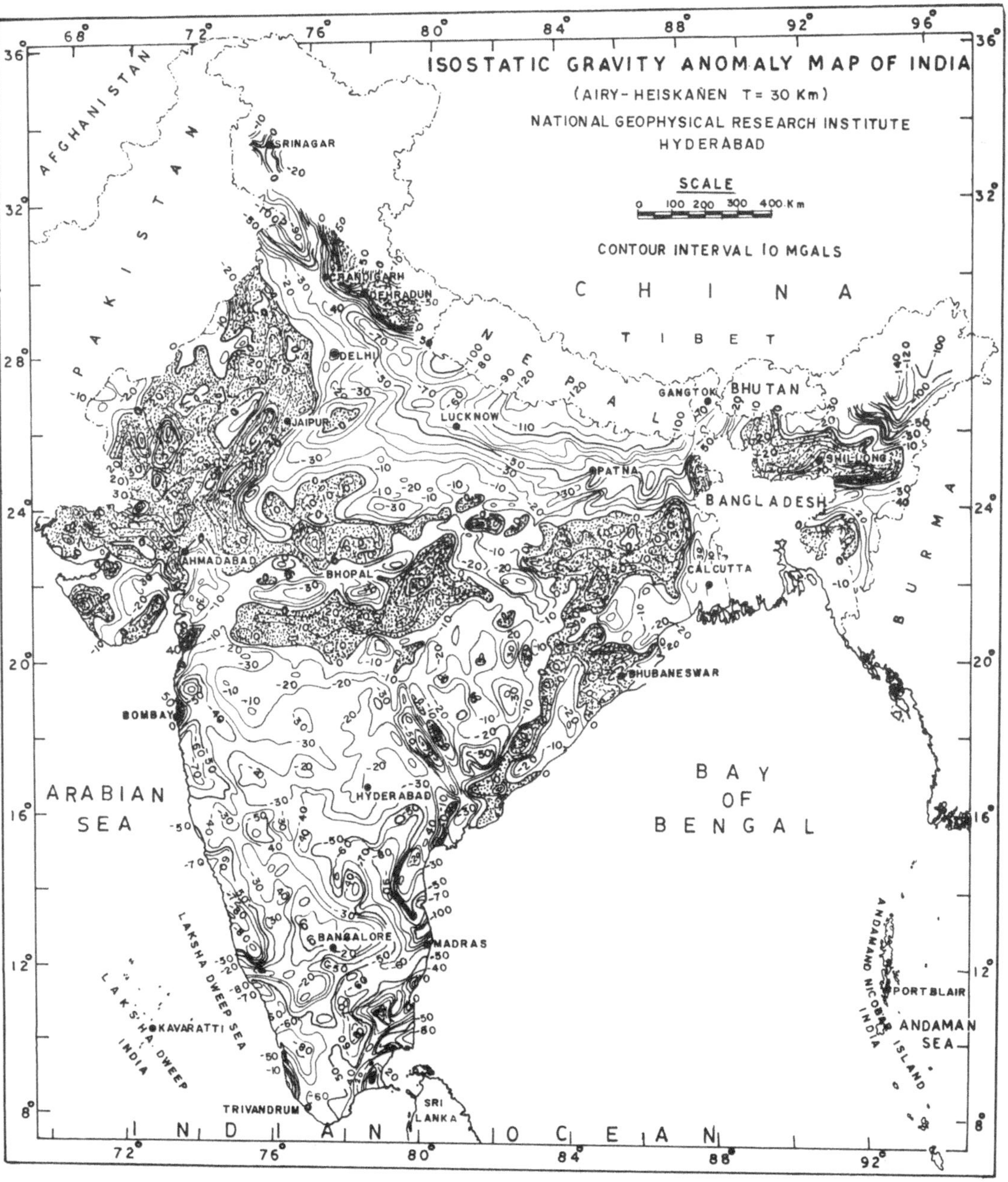

Fig. 8.6. Airy isostatic anomaly map of India for $T = 30$ km prepared by National Geophysical Research Institute, Hyderabad, India. (Reproduced with permission of Director, N.G.R.I.).

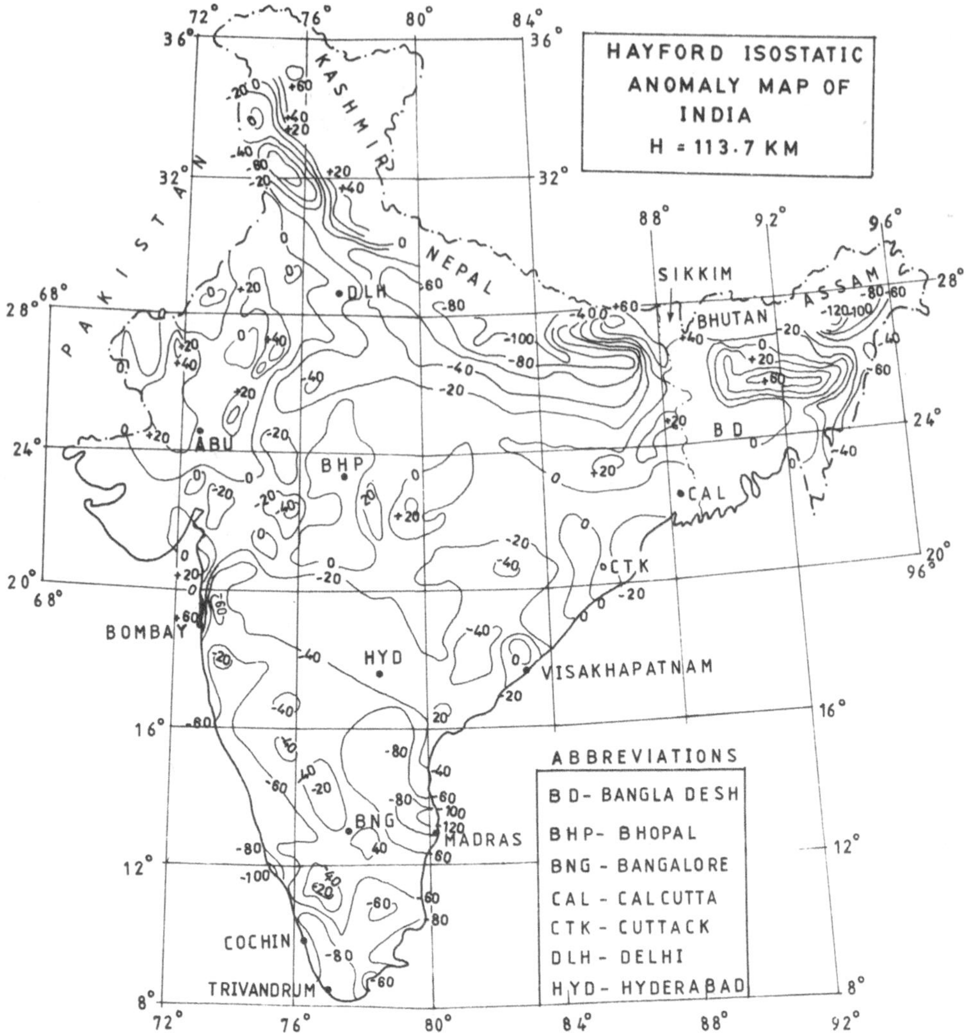

Fig. 8.7. Hayford isostatic anomaly map of India, $H = 113.7$ km, after Survey of India.

the latter oscillates around zero value.

The above mentioned features of the gravity field suggest that these mountains are not locally compensated. It appears that isostatic compensation prevails on a regional scale.

In southern peninsula, the Deccan plateau as well as the Mysore plateau appear to be in isostatic equilibrium, since the isostatic anomalies are of the order of −20 to −30 mgal and 0 to −20 mgal respectively in these areas. In plain areas, close to sea level, isostatic correction is very small in general (unless the area is close to mountains). Hence the difference between the isostatic anomalies and Bouguer anomalies is very small. In such areas, the isostatic anomalies like Bouguer anomalies reflect the effects of underlying geological masses rather than topography. Typical example of these are the Godavari Valley and the area near Bombay. In the former case the negative values are clearly due to the low density Gondwana sediments (Qureshy *et al.*, 1968), and in the latter case the positive anomalies are attributed to the high density magmatic mass which has been differentiated from a basic magma (Takin, 1966).

The Shillong Plateau in north-eastern India is one of the most conspicuous areas in India from the point of view of gravity. Here isostatic anomaly is strongly positive, maximum value reaching about +80 mgal. Bouguer anomalies (see Chapters 3 and 11) are also positive and do not show the inverse relationship to elevation expected for an elevated area. These features suggest that the Shillong Plateau is not locally compensated and may be composed of high density rocks. The nature of anomalies over the plateau will be discussed in detail in Chapter 11.

The Cuddapah basin in the southern peninsula of India, is characterized by large isostatic anomalies, varying from −30 mgal in the western margin to −90 mgal in the east. The anomalies can be explained in terms of a crustal downwarp in the eastern part of the Cuddapah Basin as revealed by D.S.S. studies (Kalia, 1982). The downwarp is probably related to the nature of the boundary between the Dharwar craton to the west and the Eastern Ghats to the east (see discussion at the end of Chapter 6).

The state of isostatic compensation in the southern peninsula of India can be further understood from study of the nature of Bouguer, free-air and isostatic anomalies prevailing over the southern hill masses such as the Western Ghats, the Nilgiri Hills, and the Palni-Cardamom Hills etc. Figure 8.8 shows a regional elevation map of the southern peninsula south of 16° N. The regional elevation contours clearly bring out the Western Ghats, the Nilgiri and Shevroy Hills, the Palni-Cardamom group of hills and the Palghat Gap, separating the Palnis from the Shevroys. The nature of gravity field prevailing over the hill masses is considered along three profiles AA′ A″, BB′, and CC′ (after Subrahmanyam, 1978) shown in Figure. 8.8.

8.4.1. PROFILES ACROSS SOUTHERN HILL MASSES

8.4.1.1. Profile AA′ A″

Figure 8.9a shows the nature of free-air, Bouguer and isostatic anomalies (Airy, for T= 30 km) as well as the nature of topography along the profile AA′ A″. Various numbers along the profile refer to gravity stations. The profile starts from the low lying coastal areas west of station 492 and passes through the Western Ghats, maintaining a steady elevation of 800 to 1000 m east of station 488. Bouguer anomalies show a decreasing trend from −40 mgal to −120 mgal along the profile. It is interesting to note that the Bouguer anomaly shows a steady decrease in values from −40 to −100 mgal along the coastal areas from station 56 to 492, although there is no corresponding increase in elevation between these stations. There is no steep fall in Bouguer anomaly from station 492 to 488 inspite of a sharp increase in elevation by about 800 m between them. Instead, Bouguer anomaly shows a gradual decreasing trend. Variations of Bouguer anomaly from regional values can be explained in terms of local geological factors. Free-air anomaly follows the Bouguer anomaly in the coastal areas as expected on account of low elevations. Further east, the anomaly rises sharply from station 492 to 488, in response to rise in topogrpahy and then maintain values between 0 and −20 mgal. Isostatic anomaly is −40 mgal near the west coast, shows a slight decrease in values and then rises to about −20 mgal over the hill areas. Variations in isostatic anomaly over the Western Ghats can be explained in terms of underlying geological formations. Total lack in response of Bouguer anomalies to the abrupt rise in topography, the tendency of isostatic anomalies to reach near zero values and sharp response of free-air anomaly to elevation over the Ghats suggests that regional compensation prevails over the area.

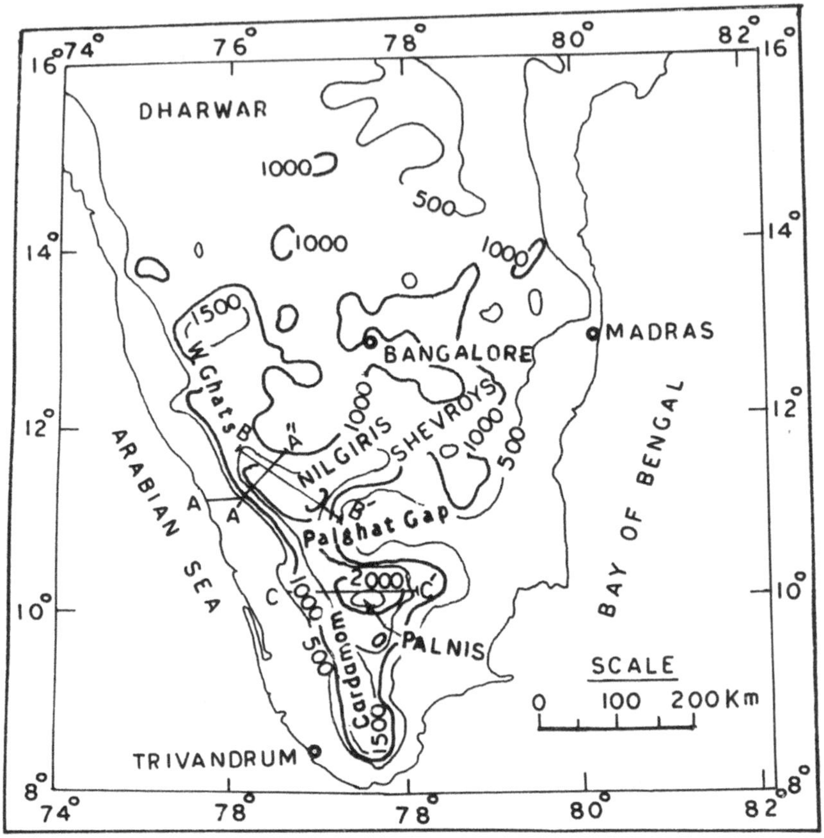

Fig. 8.8. Regional elevation contours for topographic hill masses of S. India. Locations of profiles AA′ A″ and BB′ across Western Ghats, Nilgiri Hills and profile CC′ across Palni-Cardamom Hills discussed in the text are shown in the figure.

8.4.1.2. Profile BB′

This profile cuts across the Nilgiri Hill mass in a general NW–SE direction. Topography and gravity anomalies along the profile are shown in Figure. 8.9b. Nilgiris are indicated by highly elevated topography between stations 475 to 456, elevations reaching as high as 2400 m. The Nilgiri hills are the widest along the profile, being nearly 60 km across.

Free-air anomaly is near zero on the west coast, but increases sharply to −180 mgal over the Nilgiri Hills. Both Bouguer and isostatic anomalies follow the trend of free-air anomaly although to a lesser degree. Isostatic anomaly reaches values close to zero over the hills. The Bouguer anomaly however, does not show the inverse relationship to topography over the Nilgiri Hills as would be expected if there were a local root formation under the Hills. Instead, the anomaly becomes more positive over the hills. The positive nature of the gravity field is probably due to the high density rocks composing the hills (discussed in Chapter 6). The regional Bouguer anomaly along the profile is at a general level of about −120 mgal, which can be explained only on the basis of a regional root formation under the hill masses. There appears to be no local root formation under the Nilgiris.

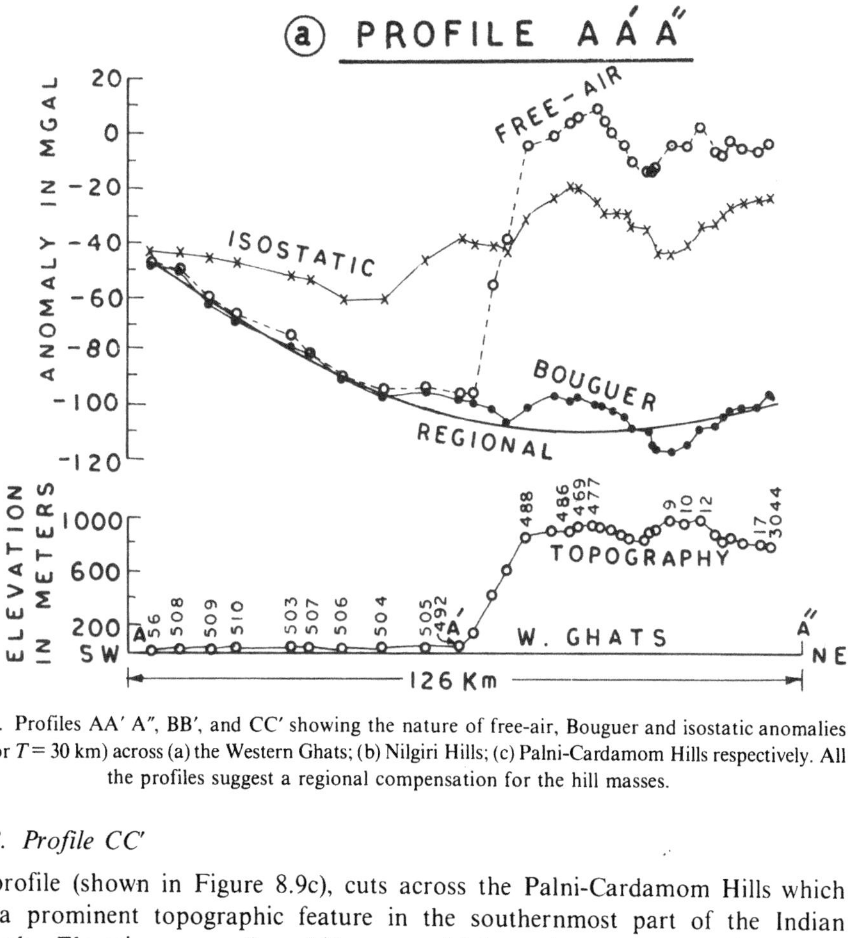

Fig. 8.9. Profiles AA′ A″, BB′, and CC′ showing the nature of free-air, Bouguer and isostatic anomalies (Airy, for $T = 30$ km) across (a) the Western Ghats; (b) Nilgiri Hills; (c) Palni-Cardamom Hills respectively. All the profiles suggest a regional compensation for the hill masses.

8.4.1.3. Profile CC′

This profile (shown in Figure 8.9c), cuts across the Palni-Cardamom Hills which form a prominent topographic feature in the southernmost part of the Indian Peninsula. Elevations over these hills are as high as those over the Nilgiris. An interesting feature is that the topographic gradients encountered in the western part are very gentle, while the eastern margin is characterised by a very sudden drop. Free-air anomalies systematically rise over the hills, responding to topographic elevations as in earlier cases. Bouguer anomaly shows a rather mild decrease with rise of topography without any abrupt change. Even near the coastal areas the anomalies are of the order of −80 to −100 mgal, suggesting that these are influenced by regional factors. Isostatic anomaly maintains a rather uniform level at about −70 mgal.

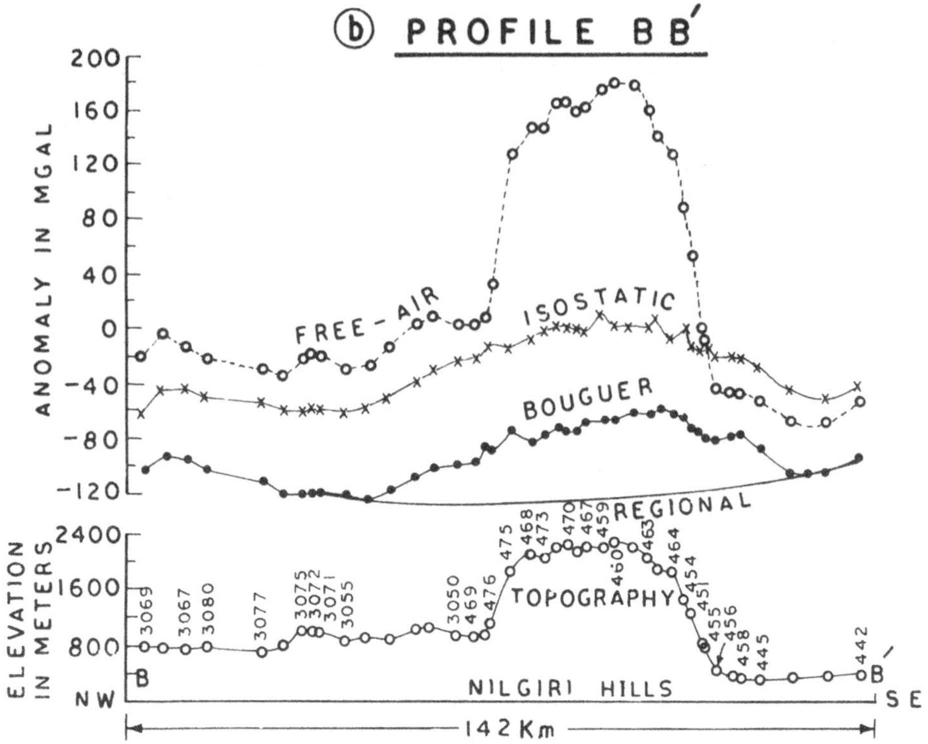

(b) PROFILE BB′
ANOMALY IN MGAL
FREE-AIR
ISOSTATIC
BOUGUER
REGIONAL
ELEVATION IN METERS
TOPOGRAPHY
NILGIRI HILLS
NW
SE
B
B′
142 Km

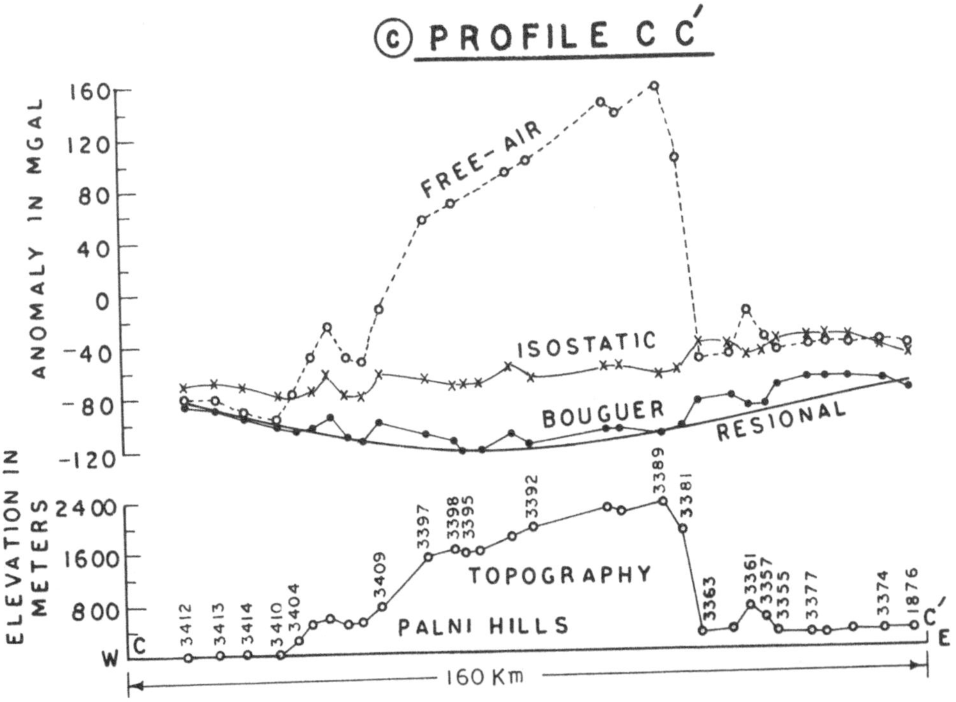

(c) PROFILE CC′
ANOMALY IN MGAL
FREE-AIR
ISOSTATIC
BOUGUER
RESIONAL
ELEVATION IN METERS
TOPOGRAPHY
PALNI HILLS
W
E
C
C′
160 Km

The nature of gravity anomalies along the three S. Indian profiles shows that regional compensation prevails in the area. In other words the continental lithosphere is able to withstand the load of the hill masses through a process of regional compensation.

8.5. Depth of Compensation

Airy and Pratt's hypotheses assume in some way or the other that isostatic compensation is local, since either a mass deficiency or the root underlying a topographic high should be proportional to elevation. Thus local compensation assumes movement of material horizontally or vertically in order to achieve compensation. It is generally believed that isostatic compensation is achieved at the level of Moho which bears an inverse relationship to topographic elevations. Indeed a general correlation between topographic elevation and Moho is found over continental areas (Heiskanen and Vening Meinesz, 1958; Woollard, 1962). However, when examined in the light of Deep Seismic Soundings (D.S.S.), it is found that Pratt and Airy's hypotheses are not always supported by observations of crustal thickness (Pakiser and Steinhart, 1964; Kaila, 1981).

Figure 8.6 shows that appreciably negative isostatic anomalies (of the order of −60 to −80 mgal) prevail over large parts of southern peninsula of India. The anomalies do not disappear even if calculated on the basis of Pratt's hypothesis for depth-of-compensation, $H = 113.7$ km as seen in Figure 8.7. The isostatic anomalies in southern peninsula appear to be influenced by the Indian Ocean gravity low (Kahle and Talwani, 1973) which extends over large parts of the Indian Ocean (see Figure 10.3, Chapter 10). The gravity low has a wave length of about 2000 km and appears to be related to a major mass deficiency existing in the upper mantle at a depth of few hundred kilometers (discussed in detail in Chapter 10). It appears therefore that part of the isostatic compensation for the hill masses in Indian peninsula is achieved at depths exceeding a hundred kilometers.

A recent approach to the study of isostatic mechanism has been that of Dorman and Lewis (1970, 1972), Lewis and Dorman (1970). They have analysed the gravity data for U.S.A. in frequency domain and show that isostatic response varies for different wave length. According to them positive as well as negative density differences are required in order to achieve complete isostatic compensation of topographic masses. Such differences may exist up to depths of the order of 400 to 500 km.

Banks *et al.* (1977) have also studied isostatic response as a function of wave length and show that a major change in density occurs at a depth of about 40 km. They have proposed a model involving flexure of the Moho as well as of the lithosphere down to a depth of about 100 km for purposes of accommodating the overlying masses. According to their model isostatic compensation is achieved by outflow of material in the underlying asthenosphere but the nature of gravity field is predominantly determined by the warping of the interfaces at the Moho, where the density contrast is maximum. In such a model Moho will have, in general, an inverse relationship to the topogrpahy but isostatic compensation will not be complete at this level. The bending of lithosphere will play some part and contribute to long wave length anomalies. This appears to be the case for isostatic compensation of topogrpahic masses in southern Peninsular Shield of India (discussed earlier).

Now we may look at isostasy on the basis of theory of plate tectonics. This theory is based on the concept of a strong lithosphere overlying a weak asthensophere. One of the well studied loads in the oceanic lithosphere is that of Hawaiian-Emperor Seamount chain in the Pacific Ocean. Vening Meinesz (1941) had suggested that this load was not compensated according to Airy's model but was regionally compensated. Walcott (1970), Watts and Cochran (1974) have constructed theoretical models to explain the nature of compensation of the Hawaiian chain. They have shown that the gravity data are consistent with a simple model in which the lithosphere supports the load in a manner similar to a thin plate overlying a weak fluid. The flexural rigidity of the lithospheric plate is found to be of the order of 2×10^{30} dyne −cm. The lithospheric model constructed by them can explain the nature of gravity anomalies observed over the Hawaiian Islands.

It appears that in most cases the lithosphere bends or adjusts itself to overlying load so as to achieve regional compensation for the overlying masses such as mountains or sedimentary basins.

CHAPTER 9

GRAVITY FIELD AND ISOSTASY IN THE HIMALAYAS

Most of the information concerning the phenomenon of isostasy prevailing in the Himalayas has come from the geodetic measurements as well as from the study of the gravity field. However, before discussing these evidences it is useful to briefly outline the geology and tectonics of the Himalayas, since the gravity field is considerably influenced by the tectonic history of the highest mountains in the world.

9.1. Geology and Tectonics of the Himalaya

The oldest rocks present in the Himalayas are the metamorphosed sediments, intrusive granites, granulites, ortho-gneisses, phyllites and schists of Archaean age. Most of the high ranges of the Himalaya forming the Central Axis are composed of these rocks (Wadia, 1953, p. 88). Three kinds of granites have been recognised in the Archaean Complex. These include; biotite granite, hornblende granite and tourmaline granite. The first one of these is the most abundant, extending from Kashmir up to Assam. It is quite reasonable to assume that the Precambrian rocks of the Himalaya once formed a part of the Indian peninsula.

According to Wadia (1953), the Himalaya have had their origin in a series of earth movements, which originated from outside India. The most dominant of these movements are referred to as powerful thrusts, acting from north of Tibet and directed towards the peninsula of India. The peninsula has limited the extent and also to some extent controlled the form of the mountain ranges. The arcuate form of the Himalaya is to some extent due to the resistance offered by the peninsular foreland to the advance of these movements or crustal waves (Wadia, 1953).

The most prominent of the thrusts is the innermost one, which can be traced from the Kashmir Himalaya, all through its entire length to their extremity in eastern Assam. This great fault is known as the Main Boundary Fault (MBF). This low angle thrust fault dips towards north, and separates the Upper Paleozoic (or older) sedimentary formations on the north from the Upper Tertiary sedimentary formations (Siwalik) towards the south. Figures 9.1 and 9.8a show geological maps of the northwestern Himalaya (including Kashmir, Panjab and Kumaun) and the Nepal Himalaya respectively. Another equally important thrust in the Himalaya is the Main Central Thrust (MCT). This thrust also dips gently towards north, across which the Precambrian crystalline and metamorphic formations of the north thrust over the Upper Paleozoic formations to the south. The MCT is also a very dominant feature of the Himalaya and extends from Assam in the east to Panjab in the west (Gansser, 1964).

The Himalayan chain is also classified into three parallel zones, differing from one

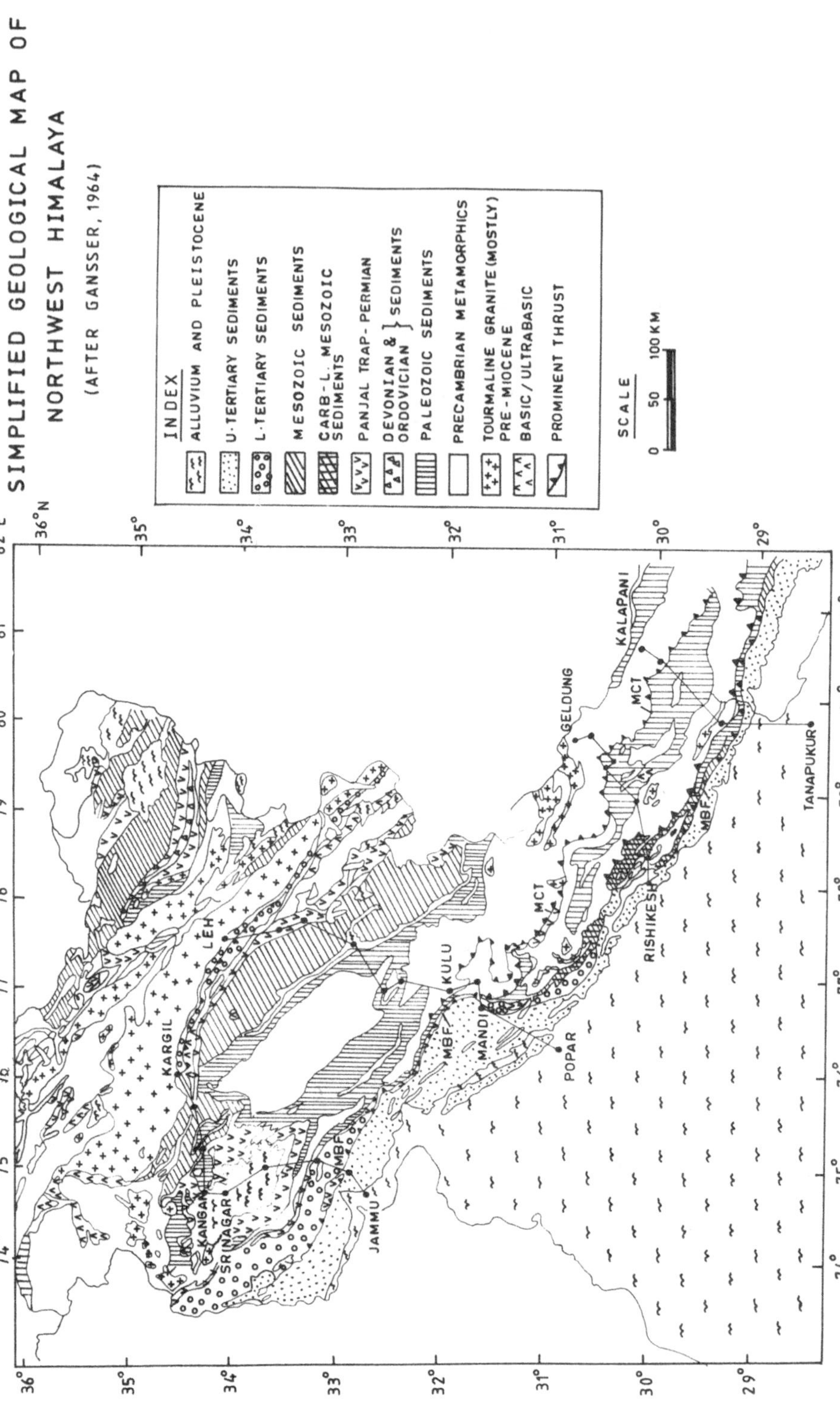

Fig. 9.1. Geological map of NW Himalaya (after Gansser, 1964) showing the location of traverses across Kashmir, Panjab and Kumaun Himalaya

another in well marked orographical features. These are the following.

(1) The Foothills or the Sub-Himalaya. These ranges lie in general between the Indo-Gangetic plains and the MBF and have an average elevation of about 1000 m. The ranges are composed mostly of Upper Tertiary river deposits. (2) The Lesser or Lower Himalaya. These lie mostly north of the MBF and extend in general up to MCT. Their elevation lies between 1000 and 4000 m. These ranges are composed mainly of unfossiliferous Precambrian sediments, crystalline and metamorphic rocks of uncertain ages. (3) The Higher Himalaya. These lie north of the Main Central Thrust and are mostly composed of crystalline metamorphics of Precambrian age as well as crystalline granitic intrusives of Tertiary and older ages. Their average elevation is more than 4000 m. (4) The Tibetan/Tethys Himalaya. These consist of a full sequence of highly fossiliferous sedimentary deposits ranging in age from Cambrian to Lower Eocene (Gansser, 1964). Many of these are marine deposits with inclusions of 'exotic blocks'. These beds are believed to have been deposited in an environment completely different from rest of the Himalaya. The average elevation of the mountains in the Tibetan Himalaya is of the order of 6000 to 6500m.

9.2. Geodetic Evidence for Prevalence of Isostasy

Survey of India (SOI) has done pioneering work in the field of isostasy by studying plumb line deflections in the Himalayas and in the Indo-Gangetic basin in late 19th and early 20th centuries. Burrad (1901, 1912, 1918) was one of the earliest workers in the field of geodesy to suggest that the folding of the Himalayan belt is due to subcrustal flow of mass from underneath the Indian shield towards the Eurasian side, which implies underthrusting of the Indian subcontinent. He ascribed the gravity low over the Ganga basin to the low density sediments deposited in a rift valley and the gravity high over the region of zero deflections over Central India to a subcrustal 'hidden range' composed of high density material. Oldham (1917) found Burrard's idea of a rift valley unlikely and inferred a gradually northward sloping floor underneath the Ganga basin. Burrard (1918), however attributed the positive isostatic anomaly over the great Himalaya to incomplete compensation. Chugh (1978) has discussed the results of detailed geodetic surveys carried out by SOI in Kashmir, Ladakh, Panjab, and Kumaun Himalaya. Figure 9.2 shows the magnitude and direction of the resultant astronomical plumb line deflections with respect to the international spheroid in the Himalayan region.

The deviations depend upon the definition of the spheroid and the actual form of geoid as brought out by earth's irregularities of form and density. It is interesting to note that from Foothills, the deflections point towards the Himalaya in a general northeasterly direction. Further north towards Leh, where altitudes increase to about 5000 m, the deflections decrease in magnitude and finally get reversed, pointing southeast towards the High Himalayan Ranges. Deflections beyond the Ladakh range seem to point towards the Karakoram Range. These deflections by and large suggest the presence of high density masses underneath the Higher Himalayan Ranges. SOI has also carried out computations of isostatic deflection anomalies assuming that isostatic compensation takes place under the Himalaya, according to Hayford system with depth of compensation equal to 113.7 km. The difference between the astronomical

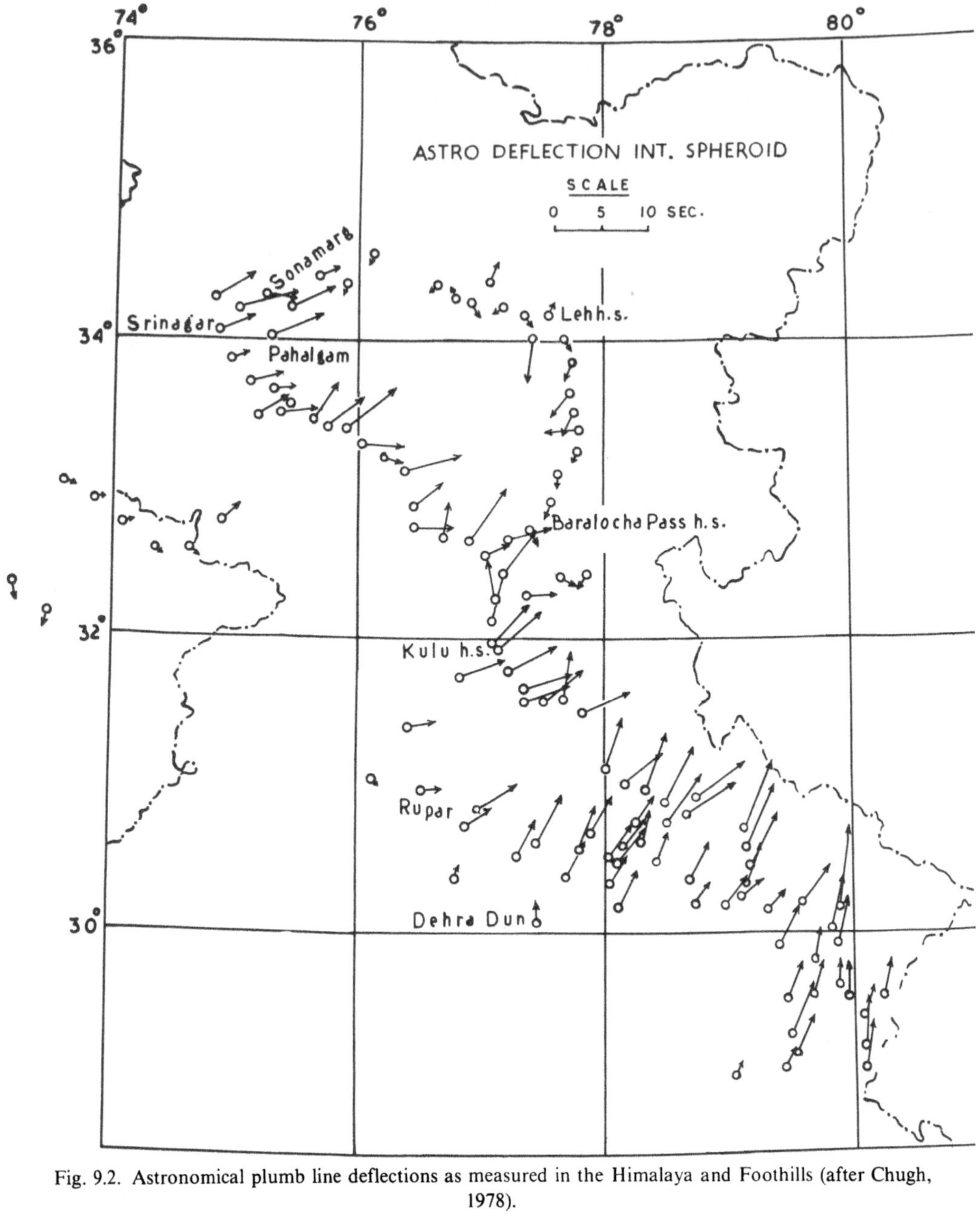

Fig. 9.2. Astronomical plumb line deflections as measured in the Himalaya and Foothills (after Chugh, 1978).

deflections and Hayford deflections are known as Hayford deflection anomalies, and are shown in Figure 9.3. The mean Hayford deflection anomaly for 49 stations north of 32° N latitude is: meridional +4.0″ (maximum range, −24.9″ to 29.0″), prime vertical, −1.9″ (maximum range, −28.8″ to +29.0″) and for 34 stations south of 32° N latitude, is meridional −5.6″ (range 42″) and prime vertical −2.1″ (range 16″), (Chugh, 1978). The

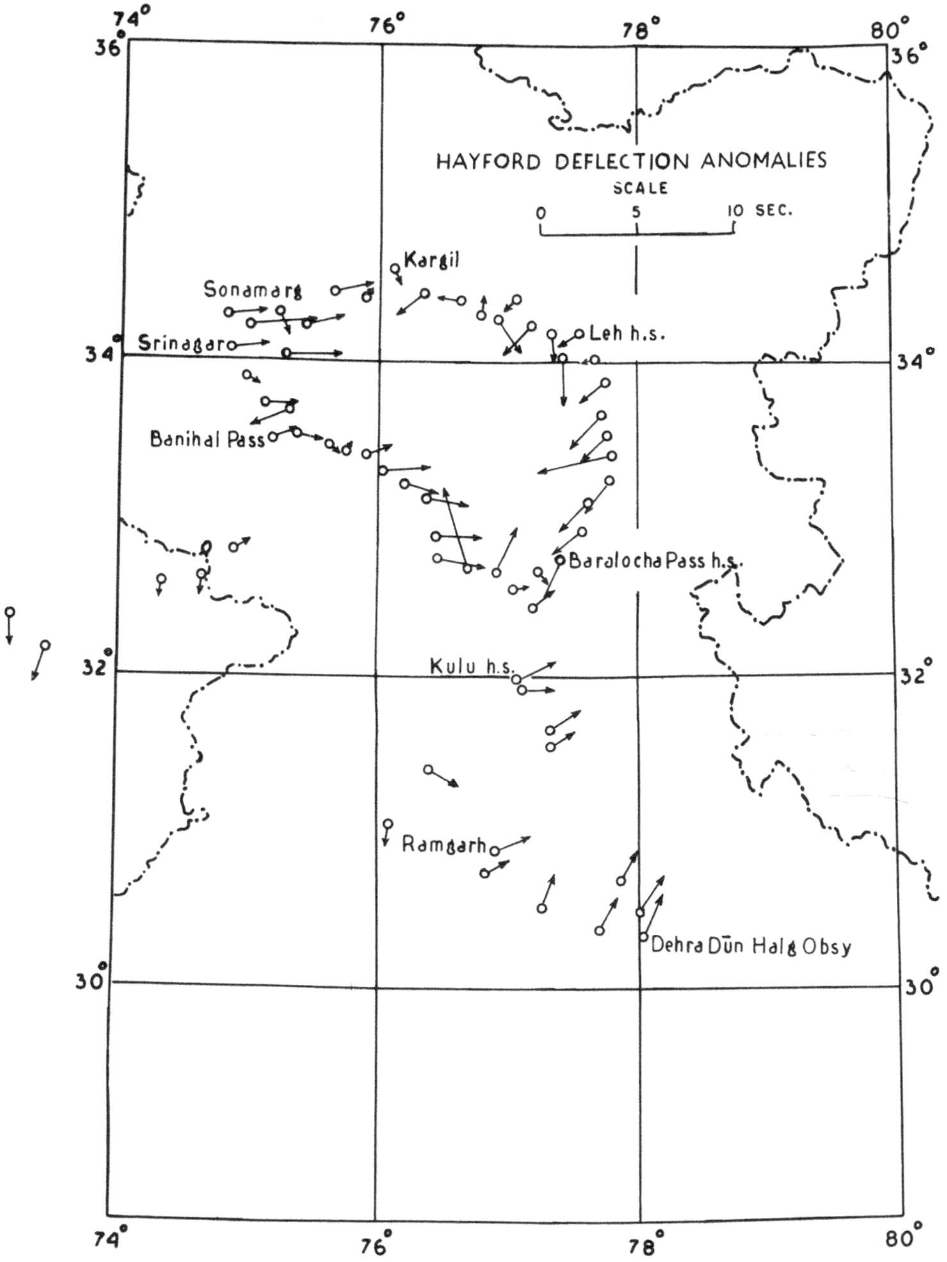

Fig. 9.3. Hayford deflection anomalies in the Himalaya and the Foot-hills (after Chugh, 1978).

resultant of all deflections is close to zero, which lends support to the idea that isostasy is generally in existence in the Himalaya, as a whole.

9.3. Gravity Field in the Himalaya

Although the studies of plumb line deflections throw some light on the nature of isostasy, the studies of gravity anomalies present in the Himalaya give the most useful information regarding the nature of subterranean mass distributions. The two most important anomalies which give information regarding the root formation underneath the Himalaya are the Bouguer and isostatic (Airy or Hayford) anomalies. Figure 9.4 shows Bouguer anomaly map of north-western Himalaya including the Kashmir and Kumaun Himalaya. It may be noted that in the Foothills of Himalaya, where the elevation is less than 300 m, the Bouguer anomalies are of the order of −100 to −120 mgal. It is a characteristic feature of the gravity field in northern India that before entering the Foothills of the Himalaya (which are composed predominantly of Middle and Upper Tertiary sedimentary formations), the Bouguer anomalies over the Indo-Gangetic plains are of this order of magnitude. Such large negative anomalies can be explained partly in terms of large thickness of sediments belonging to the Indo-Gangetic alluvium (Pleistocene) and Upper Tertiary formations. As we enter the Lower Himalaya through Foothills, the Bouguer anomalies become increasingly negative, reaching a value of the order of −175 to −200 mgal. This trend continues towards the Higher Himalaya and the Tibetan Plateau (average elevation approx. 5000 m), where the anomalies reach values of the order of −450 to −500 mgal. The continuous decrease of Bouguer anomaly with elevation is a clear indication of the presence of root formation under the Himalaya.

The Airy isostatic anomaly map of the north-western Himalaya after Chugh (1978) is shown in Figure 9.5. A noteworthy feature of the map is that the anomalies follow the trend of the Himalaya, are appreciably negative (minimum value, −80 mgal) over the Foothills and the Lower Himalaya, become increasingly positive in the Higher Himalaya (maximum value, about +100 mgal) and then decrease to near zero value over the Tibetan Plateau. Thus isostatic anomalies have a definite correlation with elevation in the Himalaya and it appears that significant departures from isostasy may prevail under the Himalaya, at least locally. In the succeeding section are discussed the results of four geotraverses taken by SOI across the Himalaya (Chugh, 1978), with particular emphasis on the state of isostatic compensation.

9.4. Gravity Traverses across the Himalaya

Figure 9.1 shows the location of four gravity traverses taken across the Kashmir, Panjab and Kumaun Himalaya.

9.4.1. JAMMU TO KARGIL VIA SRINAGAR

The relationship of various gravity anomalies to elevation along this traverse is shown in Figure 9.6a. The traverse starts from Jammu located in the Foothills of the Himalaya, where the Bouguer anomaly is of the order of −200 mgal and free-air about −160 mgal. Proceeding northeast the traverse passes through the Foothills underlain

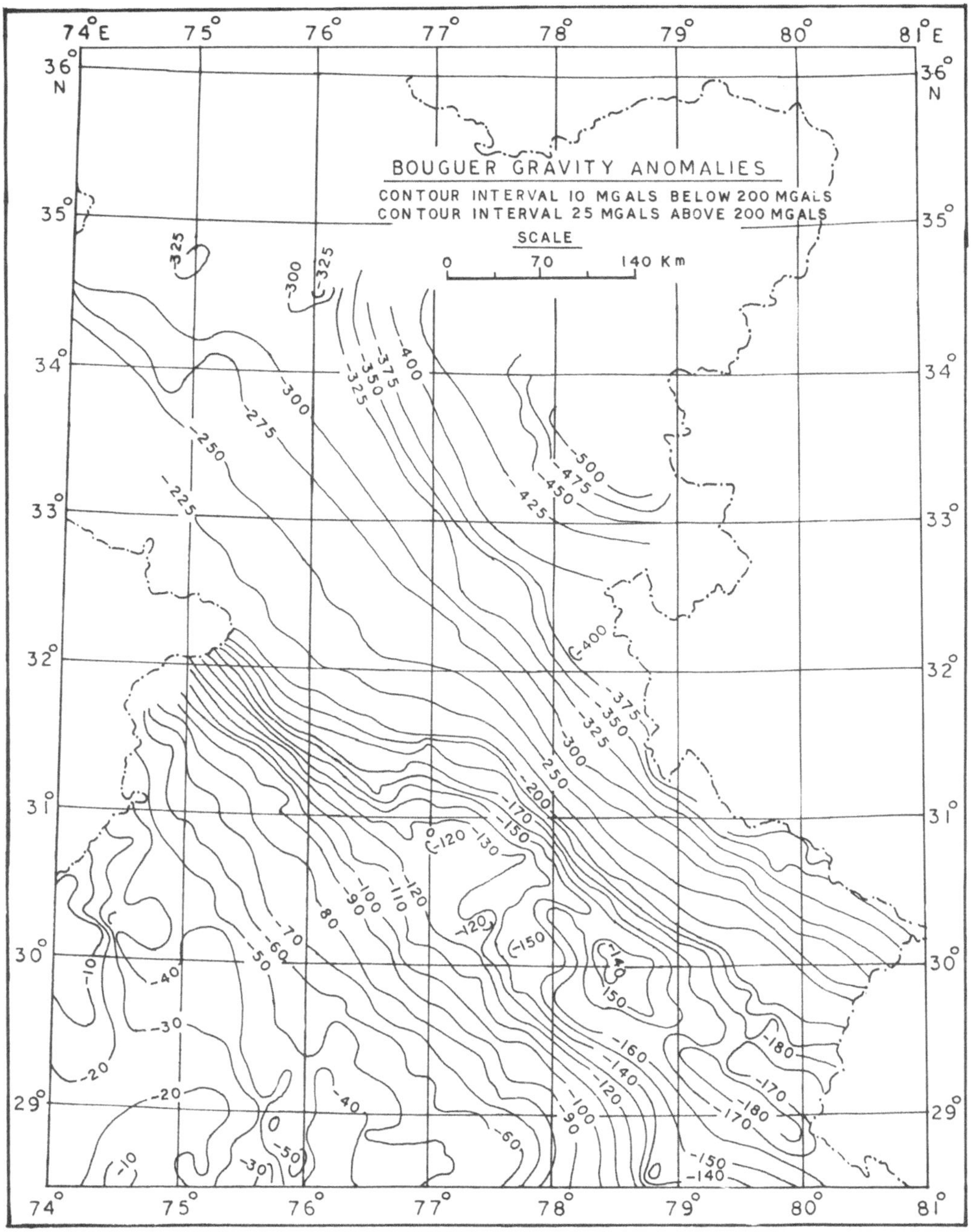

Fig. 9.4. Bouguer anomaly map of NW Himalaya (After Chugh, 1978). Reproduced with permission of R. S. Chugh.

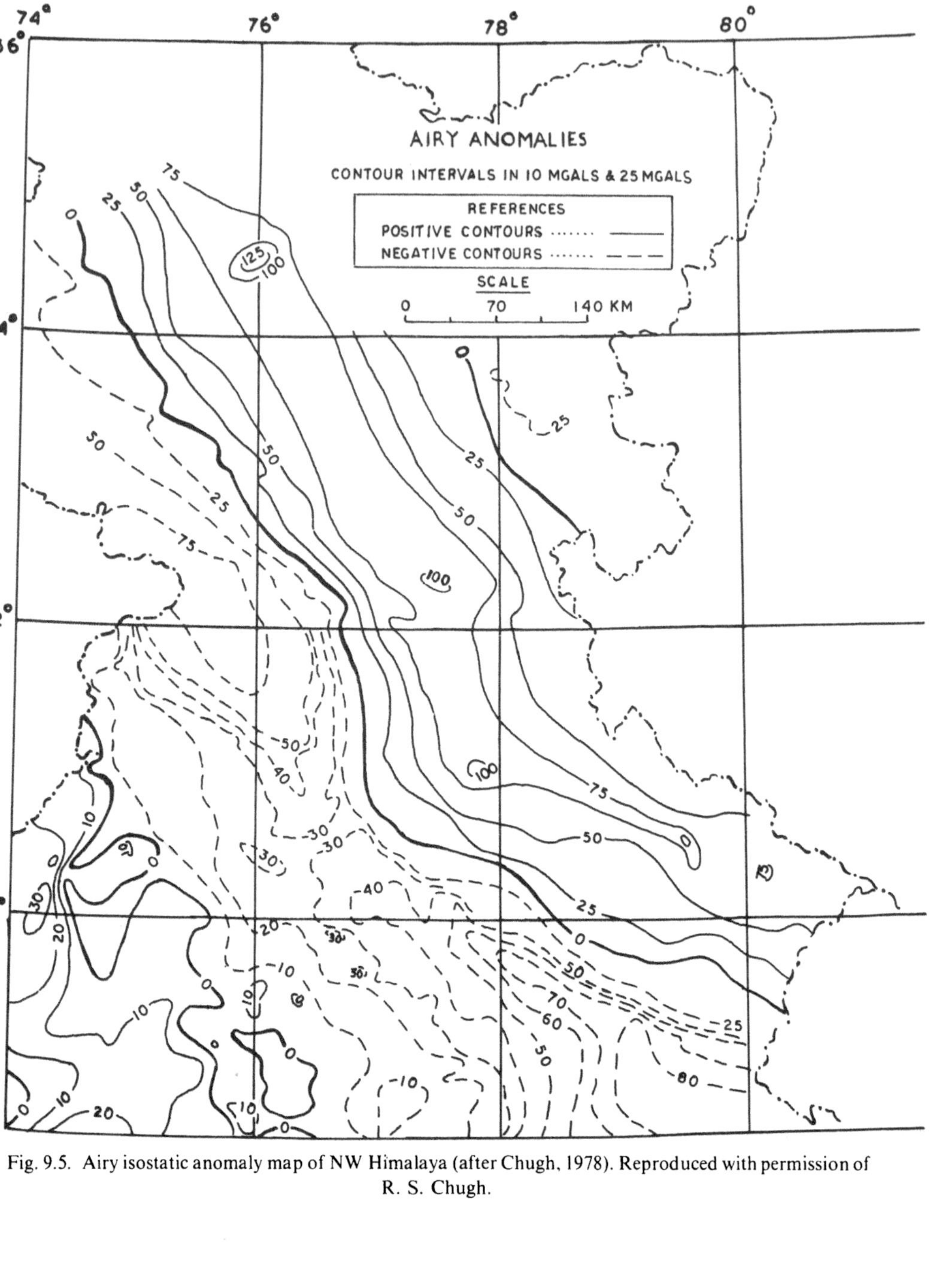

Fig. 9.5. Airy isostatic anomaly map of NW Himalaya (after Chugh, 1978). Reproduced with permission of R. S. Chugh.

by Tertiary formations and crosses the MBF near Udhampur. Further north it passes through the Lower Himalaya underlain by Precambrian, Paleozoic and Mesozoic formations, before entering the Kashmir valley through Banihal pass. Elevations in the Kashmir valley are of the order of 1500 m, while in the Lower Himalaya the fluctuations in elevation are rather large. However the Bouguer anomalies consistently decrease across the ranges and the valley. The free-air anomaly follows the trend of elevations responding to large fluctuations in elevations. However, the values reach close to zero over the Higher Himalaya.

Beyond Kangan in Kashmir valley, the traverse direction changes to northeast towards Kargil. The profile passes through Mesozoic sedimentary formations (Tethys Himalaya) before ending over Tertiary granites near Kargil. The general elevation in the Higher Himalaya is of the order of 2800 to 3000 m. After crossing the Kashmir valley the elevation increases suddenly. However, the Bouguer anomaly does not show a sudden decrease in values, as would be expected if the Himalayas were locally compensated. On the other hand the Bouguer anomaly shows a general negative (sometimes positive) trend. The Airy isostatic anomaly shows an increasing trend, becomes nearly zero in the Kashmir valley and reaches high positive values (maximum, +100 mgal) over the Higher Himalaya.

9.4.2. ROPAR TO LEH VIA KULU

This profile is almost perpendicular to the trend of the major geological formations in the area (Figure 9.1). The profile starts from Ropar at an elevation of about 300 m, where the Bouguer anomaly is about −135 mgal, the free-air anomaly is nearly −100 mgal and isostatic anomaly is about −40 mgal (Figure 9.6b). Along the traverse, the Foothills are composed of Tertiary formations, the Lower Paleozoic are underlain by Precambrian-Paleozoic formations and the Higher Himalaya mostly consist of Paleo-Mesozoic formations as well as Tertiary granites. The traverse ends near Leh, where Tertiary granites are dominant. The free-air anomaly is nearly the same along the traverse up to Manali, beyond which it increases sharply on account of sudden increase in elevation (2000 to 5000 m) from Lower Himalaya to Higher Himalaya. The Bouguer anomaly along the traverse shows a steady decrease (maximum amplitude exceeding −320 mgal), without any jump due to increase in elevation. This supports the validity of regional compensation underneath the Himalaya. The Airy isostatic anomaly becomes nearly zero in the Lower Himalaya, reaching a maximum value of +75 mgal along the traverse.

9.4.3. RISHIKESH TO GELDUNG

Located in the western part of the Kumaun Himalaya, this traverse is underlain mostly by the Precambrian formations of the Himalayas and ends near Geldung, where Mesozoic formations are found. The major features of the various gravity anomalies are repeated along this profile (Figure 9.7a) as well. Bouguer anomaly maintains a steady decreasing trend along the traverse. The free-air anomalies show slight fluctuations but are fairly negative up to about 2500 m, then start rising sharply and follow the elevations up to the end of the profile. The Airy isostatic anomalies are fairly positive along the entire profile, reaching a maximum value of the order of +75 mgal over the Higher Himalaya.

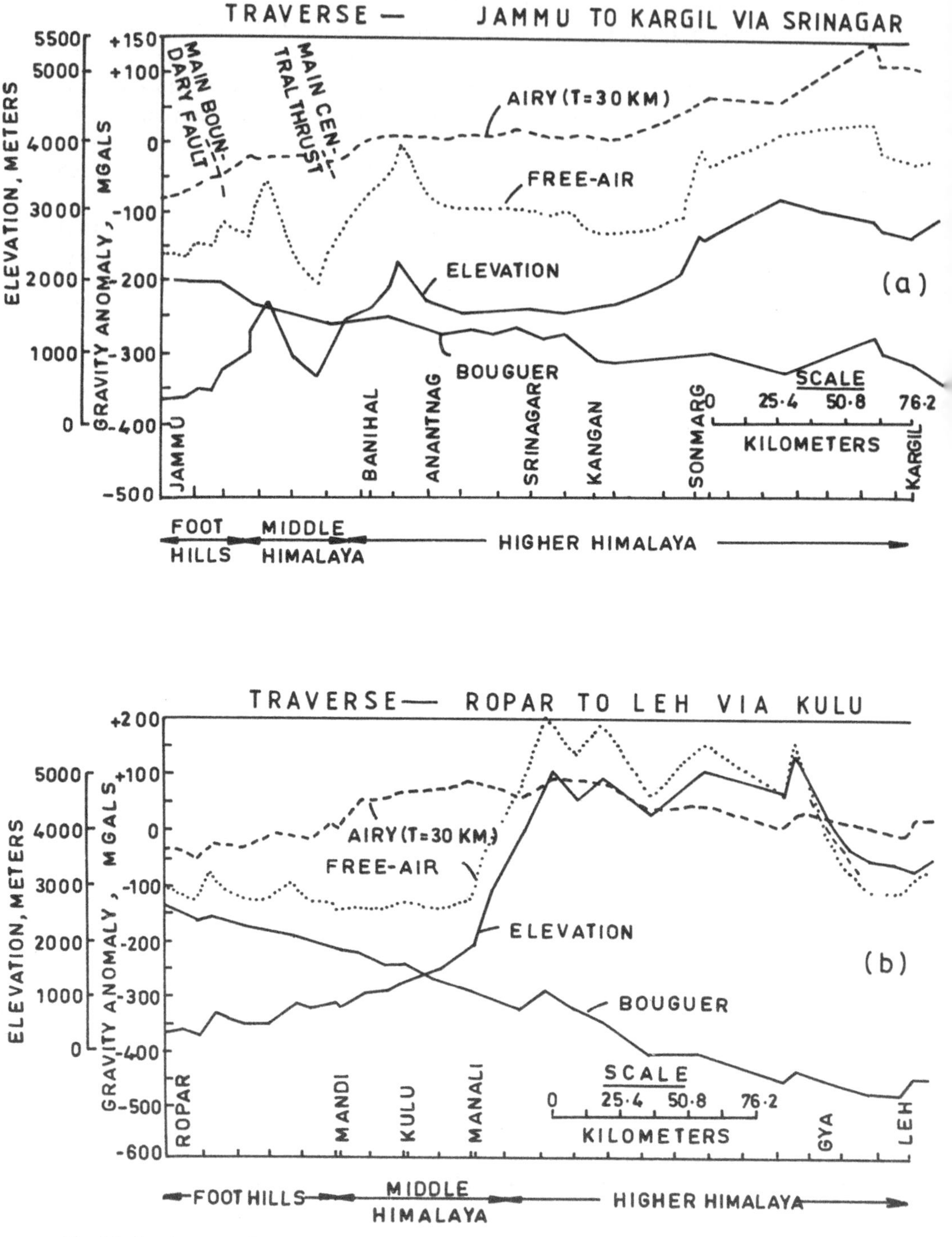

Fig. 9.6. Gravity anomalies and elevation along profiles: (a) Jammu to Kargil; (b) Ropar to Leh in Kashmir and Panjab Himalaya respectively. (after Chugh, 1978). Reproduced with permission of R. S. Chugh.

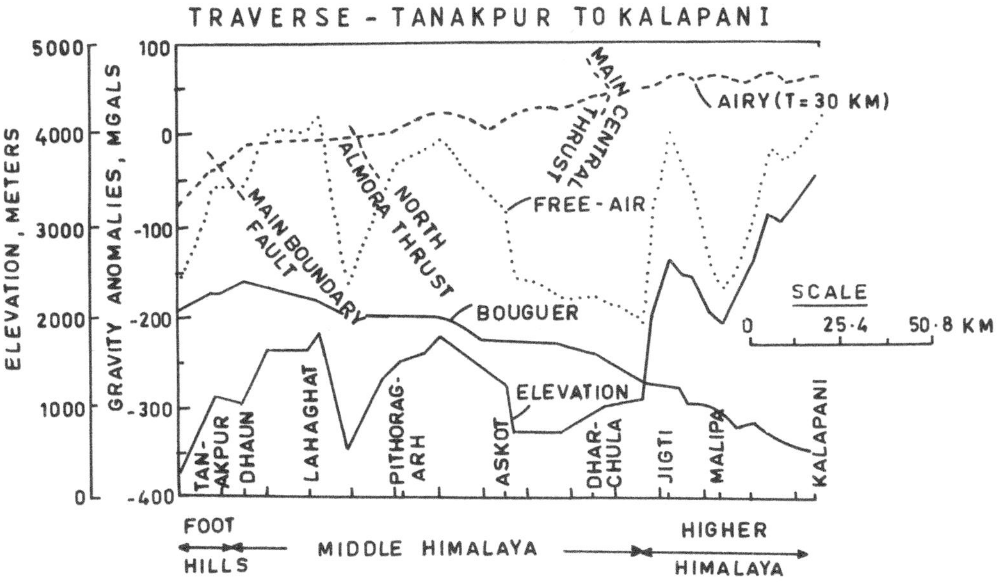

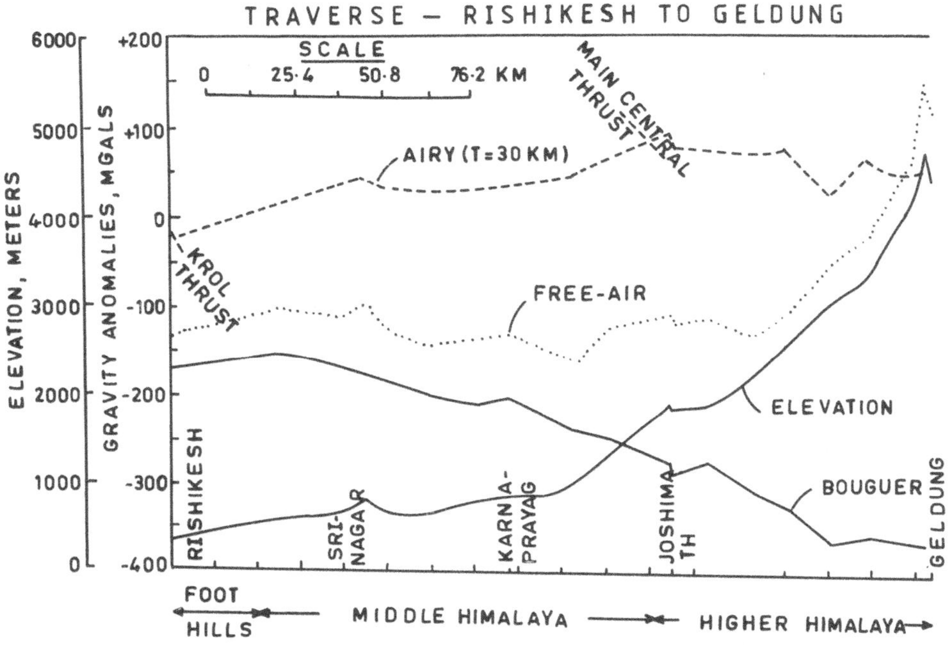

Fig. 9.7. Gravity anomalies and elevation along profiles—Rishikesh to Geldung, b—Tanakpur to Kalapani, in Kumaun Himalaya (After Chugh, 1978). Reproduced with permission of R. S. Chugh.

9.4.4. TANAKPUR TO KALAPANI

This traverse is located close to the Nepal Himalayas. The behaviour of various gravity anomalies is shown in Figure 9.7b and is similar to that discussed for earlier traverses. Here the free-air anomaly has almost a direct correlation with elevation throughout the length of the profile. The Airy isostatic anomalies are negative over the Foothills but are mildly positive over the Middle Himalaya, and maintain a continuously rising trend over the Higher Himalaya. The regional Bouguer anomaly along the profile shows a similar steady fall with amplitude of about 140 mgal.

9.4.5. KATHMANDU TO MOUNT EVEREST

Kono (1974) has established about 145 gravity stations in the Nepal Himalaya (Lower and Higher). The location of the station is shown in Figure 9.8a, along with generalised geological map of the area. Most of the stations are located over the Lower Paleozoic and Precambrian formations. All the stations were projected on two profiles in a direction N 95° E (from Kathmandu to Chokan Dala) and N 10° E (from Chokan Dala to Mount Everest Base Camp). The values of various gravity anomalies as well as elevation are plotted in Figure 9.8b.

It is interesting to observe that the Bouguer anomaly shows very little variation within the Lower Himalaya (being close to −225 mgal), and significant changes occur only within the Higher Himalaya. But the free-air anomaly shows good correlation with elevation and Airy anomaly ($T = 30$ km) shows significant changes from −65 mgal in the Lower Himalaya to +68 mgal in the Higher Himalaya. The remarkably small variations of the Bouguer anomaly within the Lower Himalaya are due to the fact that these observations have been taken almost along the strike of the Himalaya and Bouguer anomaly is sensitive only to regional rather than local elevations. The positive nature of the Airy isostatic anomaly in the Higher Himalaya is consistent with observations made in the northwestern Himalaya discussed earlier.

9.5. Analysis of Results

Since Bouguer anomaly decreases with elevation, primarily on account of root formation under the mountains, the behaviour of this anomaly with elevation in the Himalayas has been the subject of investigation of several workers (Choudhury, 1975; Chugh 1977, 1978; Kono, 1974; Qureshy, 1971; Qureshy *et al.*, 1974; Woollard, 1962, 1969). These studies have thrown some light on the nature ot root formation, the thickness of the crust as well as the degree of compensation prevailing under the Himalayas.

Qureshy (1971) on the basis of the gravity data available up to 1971 for the Kumaun Himalaya, suggested the following relationship between the Bouguer anomaly and the elevation.

$$\text{B.A.} = -176.3 - 0.033H, \text{ above 700 m}$$
$$\text{B.A.} = -154.0 - 0.038H, \text{ below 700 m},$$

where H is elevatgion in meters.

According to these relationships, Bouguer anomaly decreases with elevation at the rate of 33 to 38 mgal km^{-1}. Further, dividing the Himalaya into three different zones;

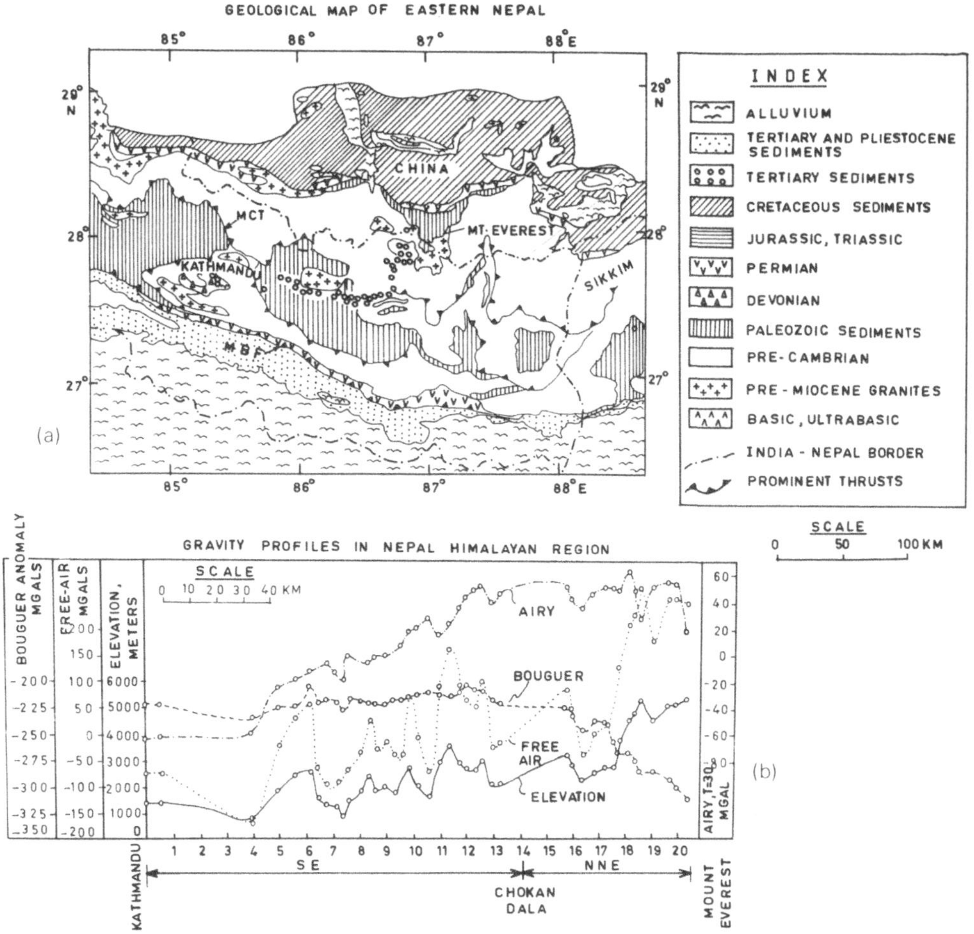

Fig. 9.8. (a) Geological map of a part of Nepal Himalaya, showing the location of gravity stations. (b) Gravity anomalies along Nepal Himalaya (data after Kono, 1974).

(1) the Foot-Hills (south of MBF), (2) the Lower Himalaya (generally between MBF and MCT) and (3) Higher Himalaya (north of MCT), Kono (1974) and Chugh (1978) have obtained different relationships for different parts of the Himalayas as given in Table 9.1.

These relationships suggest that the rate of decrease of Bouguer anomaly with elevation in the Himalayas in general is of the order of −20 to −40 mgal km^{-1}. In the Foothills of the Himalaya, the rate varies from about + 12 to −38 mgal km^{-1}. In the Lower Himalaya, the rate appears to vary considerably in different parts. In Kashmir and Panjab Himalaya, the rate is approximately −141 mgal km^{-}1, while in Nepal Himalaya, the rate is only about +3 mgal km^{-1}. Such differences can be explained mostly in terms of the geological effect of the underlying rock formations. For instance in Panjab and Kashmir Himalaya, the large thickness of relatively low density sediments present in the Lower Himalaya contributes appreciably to the variations of

TABLE 9.1

The relationship of Bouguer anomaly to elevation for Nepal, Kumaun and Kashmir-Panjab Himalaya

H is in meters and Bouguer anomaly in mgal.

Region	Nepal Himalaya B.A. =	Kumaun Himalaya B.A. =	Kashmir and Panjab B.A. =
Foot Hills	$-181.6+0.012H$	$-143.6-0.038H$	$-140.7\ -0.0204H$
Lower Himalaya	$-223.6+0.003H$	$-161.1-0.020H$	$-\ 18.98-0.14H$
Higher Himalaya	$-195.6-0.019H$	$-239.3-0.023H$	$-248.0-0.021H$
Reference	Kono (1974)	Chugh (1978)	Chugh (1978)

Bouguer anomaly. In Nepal Himalaya, the gravity observations have been taken mostly along the strike of the Lower Himalaya (ESE direction) where mostly Paleozoic and Precambrian rocks are exposed, and since the average elevation along the profile shows little variation, so does the Bouguer anomaly. It may be noted that for the gravity observations taken along the NNE direction over the Higher Himalaya, the Bouguer anomaly shows a variation of about -20 mgal km^{-1}. This rate is nearly the same as in Kashmir/Panjab as well as Kumaun Himalaya.

Average Bouguer anomaly vs average elevation relationship for $1° \times 1°$ areas of the North-Western Himalaya is shown in Figure 9.9a. The figure shows that on the whole Bouguer anomaly decreases with elevation at the rate of about 50 mgal km^{-1}. The presence of large Bouguer anomaly reaching upto -500 mgal close to Tibetan plateau is an indication of the presence of roots and therefore an evidence in support of isostasy. However, it may be noted that Bouguer anomaly does not give a correct picture of the state of isostatic compensation. Besides Bouguer, the nature of isostatic anomaly should also be considered.

For perfect isostatic equilibrium, the isostatic anomalies should approach near zero values. However, this is not so in the case of the Himalayas. At low elevations (Foot-Hills and Lower Himalaya) average isostatic anomaly is appreciably negative and at high elevations (Higher Himalaya) it is appreciably positive, as shown in Figure 9.9b for $1° \times 1°$ areas in the Northwestern Himalaya. Similar is the case for other parts of the Himalaya as well. The parameters for anomaly vs elevation curves are given in Table 9.2. This suggests that the Lower Himalaya are under a state of over-compensation (excess root formation), while under Higher Himalaya a state of under-compensation prevails. It appears that certain forces are still active in the Himalayas which are preventing it from reaching a state of perfect isostatic equilibrium. These forces could be due to the Indian plate pushing northward against the Tibetan landmass (Molnar *et al.*, 1977).

9.6. Model for Isostatic Compensation

In order to build models for isostatic compensation of the Himalayas, the nature of Bouguer as well as isostatic anomalies needs to be considered. Qureshy (1969) has built a model for the Himalayas showing downbuckling of the crust near the MBF and

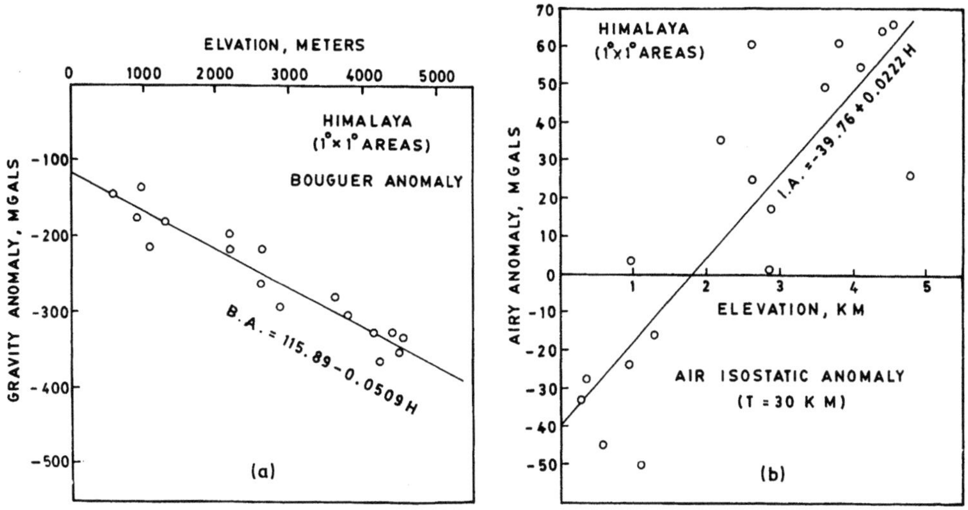

Fig. 9.9. (a) Relationship between Bouguer anomalies and elevation for the NW Himalaya, for 1° × 1° areas. (b) Airy isostatic anomaly vs elevation-relationship for 1° × 1° areas in NW Himalayas. Parameters for the curves are given in Table 9.2.

TABLE 9.2

The slope, intercept and other parameters of the curve, showing relationship between gravity anomalies and elevation for the Himalaya (1° × 1° areas)

Anomaly al)	Intercept	Slope	Correlation Coefficient	Mean Anomaly (mgal)
Bouguer (1° × 1° areas)	−115.9	−50.9	0.95	−256.1
Airy-Isostatic (1° × 1° areas)	− 39.7	+22.2	0.86	+ 14.64

explained the positive isostatic anomalies prevailing in the Middle to Higher Himalaya in terms of thickening of the basaltic layer. He has inferred a crustal thickness of nearly 82 km assuming isostatic compensation underneath the Middle Himalaya. However, there has been no confirmation of this from any independent source.

Kono (1974) has constructed two dimensional models for the crust and mantle configuration in order to explain the nature of Bouguer anomaly in Nepal Himalaya. On the basis of these models he has concluded that the crustal thickness reaches a maximum value of 75 km underneath the Tibetan Himalaya, about 250 km away from the Main Boundary Fault (MBF), instead of reaching its maximum value under the Higher Himalaya nearly 150 km away from MBF, as would be expected under Airy's scheme of compensation. On the basis of these results, he concluded that the Himalayas are not in complete isostatic equilibrium. According to him the excess load of the Himalayas is probably supported by tectonic forces by which the Indian subcontinent is being pushed towards Eurasia. Alternately the strength of the Indian lithosphere as it slides beneath the Himalayas may prevent isostatic compensation from taking place locally (Molnar *et al.*, 1977).

The negative Bouguer anomalies over the Himalayas have been explained by Warsi and Molnar (1977) in terms of plate tectonics model involving underthrusting of the Indian plate beneath the Tibetan landmass (as shown in Figure 9.10). For construction of their model a density contrast of 0.55 g cm^{-3} has been assumed between the crust and the mantle. The model constraints the dip of MBF and MCT to $10° \pm 5°$ as shown in the figure. At present the seismic data in the Himalaya is not precise enough to verify this inference. If we assume that isostatic compensation prevails in the Tibetan Himalaya and the Tibetan plateau, we should expect a crustal thickness of the order of 70–75 km in this region (Chang and Cheng, 1973; Choudhury, 1975; Kono, 1974; Marussi, 1964). This needs to be confirmed along various parts of the Himalaya through Deep Seismic Sounding profiles. It may be remarked here that the crustal models based on gravity data alone suffer from a basic uncertainty of density values for the crustal layers, their thickness and the density contrast between the crust and the mantle. By variations of these parameters several models are possible which may satisfy the gravity data.

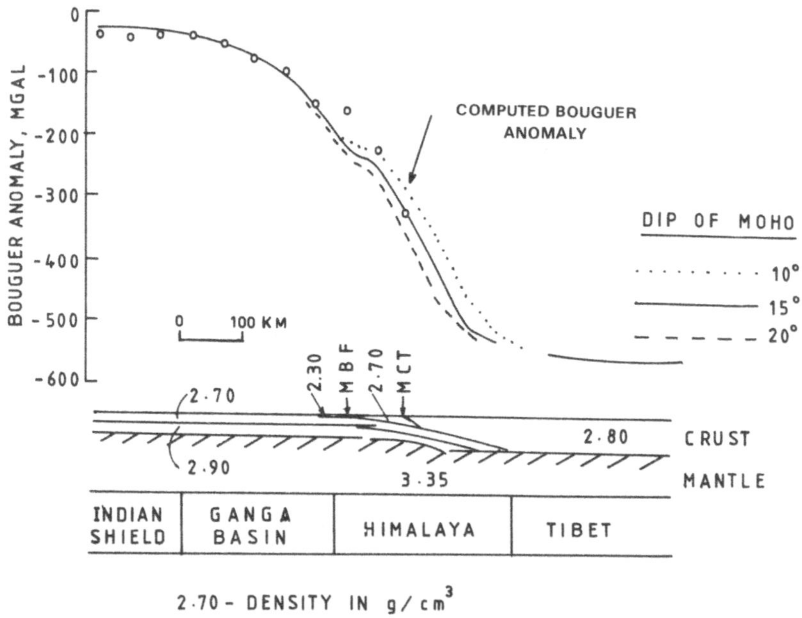

Fig. 9.10. Warsi and Molnar's (1977) model for the Himalayas showing underthrusting of the Indian plate beneath the Tibetan landmass to explain the nature of Bouguer anomaly observed in the Himalayas. Reproduced with permission of P. Molnar and W. E. K. Warsi.

9.7. DSS Profile Across NW Himalaya

An interesting light on the nature of the Moho underlying the Northwestern Himalaya is thrown by Deep Seismic Sounding (DSS) profile shot by the National Geophysical

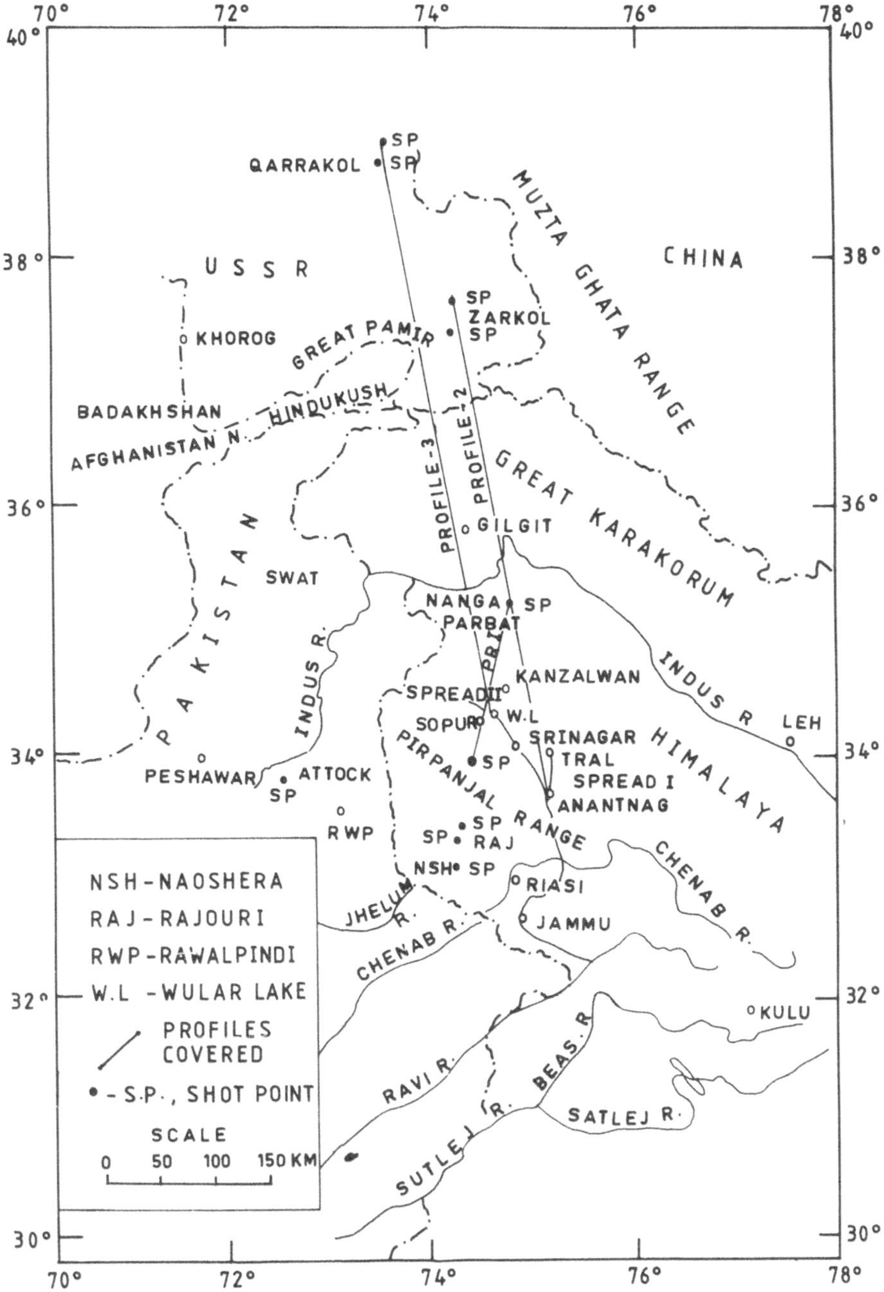

Fig. 9.11. Location of Deep Seismic Sounding (DSS) profiles shot by National Geophysical Research Institute across Himalayas towards Badakshan-Pamir for deep crustal studies. Reproduced with permission of Dr. K. L. Kaila.

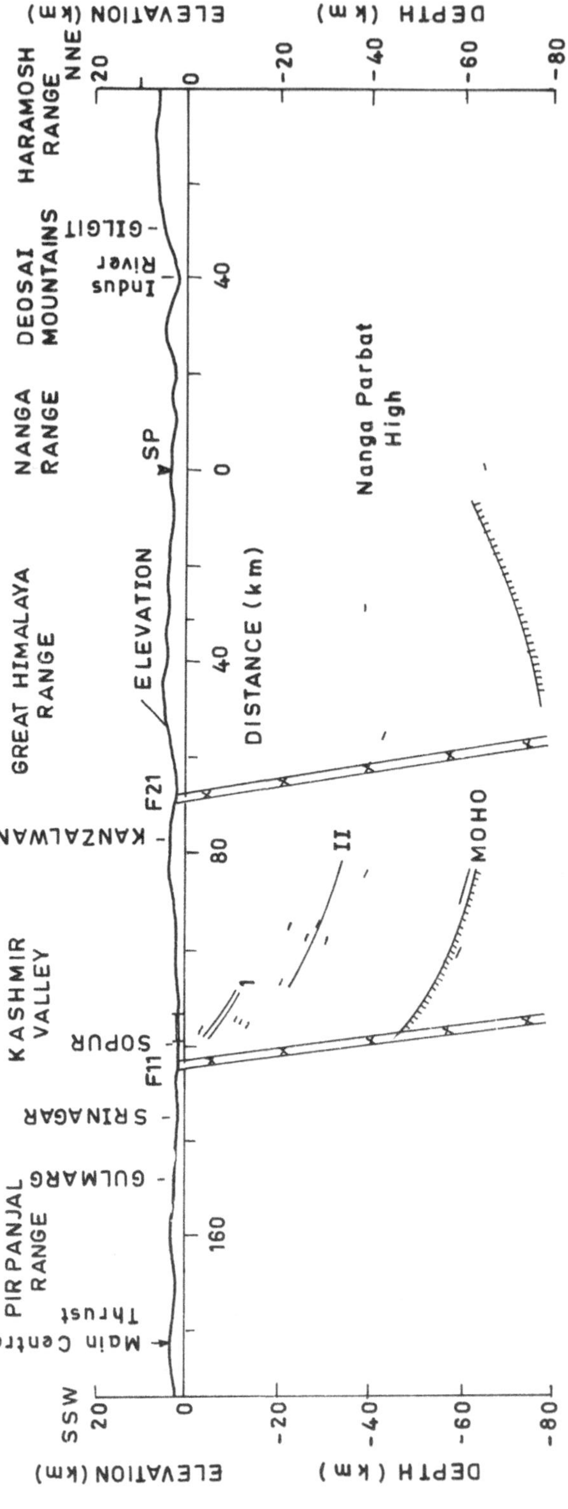

Fig. 9.12. Crustal depth section along Nanga Parbat–Wular Lake profile obtained from DSS (location shown in Figure 9.11). Reproduced with permission of K.L. Kaila.

Research Institute (NGRI), Hyderabad, in collaboration with USSR Scientists during 1974-75 (Kaila, 1981). The shots for this profile were fired in N. Pamir near Lake Qarrakol, Badakshan mountains in Lake Zarkol, near Attock in Pakistan and Naoshera, Rajouri, Gulmarg, Wular Lake and Nanga Parbat in Kashmir, India. The seismic spreads were located at two locations, few km southeast of Srinagar and near Sopur as shown in Figure 9.11. The shot-detector distance varied from about 115 to nearly 600 km. The reflections recorded from the crust and the Moho along profile 2 covering Kashmir Himalaya are shown in Figure 9.12 (after Kaila, 1981).

The DSS profile has brought out several interesting features concerning the Northwestern Himalayas. It is noticed from Figure 9.12 that the region between the Kashmir Valley and the Great Himalayan Range to the north is bounded by two steeply dipping faults F_{11} and F_{21} reaching up to a depth of about 80 km in the upper mantle. Thickness of the crust is nearly 42 km near Sopor in Kashmir Valley and increases to nearly 60 km south of Kanzalwan, where the elevation is of the order of 3 km. Dip of Moho towards north is expected on the basis of Airy's scheme of compensation, since the elevation of the mountains increase towards north. However, the most unexpected result is an upward dip of Moho from fault F_{21} towards Nanga Parbat, where the thickness of crust decreases from nearly 76 km, north of Kanzalwan to 62 km near Nanga Parbat. Kaila (1981) has suggested that this is due to the uplift of Nanga Parbat massif which has possibly taken place during the early phases of the Himalayan orogeny.

Recently the French and Chinese geoscientists have carried out wide angle refraction profiles in parts of Tibet. Their results have thrown a new light on the nature of crust and mantle relationships underlying the Himalayas and southern part of Tibet (Allègre *et al.*, 1984, Hirn *et al.*, 1984). It has been found that in this region the Moho has two or more layers. At several places the Moho has a sharp discontinuity rising from nearly 70 to 50 km. An unexpected feature of the Moho is its updip underneath the Higher Himalayas, where elevations are the highest (see Figure 12.11, Chapter 12).

The D.S.S. observations taken by the Indian and French-Chinese geoscientists show that isostatic compensation underneath Himalayas and Tibet is not limited to the depth of Moho. Thus Airy's scheme of compensation as conventionally understood is not strictly valid for this area. Density contrast may exist at deeper levels (perhaps up to 100 km depth) contributing to the process of isostatic compensation. Apart from this the bending of the lithosphere may play an important part as suggested by Banks, Parker and Huestis (1977), (see discussion at the end of Chapter 8).

CHAPTER 10

SATELLITE GRAVITY FIELD OVER INDIA AND THE SURROUNDING REGIONS

10.1. Introduction

At points rotating with the earth the gravity field is considered to be composed of a vector sum of gravitational attraction of masses of the earth and centrifugal force due to its rotation. For points not rotating with the earth (such as a satellite) the gravitational field acting is only due to masses inside the earth and the effect of the atmosphere. The path of a satellite is influenced by the shape of the earth as well as masses lying inside at various depths and hence analysis of exact path of a satellite may give us considerable information regarding the distribution of masses (inhomogeneities) as well as the shape of the earth. However, the inverse potential problem is not unique.

For most practical purposes the reference equipotential surface of earth's gravity field is the geoid which lies close to the ellipsoid but does not coincide with it. A major task of geodesy has been to determine undulations of the geoid 'N' using terrestrial gravity data (see Chapter 1). The undulation or geoid height can be computed by using Stokes theorem which relates the free-air anomaly Δg of an element of surface da to geoid height N by using the equation (Heiskanen and Vening Meinesz, 1958).

$$N = \frac{1}{4\pi g R} \int_0^A S(\psi)\, \Delta g\, da \tag{10.1}$$

where

R mean radius over the geoid

g mean value of gravity over the geoid

A surface area of the sphere with mean radius R of the geoid

ψ angular distance of da from point P at which N is to be calculated.

$$S(\psi) = \frac{2}{\sin(\psi)} F(\psi) \tag{10.2}$$

and

$$F(\psi) = \cos\frac{\psi}{2} + \frac{1}{2}\sin\psi \left[1-5\cos\psi - 6\sin\psi/2 - 3\cos\psi \ln(\sin\psi/2 + \sin^2\psi 2) \right].$$

The surface integral has to be extended over the whole surface of the earth. For this purpose we use mean free-air anomaly values over $5° \times 5°$ or smaller areas. Now through the satellites the figure of the earth and distribution of masses inside can be studied much more precisely.

The satellite tracking program was initially conceived by the Smithsonian Astrophysical Observatory (SAO), U.S.A., and coordinated by this institution for a number of years. The satellites are tracked by means of Baker-Nunn Cameras installed by SAO at several observatories distributed around the world. In order to improve the accuracy of observations Satellite Laser Ranging (SLR) and Satellite Altimetry have been used (GEOS-3, 1979; SEASAT, 1982; Lerch, 1983). The modern techniques have considerably improved our knowledge of the geopotential.

10.2. Theoretical Background

The gravity field of the earth or its geopotential is very irregular. It can be represented in the form of an infinite series of spherical harmonics of the form,

$$V(r, \phi, \lambda) = \frac{GM}{r}\left[1 + \sum_{l=2}^{\infty}\left[\frac{a_e}{r}\right]^l \sum_{m=0}^{l} (C_{lm} \cos m\lambda + S_{lm} \sin m\lambda) P_{lm} (\sin \phi)\right]. \tag{10.4}$$

Here G is gravitational constant of the earth; M is mass of the earth; r, ϕ, λ are geocentric coordinates of a point, distance to the center of mass, latitude and longitude; a_e, is equatorial radius of the earth; $P_{lm}(\sin\phi)$ are normalised Legendre function of degree 1 and order m; C_{lm}, S_{lm} are called tesseral harmonics.

The average gravity anomalies Δg are given as,

$$\Delta g = \frac{GM}{r^2}\left[\sum_{l=2}^{\infty} (l-1)\left(\frac{a_e}{r}\right)^l \sum_{m=0}^{l} (C_{lm} \cos m\lambda + S_{lm} \sin m\lambda) P_{lm} (\sin \phi)\right]. \tag{10.5}$$

It is apparent from this equation that the gravity field is relatively more influenced by C_{lm}, S_{lm} coefficients of high degree because of $(l - 1)$ term in the expression. Terrestrial values of gravity anomalies (Δg) are therefore useful in determination of tesseral harmonics of high degree.

The SAO has determined a number of Standard Earth models based on satellite tracking data (Gaposchkin and Lambeck, 1970; Gaposchkin, 1973, 1974). The standard earth model consists of a set of geocentric coordinates for 39 stations observing satellites and a set of spherical harmonics representing the gravity field complete through degree and order 16. The parameters adopted by SAO for the Standard Earth are,

$$\begin{aligned}
J_2 &= 1082.637\\
a_e &= 6.378\ 140 \times 10^8 \text{ cm}\\
GM &= 3.986\ 013 \times 10^{20} \text{ cm}^3 \text{ sec}^{-2}\\
\omega_e &= 7.292\ 115\ 085 \times 10^{-5} \text{ sec}\\
\frac{1}{f} &= 298.256.
\end{aligned}$$

Here, J_2 is zonal harmonic of 2nd degree, ω_e is the angular velocity of the earth and $1/f$ is the ellipticity. Zonal harmonics of gravity field have been derived from the satellite perturbation data. Tesseral harmonics have been computed by combining satellite gravity data using a set of 19, 328 values of 1° × 1° mean free-air anomalies over continents and oceans supplied by Aeronautical Chart Information Center, U.S.A. The coefficients of zonal and Tesseral harmonics determined by SAO have been listed by Gaposchkin (1973, 1974). The gravity field derived from computations of satellite + terrestrial gravity data has been compared with observed gravity field over land and oceans (Kahle and Talwani, 1973; Gaposchkin, 1974). In most cases the agreement has been found to be reasonably good. The geoid heights and free-air anomalies determined from satellite gravity data bring out large wave length anomalies over continents and oceans. For instance the geoid height map reveals large negative anomaly over the Indian Ocean, western part of North Atlantic, eastern part of Northern Pacific and Antarctica. Large positive heights are found over western Pacific, south of Africa and central part of North Atlantic.

Recently there has been a considerable improvement in the field of satellite geodesy with the introduction of satellites (such as GEOS-3, GEM-9, SEASAT) carrying radar altimeter and improvement in terrestrial data collection. The altimeter makes it possible to measure directly the elevation of the satellite over ocean surface. When combined with precise orbital tracking one obtains the height of the ocean surface relative to the earth's ellipsoid. The sea surface is not at constant gravitational potential since non-gravitational forces such as ocean currents cause slight deviations. Making corrections for these one can obtain the geoid height from the satellite track.

10.3. Geoid over India and the Surrounding Regions

Geoid undulation map for India and the surrounding regions up to degree $n = 20$ based on the GEM-9 potential coefficients (Lerch *et al.*, 1979) prepared by Rapp (personal communication) is shown in Figure 10.1. The map is based on flattening of 1/298.25722 and equatorial radius of 6378136 m. The map presents very long wave length features of the gravity field and shows a negative anomaly with a minimum of −104 m centred near 5° N, 80° E, south of Indian peninsula. The geoid height increases over Indian peninsula and reaches a value of −38 m over the Himalayan arc, where it has a NW–SE trend. Further northeast the geoid height becomes increasingly negative. It may be mentioned here that this geoid anomaly which has a wave length of nearly 10000 km is one of the most prominent ones observed over the surface of the whole earth, land as well as the oceans as shown by Gaposchkin (1973).

Geoid undulation map for India and the surrounding regions prepared by Rapp (personal communication) using potential coefficients up to degree 180 is shown in Figure 10.2. The coefficients have been based on a combination of satellite information and terrestrial gravity data for 1° × 1° areas. The figure shows several small wave length features which are not apparent in Figure 10.1. The minimum of geoid is again centered at the same place but has a value of −104 m. The geoid rises sharply to the east attaining a value of −8 m over Sumatra. It rises to a value of −56 m in the north over Indian peninsula. The geoid has a major depression along Ganga-Brahmaputra (GB) foredeep which stretches from Panjab in the west to Assam Valley in the east, the values ranging from −52 to −60 m. The undulation takes a sharp turn near 33° N, 72° E

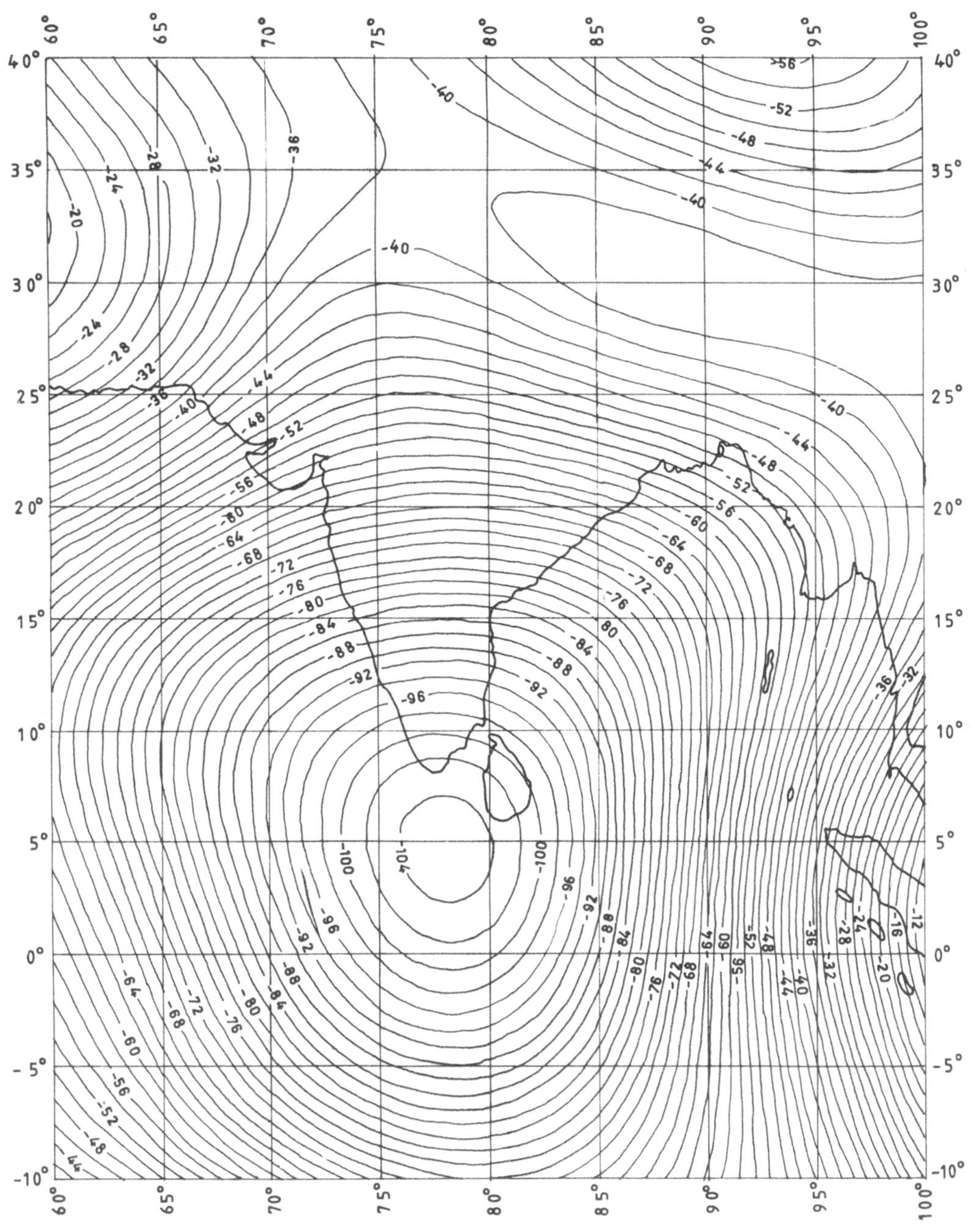

Fig. 10.1. Geoid undulation map for India and surrounding regions up to degree $n = 20$ based on GEM-9 altimeter data, prepared by Rapp (personal communication). Contour interval 2 m. The map shows a major geoid anomaly covering large parts of the Indian peninsula and the Indian Ocean. Reproduced with the permission of Prof. R. H. Rapp.

and strikes in N–S direction in accordance with the trend of Indus basin (IB). The geoid rises again in the northwest over Afghanistan and over Tibetan plateau to the east. The geoid anomalies have a close correlation with free-air anomalies as discussed later.

The negative geoid height over a large part of the Indian ocean represents a major mass deficiency present in the upper mantle. The undulation map has been prepared by several workers using satellite and terrestrial gravity data from different sources (Gaposchkin and Lambeck, 1971, Kahle and Talwani, 1973, Kahle *et al.*, 1978; Rapp, 1981, 1983). All the maps show that this major geoidal anomaly extends towards Australia on the southeast, Somali basin on the southwest and appears to be related to the formation of the Indian Ocean as a result of separation of the Indian subcontinent from Antarctica and Australia which formed the Gondwana Land during Mesozoic times (McElhinny, 1973).

10.4. Free-Air Anomaly over India and the Surrounding Regions

The free-air anomanly map prepared by Rapp (personal communication) using a spherical harmonic expression to degree 180 is shown in Figure 10.3. The map shows several interesting features concerning Indian peninsula, the Himalayas and the surrounding regions. These are discussed below.

(1) Large parts of the Indian Ocean are characterised by a negative free-air anomaly of the order of −40 mgal. However, the Laccadive ridge (LR) and Ninety East ridge are relatively positive as compared to the surroundings. To the east of Ninety East ridge, a long belt of negative free-air anomaly follows the west coast of Sumatra (near 5° N, 95° E), passes to the east of Andaman Islands (AI) and reaches up to Bengal basin near 24° N, 92° E.

(2) Appreciably negative anomalies of the order of −40 to −60 mgal prevail over parts of southern peninsula of India (see Chapter 3 for comparison with land gravity data). Mysore plateau (MP) is characterised by nearly zero free-air anomaly, while the Eastern Ghats (E.G.) are relatively positive (+20 mgal).

(3) Over northern peninsula of India between 20° to 25° N, 75°–85° E and Deccan trap (D.T.) areas, free-air anomaly has near zero value. To the west of this area the Aravalli mountains (AR) are characterised by positive free-air anomaly trending in NE–SW direction and reaching up to +40 mgal. To the east over Shillong plateau the anomaly is appreciably positive, (+40 to +60 mgal).

(4) North of Indian peninsula a major gravity low prevails over the Ganga–Brahmaputra basin. The low has a NW–SE to E–W trend conforming to the trend of the Himalayas and extends from 72° E to 95° E. Further west the low takes a sharp turn to N–S direction between 27°–35° N, 70–72° E. This major low is attributable to the large thickness of Upper Tertiary sediments (of the order of 5–6 km) deposited in Indo-Gangetic foredeep as the Himalayas and the Suleiman–Kirthar ranges were formed during Miocene–Pliocene times.

(5) Over the Higher Himalayas between 75° to 95° E, 27° to 35° N, the free-air values are appreciably positive reaching up to +160 mgal. The anomalies are undoubtedly related to the high elevation of the Himalayas. However, it is interesting to note that between 30° to 35° N, 80°–95° E, over Tibetan plateau the anomaly has much lower value, of the order of 0 to +20 mgal.

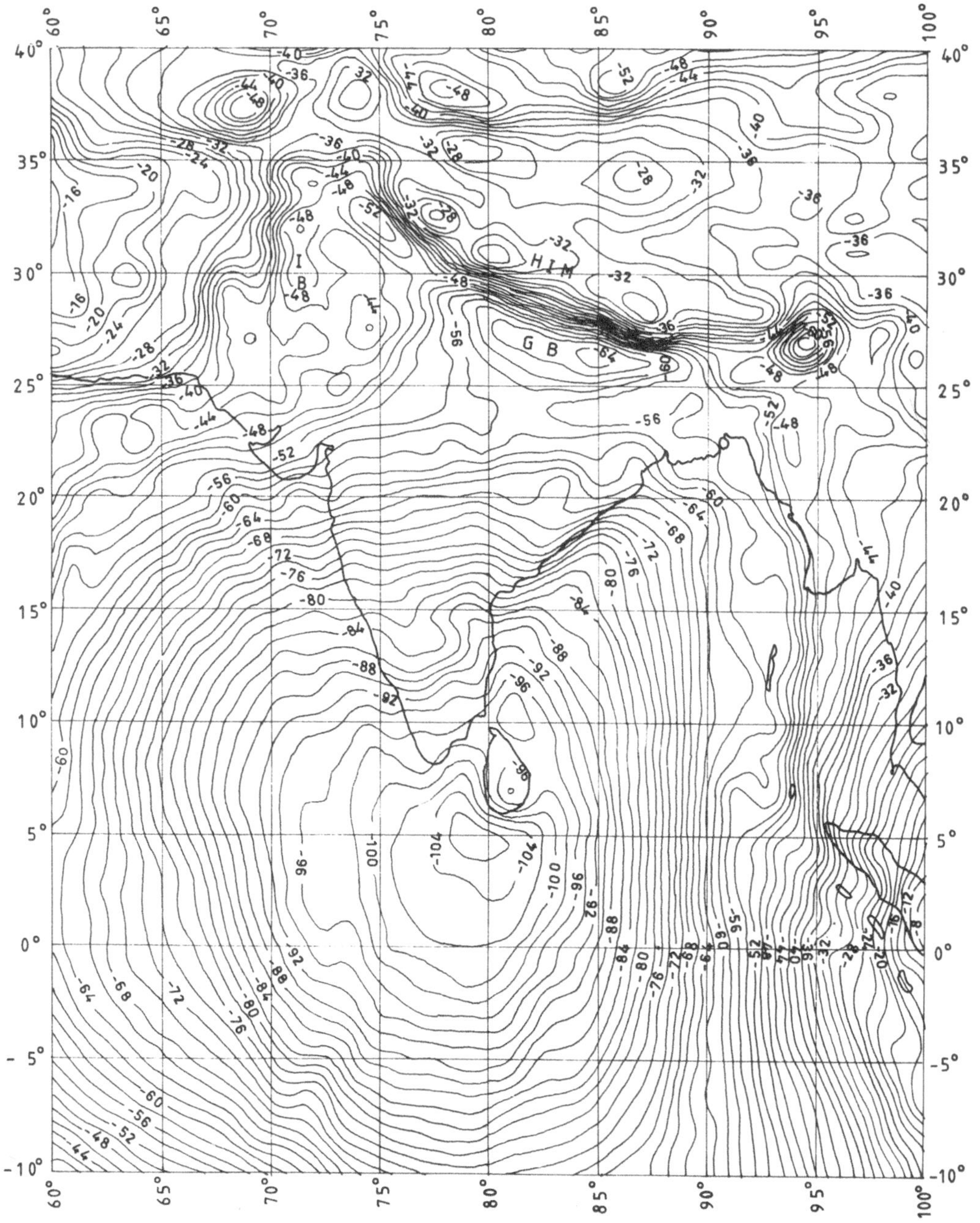

Fig. 10.2. Geoid undulation map for India prepared by Rapp (personal communication) combining satellite gravity data using potential coefficients up to degree 180 (Rapp, 1981) and terrestrial gravity data for 1° × 1° areas, contour interval 2 m. The map shows several small wave-length features not brought out in Figure 10.1. Abbreviations: (GB) Ganga Basin, (HIM) Himalayas, (IB) Indus Basin. Reproduced with the permission of Prof. R. H. Rapp.

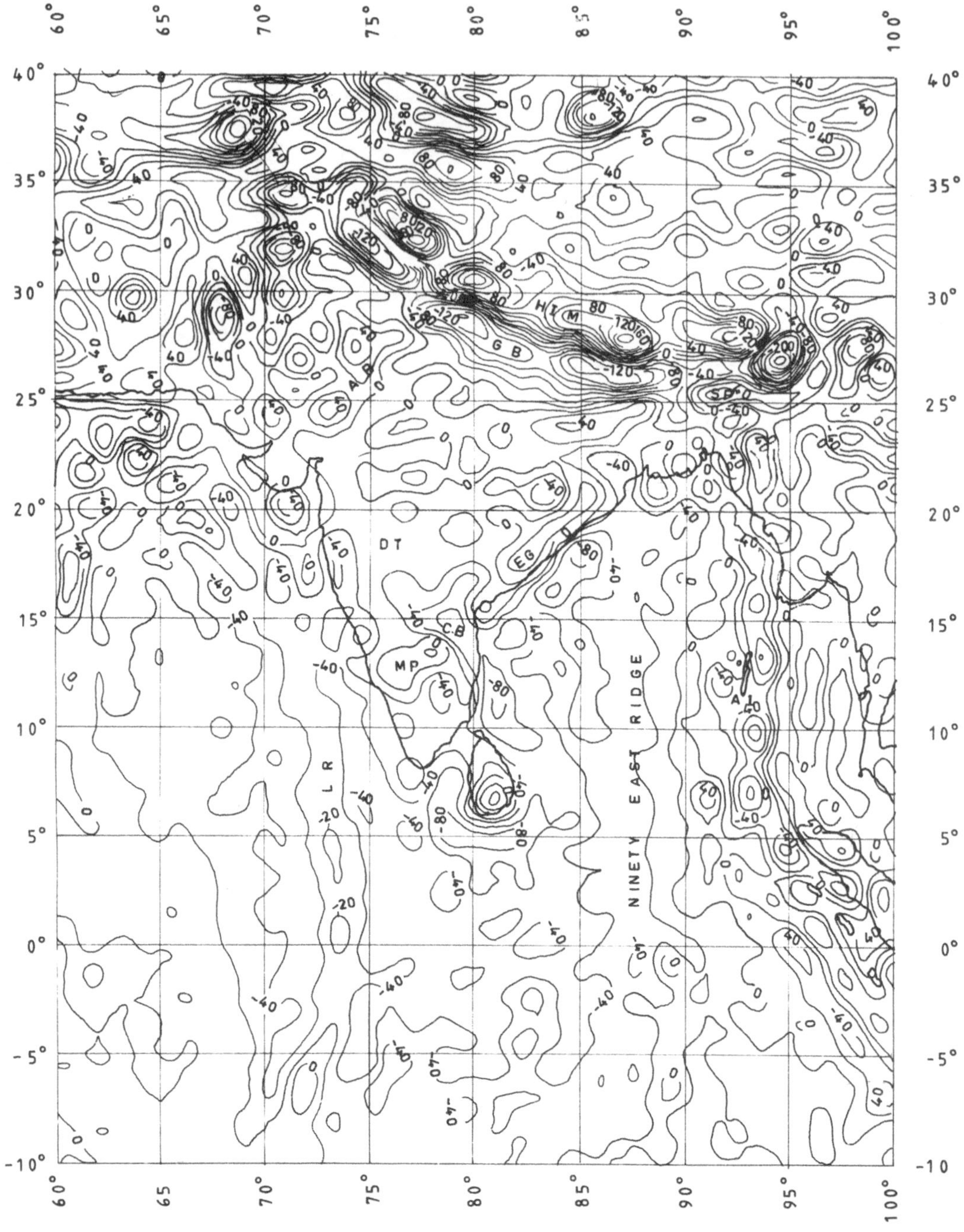

Fig. 10.3. Free-air anomaly map for India and surrounding regions prepared by Rapp (personal communication) by combining Satellite and terrestrial gravity data, contour interval 20 mgal. Abbreviations: (AI) Andaman Islands, (AR) Aravalli mountains, (CB) Cuddapah basin, (DT) Deccan trap, (E.G.) Eastern Ghats, (GB) Ganga Basin, (HIM) Himalayas, (LR) Laccadive Ridge, (MP) Mysore plateau, (SP) Shillong plateau. Reproduced with the permission of Prof. R. H. Rapp.

(6) Free-air anomaly over North Pamir and Karakorum mountains between 35 to 40° N, 68° to 80° E are appreciably positive, reaching up to +100 mgal.

10.5. General Remarks

Over the Indian Ocean the large negative anomalies are possibly related to the Indian Ocean geoid anomaly as discussed earlier. It appears that influence of this low extends partly to the southern peninsula of India. Positive anomalies over several topographic features such as the Laccadive ridge, Ninety East ridge suggest that these areas are not locally compensated. Over large topographic features which are characterised by regional isostatic compensation, free-air anomaly values are found to be very close to those of isostatic anomaly as in Rocky Mountains of N. America (Woollard, 1966). Large free-air anomalies are generally found over topographic highs which are not locally compensated. Thus large free-air anomaly values found over parts of Higher Himalaya, N. Pamir and Karakorum mountains suggest that there may be local departures from isostatic equilibrium for these mountain ranges. In comparison to above, the Tibetan plateau appears to be isostatically compensated.

CHAPTER 11

GRAVITY FIELD AND SEISMICITY IN NORTHEASTERN INDIA AND NORTHERN BURMA

11.1. Introduction

From the point of view of tectonics, northeastern India and northern Burma, comprising the area lying approximately between latitude 22° to 30° N and longitude 88° to 100° E is one of the most interesting parts of the world. Here large scale thrust movements directed from northwest and southeast have resulted in the formation of Eastern Himalaya (whose highest peak is Namche Barwa (7755 m high) in the north, Naga Hills in the south and Assam-Brahmaputra valley in the middle. According to Evans (1964) crustal shortening that has taken place in this region is of the order of few hundred kilometers. Large-scale vertical movements have taken place resulting in the uplift of the Shillong Plateau and Mikir Hills in the central part, deposition of large thickness of sediments (of the order of 12-13 km) in Surma Valley and Bengal basin and nearly 4-6 km in Assam-Brahmaputra valley. These movements have originated as a consequence of northward drift of the Indian peninsula and its collision with the Eurasian plate, in this case the Tibetan landmass to the north and Burma to the east (Dewey and Bird, 1970). All these movements have reshaped the landforms and changed the course of rivers during the Tertiary times. The effect of such movements can be studied from analysis of gravity field and seismicity of the area. The movements which have started since the initial stage of Himalayan folding phase are continuing up to Recent times as is evidenced by high seismicity of the area. The analysis of gravity field and seismicity of northeastern India and northern Burma is discussed in this chapter.

11.2. Regional Geology and Tectonics

The area comprises the following major tectonics units as shown in Figure 11.1.

(1) Eastern Himalaya.
(2) Mishmi Block.
(3) Naga Hills.
(4) Assam Valley.
(5) Shillong Plateau and Mikir Hills.
(6) Arakan-Yoma folded belt.
(7) Bengal basin and Surma Valley.

The area comprising Eastern Himalaya lies approximately between 27°–29° N, and 88°–96° E and is composed of Palaeozoic, Mesozoic and Tertiary rocks. The tectonics of these mountains is dominated by extensive thrust sheets which have moved from north towards south or southeast. The outer ranges of Eastern Himalaya consist of

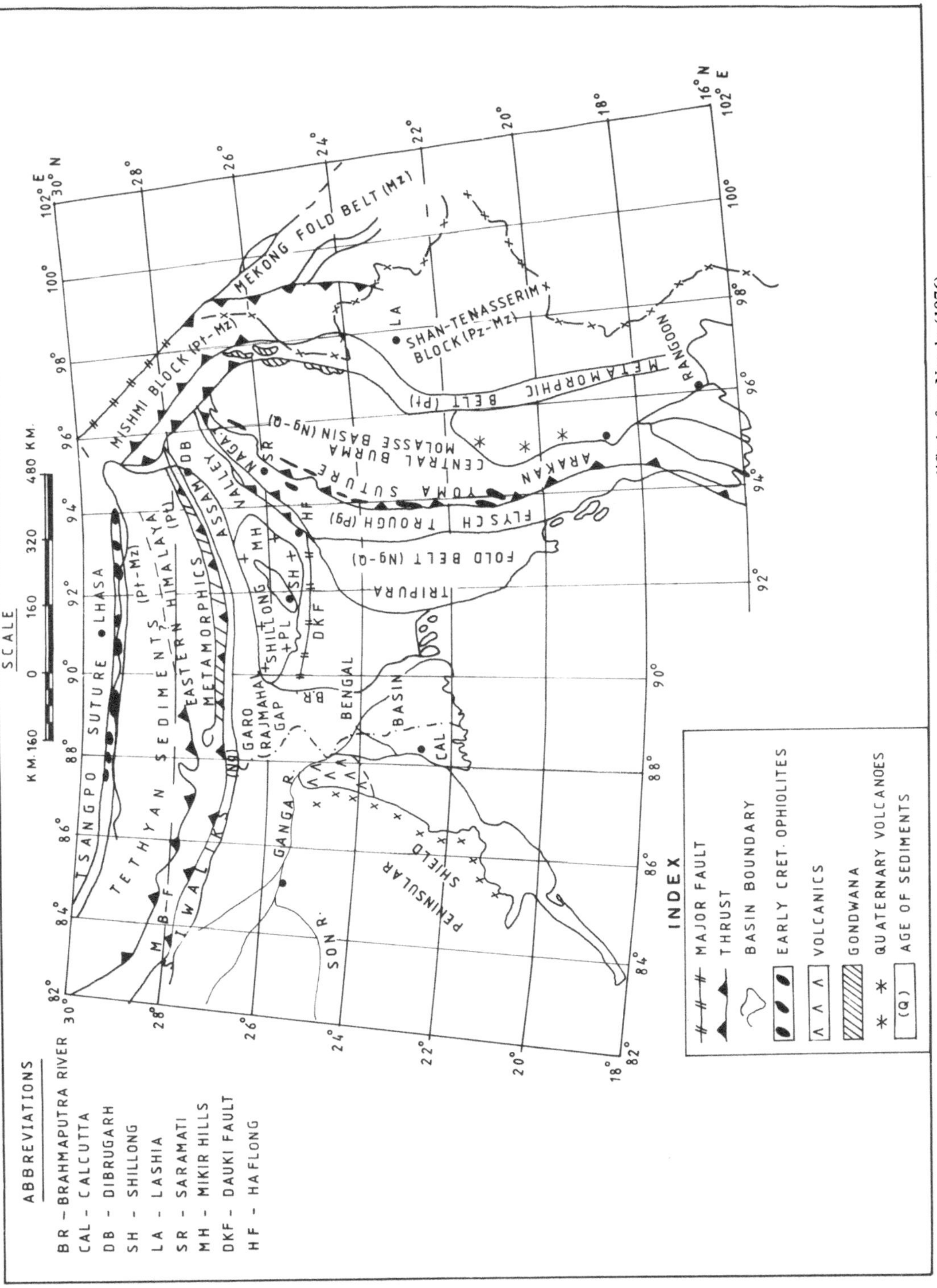

Fig. 11.1. Tectonic map of NE India and N. Burma, modified after Nandy (1976).

several overthrust masses. Lying over the Tertiaries, which form the outermost ranges, is a thrust sheet composed of Gondwana beds and over this has been thrust a complex of older crystalline rocks.

11.2.1. THE MISHMI BLOCK

The Himalaya folded belt which has a general strike of NE–SW from 94° to 96° E takes a sharp turn near 96° E and assumes a general strike of northwest to southeast. At this place often called the "Assam syntaxis", the thrusts are directed from northeast towards southwest. Near the southern part of Mishmi Hills the Himalaya branch off into two parts, Naga-Lushai hills trending NE–SW and Arakan-Yoma folded belt trending NNE–SSW in the northern part, changing to N–S in the southern part.

11.2.2. NAGA HILLS

The Naga Hills consist of an entire Tertiary succession of Eocene to Paleocene rocks (Evans, 1964). The "Belt of Schuppen" as these hills have been called, consist of several overthrusts along which the rocks have moved northwestward relative to the 'Foreland Spur', an autochthonous unit consisting of Shillong Plateau, Mikir Hills and Brahmaputra Valley as defined by Evans (1964). The thrusts form a complex pattern (convex towards northwest), one thrust overriding another. The northwestern edge of this belt closely follows the alluvial boundary of Upper Assam Valley for over 350 km, continues southwestward for another 50 km and finally ends near Haflong (25° 11′ N, 93°01°′ E). The northwestern edge of the Belt of Schuppen is called the Naga thrust.

11.2.3. THE ASSAM VALLEY

The Assam/Brahmaputra Valley lies between the Eastern Himalaya and Naga Hills in the eastern part and between Himalaya and Shillong Plateau in the western part. The valley has been formed as a result of deposition of sediments brought from the rising Himalaya to the north and Naga Hills/Shillong Plateau to the south. The alluvial coverd valley is flat, average elevation being less than 100 m. General trend of the valley is NE–SW in the eastern part. Beneath the alluvium a thick succession of Tertiary rocks (4–6 km thick) overlies the Precambrian basement complex, which is exposed in the Mikir Hills. River Brahmaputra enters at its northeastern extremity, follows all along the valley and takes a sharp turn towards south near the western margin of Shillong Plateau.

11.2.4. SHILLONG PLATEAU AND MIKIR HILLS

The Shillong Plateau and Mikir hills (together called Meghalaya Plateau) form a very prominent part of northeastern India lying between 25° to 26° N and 90° to 93° E. The plateau has an average elevation of about 1000 m and consists of Archaean gneissic complex with Proterozoic intracratonic Shillong Series. On the southern side of the plateau Sylhet traps (of probable Jurassic age) are present. During Jurassic to Cretaceous period the plateau broke up along its southern margin by the E–W trending Dauki fault through which Sylhet traps were extruded. Since Cretaceous times, there has been a history of continuous uplift of the plateau and submergence of the area lying to its north as well as south. Subsidence of northern side of the plateau led to the deposition of sediments in Lower Assam Valley up to Main Boundary Fault, while subsidence on the southern side led to the deposition of nearly 13 km of sediments in Surma Valley. To the southeast and east, the plateau terminates along the Halflong-Disang thrust. The Mikir Hills are detached from the plateau by NW–SE trending Kopili fracture zone. To the south the plateau is limited by Dauki fault, and to its western side by N–S trending Jamuna fracture/lineament. Besides above, several NE–SW trending lineaments have been found over the Shillong Plateau (Nandy, 1980).

11.2.5. ARAKAN-YOMA FOLDED BELT

The Arakan-Yoma folded belt extends from 16° N to 27° N and consists of a series of thrusts and thrust slices (Brunnschweiler, 1966). A flysch trough (Paleogene) lies on its western side, while Central Burma molasse basin (Irrawaddy) of Quaternary age lies to the east of the folded belt. Rocks in Arakan-Yoma suture zone consists of a melange of highly deformed sediments, fossiliferous (Cretaceous–Eocene) limestone, deeply exposed metamorphics (Paleozoic) with ophiolites and exotic blocks of Paleozoic sediments (Brunnschweiler, 1966).

11.2.6. BENGAL BASIN

The Bengal basin lies to the south of Shillong Plateau and is bounded by Tripura folded belt on the east and peninsular shield on the west. The basin has been formed during Eocene to Pleistocene times and is characterised by sediments having a thickness of 12–13 km (Khan and Azad, 1963), in the eastern part. In the western part Tertiary rocks are found to be resting over Gondwana formations. The eastern part of Bengal basin has suffered considerable amount of folding as a result of thrust movements directed from east towards west and is known as Tripura Folded Belt.

11.3. Bouguer Anomaly Map of Northeastern India

Bouguer anomaly map of the area prepared by incorporating data from several organisations including National Geophysical Research Institute, Survey of India, Burmah Oil Company, Oil and Natural Gas Commission is shown in Figure 11.2 (Verma and Mukhopadhyay, 1977). The map shows several distinctive features. These are discussed below.

Two major trends in the anomaly prevail over the area. One is the east-west trend, almost parallel to the structural trend of the Himalaya, and the other is the north-south trend parallel to the structural trend of the Arakan-Yoma ranges in the east and the Indian shield on the west. The east-west trend predominates over the Lower Brahmaputra Valley, the Shillong Plateau and the Surma Valley. The trend changes to northeast-southwest over the Upper Brahmaputra Valley conforming with the structural trend in the area. North-south trends predominate over the Bengal Basin, Tripura Folded Belt, and Arakan-Yoma ranges of Burma. The gravity field shows a large variation, from +44 mgal (Bouguer) over the Shillong Plateau to −255 mgal over Upper Assam Valley, a difference of 299 mgal. Whole of Upper Assam Valley is characterised by large negative anomalies varying from −180 to −255 mgal. The lowest gravity value in the area is found near North Lakhimpur (27° 15′ N, 97° 08′ E), close to the Himalayan foothills. In the Bengal Basin coastal anomalies are close to −10 mgal.

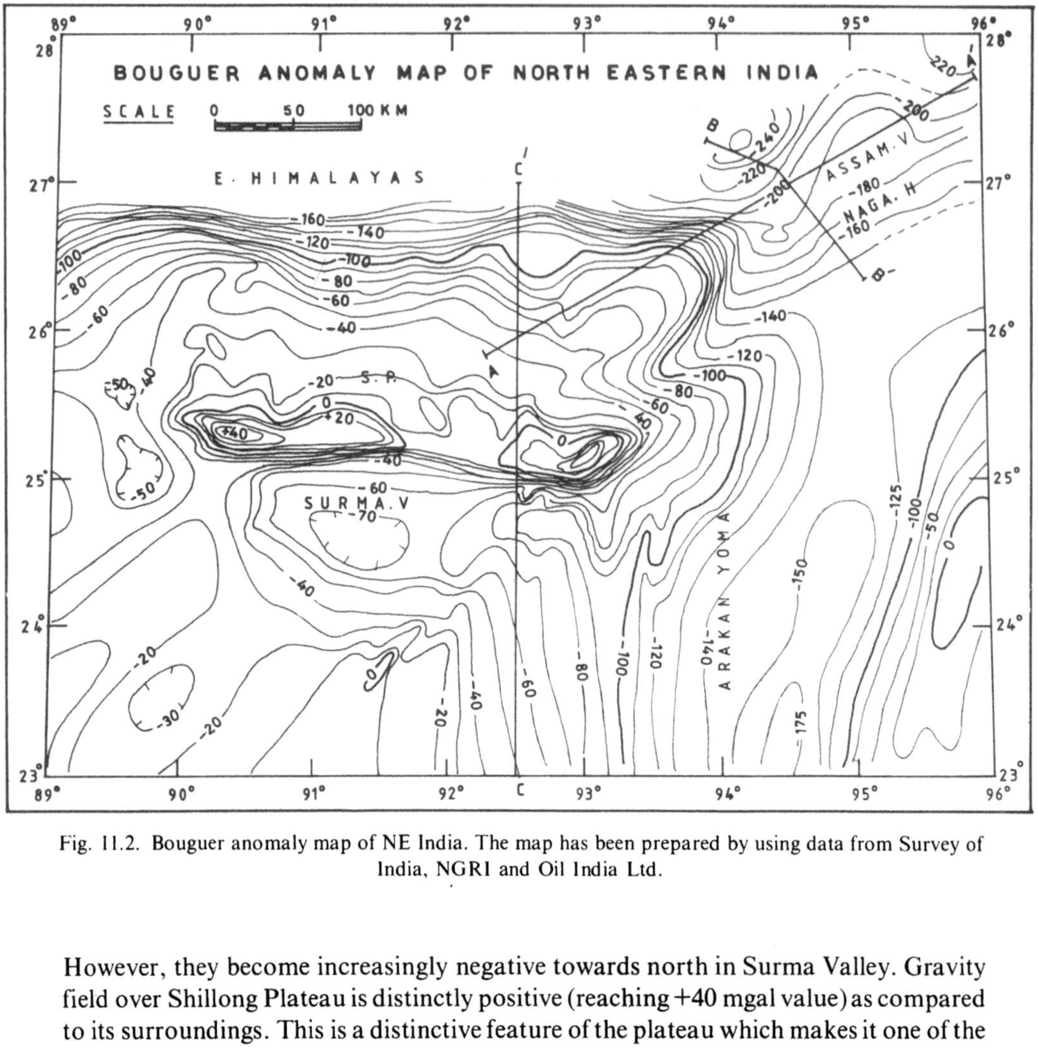

Fig. 11.2. Bouguer anomaly map of NE India. The map has been prepared by using data from Survey of India, NGRI and Oil India Ltd.

However, they become increasingly negative towards north in Surma Valley. Gravity field over Shillong Plateau is distinctly positive (reaching +40 mgal value) as compared to its surroundings. This is a distinctive feature of the plateau which makes it one of the most anomalous parts of India. Steep gravity gradients are observed over the outer edge of the Mikir Massif and the southern edge of the Shillong Plateau, near Dauki Fault.

Examining the correlation of the gravity field with geology it may be realised that the large thickness of low density sediments present in the Lower and Upper Brahmaputra Valleys as well as the Bengal Basin should have a considerable influence on the gravity field. Verma and Mukhopadhyay (1977) have examined the influence of sediments on the gravity field by applying geological correction along profiles AA′ and BB′. The location of profiles is shown in Figure 11.2. The density values for the Tertiary sediments have been taken form Evans and Crompton (1946), the geophysical wings of Oil India Ltd. as well as ONGC and are tabulated in Table 11.1. Basement density was taken as 2.67 gm cm^{-3}. Geological correction along profiles AA′ and BB′ and the interpretation of corrected anomalies is discussed below.

11.3.1. PROFILE AA′

This profile (Figure 11.3) runs from the Shillong Plateau in the southwest to the Mishmi Hills in the northeast across the Upper Assam Valley. Geological information along the profile is taken from Mathur and Evans (1964) as well as the Tectonic Map of India published by the ONGC. Density values for various rock formations were assigned as shown in Table 11.1 and respective density contrast of these with respect to the basement was determined. Gravity effect of the various layers was calculated assuming two-dimensional model. As seen in Figure 11.3 the geologic correction along

TABLE 11.1

Density values for various sedimentary formations found in Brahmaputra Valley in Assam

Age	Formation	Above 1.5 km ($gm\ cm^{-3}$)	Below 1.5 km ($gm\ cm^{-3}$)
Recent	Alluvium	2.0	2.1
Tertiary	Dihing and Dupi Tila Series	2.1	2.2
	Tipam Series	2.2–2.3	2.35–2.4
	Surma Series	2.35–2.45	2.45–2.55
	Barail Series	2.5	2.55
	Disang Series	2.55	2.60
Gondwana Formations		2.60	2.60
Upper Precambrian-Lower Paleozoic		2.62	2.62
Lower Precambrian-Basement		2.67	2.67

the profile varies from near zero to −70 mgal, the most negative value occurring over the Manabum depression in upper Assam. The observed Bouguer anomaly, the geological effect of sediments and the observed and geologically corrected (termed as g_{RA}) anomalies are shown in Figure 11.3. It may be noted from the figure that the Bouguer anomaly varies from −30 mgal at the outer edge of the Mikir Massif to −220 mgal near the Manabum depression. Geologically corrected Bouguer anomalies also show a considerable variation, from −30 mgal to −180 mgal. A structural hump near Naharkatiya is also reflected in g_{RA} anomalies.

11.3.2. PROFILE BB′

This profile runs from the Naga Hills in the southeast to the Eastern Himalaya in the northwest across the Upper Assam Valley. The underlying geological information is after Mathur and Evans (1964). The observed Bouguer anomaly, the geological correction for the sediments and the corrected Bouguer anomaly, (g_{RA}) are shown in Figure 11.4. The profile clearly shows the relationship of the Naga and Himalaya thrusts to the tectonics of Assam Valley and the sediments involved in the whole process. As seen in the profile, the sediments thicken considerably towards the Naga Hills as well as the Himalaya, where the maximum thickness of sediments is of the order of 5.5 km close to the Himalayan thrust.

The observed Bouguer anomaly ranges from −150 mgal near the Naga Hills to −260

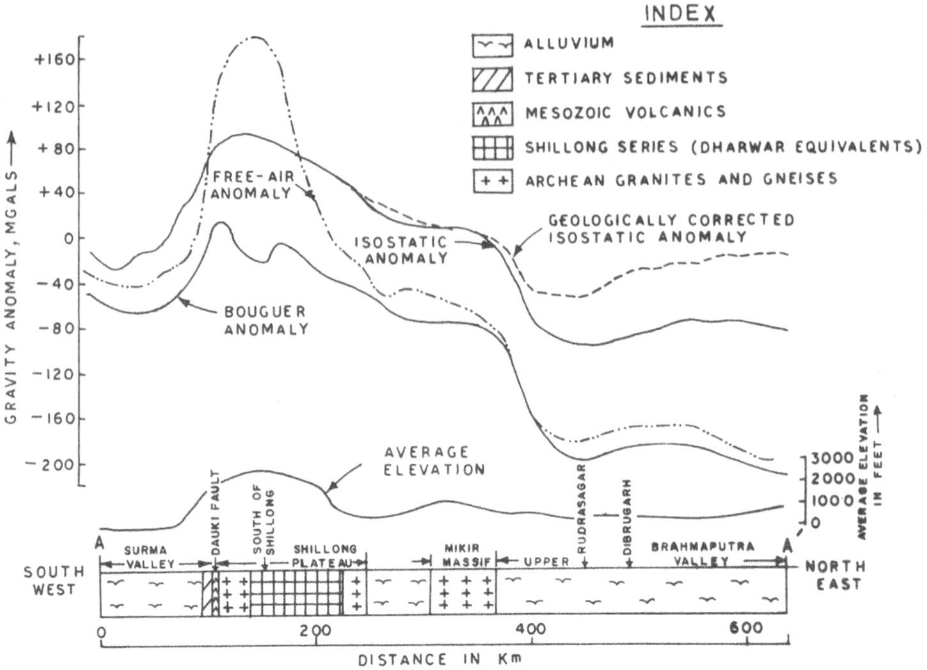

Fig. 11.3. Profile AA′ in NE–SW direction across Tripura folded belt, Shillong Plateau and Assam Valley ending near Himalaya foothills. Geologic correction shows the influence of sediments on the gravity field along the profile.

mgal near the foothills of Himalaya. The geologically corrected anomalies range from −120 mgal to −150 mgal over the valley and become increasingly negative towards the Himalaya.

It may be noted from the analysis of these profiles that the large negative anomalies observed in the Assam Valley cannot be attributed to the sediments alone. The explanation for the large g_{RA} anomalies has to be sought in terms of the crust and mantle relationships prevailing in the Assam Valley. These may also be influenced by the root formation under the Naga Hills and the Eastern Himalaya. An explanation for these will be discussed after considering the nature of the isostatic anomalies prevailing in the area.

11.4. Pratt-Hayford Isostatic Anomaly Map of Northeastern India

Figure 11.5 shows Pratt-Hayford isostatic anomaly map (for $D = 113.7$ km) of Northeastern India and the adjoining areas. The map is based upon the data from the Survey of India (Gulatee, 1956) and the map published by Evans and Crompton (1946). A maximum relief of about 230 mgal may be noted on the map, from +80 mgal over the Shillong Plateau to −130 mgal over the Upper Assam Valley.

As mentioned earlier, the Shillong Plateau is one of the most anomalous areas of India from gravity point of view. This is supported by large isostatic anomalies prevailing over the plateau reaching a maximum value of +80 mgal near Haflong

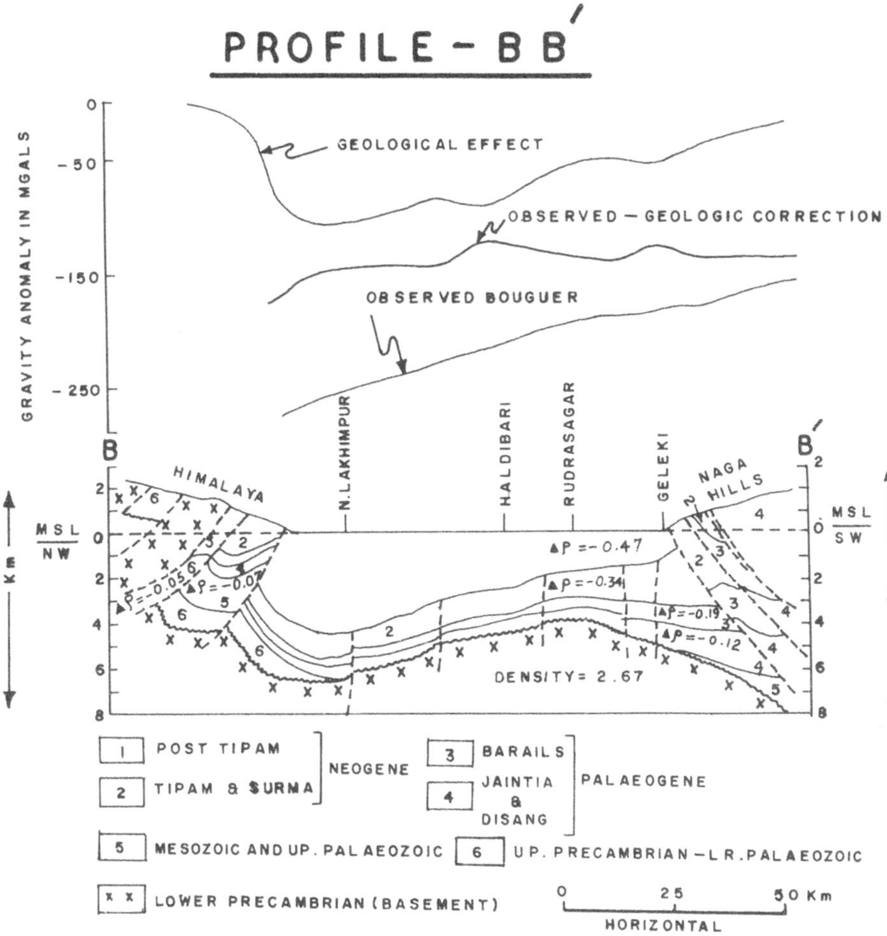

Fig. 11.4. Profile BB' in NW–SE direction from Himalaya to Naga Hills across Assam Valley. Geological correction due to sediments in Assam Valley has been computed using two dimensional model.

(25° 11′ N, 93°01′ E) which is situated at the junction of the Naga thrust and the Dauki fault. The plateau anomalies extend towards Darjeeling Himalaya in a northwesterly direction across the Lower Assam Valley. Anomalies in the Foothills of Himalaya are negative. However, a belt of positive anomalies reappears at higher elevations in the region of Darjeeling (27°03′ N, 88°03′ E) and Gangtok (27°22′ N, 88°37′ E).

The alluvial covered Brahmaputra plains as well as the eastern edge of the Indo-Gangetic plains are characterised by moderate to strongly negative issostatic anomalies, revealing a gravity minimum of about −130 mgal near North Lakhimpur (27° 14′ N, 94°06′ E) in Upper Assam. The Gangetic West Bengal and the Bengal Basin have near zero isostatic anomalies, save the Surma Valley (the northern part of the Bengal Basin, immediately to the south of the Shillong Plateau).

At the eastern portion of the area, the trend of anomalies is NE–SW, instead of east-west. This veering around of the trend starts near Jorhat (26°46′ N, 94°13′ E)—Kohima (25°40′ N, 94°07′ E) area, where a V-shaped form of anomalies is seen, the

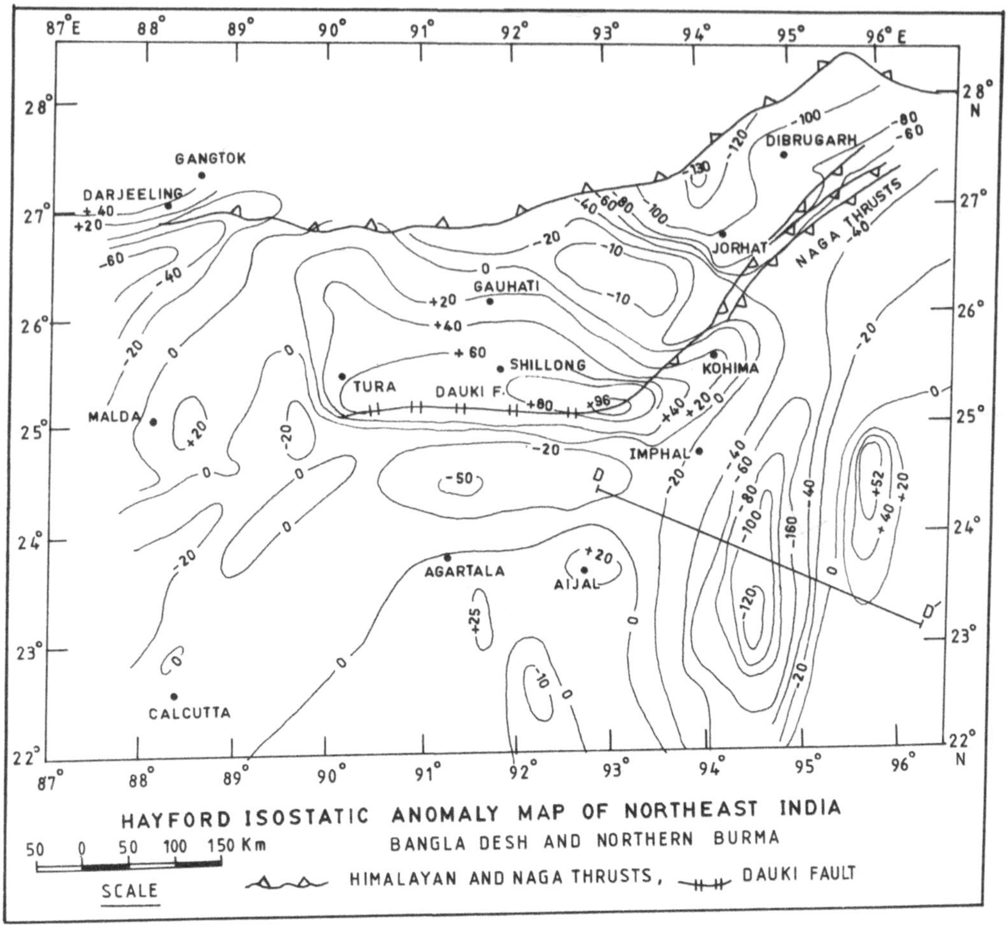

Fig. 11.5. Hayford isostatic anomaly map of NE India and N. Burma. The figure shows that the Shillong Plateau is characterised by large positive anomaly, while Assam Valley and Burmese molasses basin are significantly negative.

apex pointing towards Haflong. The nosing of the Shillong Plateau anomalies stretches out towards Kohima. However, near Imphal (24°49′ N, 93°56′ E) and further east, that is over the Burmese plains, the anomalies trend north-south. The Burmese negative strip has been described as an exceptionally anomalous area by Evans and Crompton (1946). Further east of this strip, an area of positive anomalies appears which falls over the volcanic arc of Burma.

11.4.1. SIGNIFICANCE OF ISOSTATIC ANOMALIES

In general, large isostatic anomalies may indicate appreciable departure from isostatic equilibirum. However, this may not always be the case. In areas of low elevation large isostatic anomalies may be due to the influence of local geological factors such as large thickness of sediments present, as in the Ganga basin. Application of geologic correction due to sediments enables us to remove the effect of sediments and to look for crust and mantle relationships prevailing in the area. Verma and Mukhopadhyay (1977) have examined the effect of sediments on isostatic anomalies in Assam Valley and found that geologically corrected isostatic anomalies are appreciably lesser in magnitude, being of the order of −40 to −50 mgal. Such anomalies could be explained in terms of crustal downwarp underneath Assam Valley as a result of regional compensation of Eastern Himalaya to the northwest and Naga hills to the southeast.

Analysis of gravity anomalies along profile CC′ (location shown in Figure 11.2) throws an interesting light on the nature of Shillong Plateau and the adjoining areas. Figure 11.6 shows the nature of free-air, Bouguer and Airy isostatic anomalies along with elevation along the profile which runs in a general north-south direction from Tripura folded belt, passing through Surma Valley, Shillong Plateau, Brahmaputra Valley and ends up near the Himalayan foothills. Gravity anomaly maps published by NGRI were used for this purpose. It is observed that Bouguer anomaly is fairly negative over the Tripura folded belt and reaches a minimum value of −80 mgal over Surma Valley. All the anomalies become sharply positive over Shillong Plateau after crossing the Dauki fault, and then asume a downward trend over the northern edge of the plateau and Brahmaputra Valley. Highest positive Bouguer as well as isostatic anomalies observed over Shillong Plateau are nearly +40 and +100 mgal respectively near the eastern as well as the western edge of the plateau. Near the Himalaya foothills the Bouguer anomaly is nearly −150 mgal.

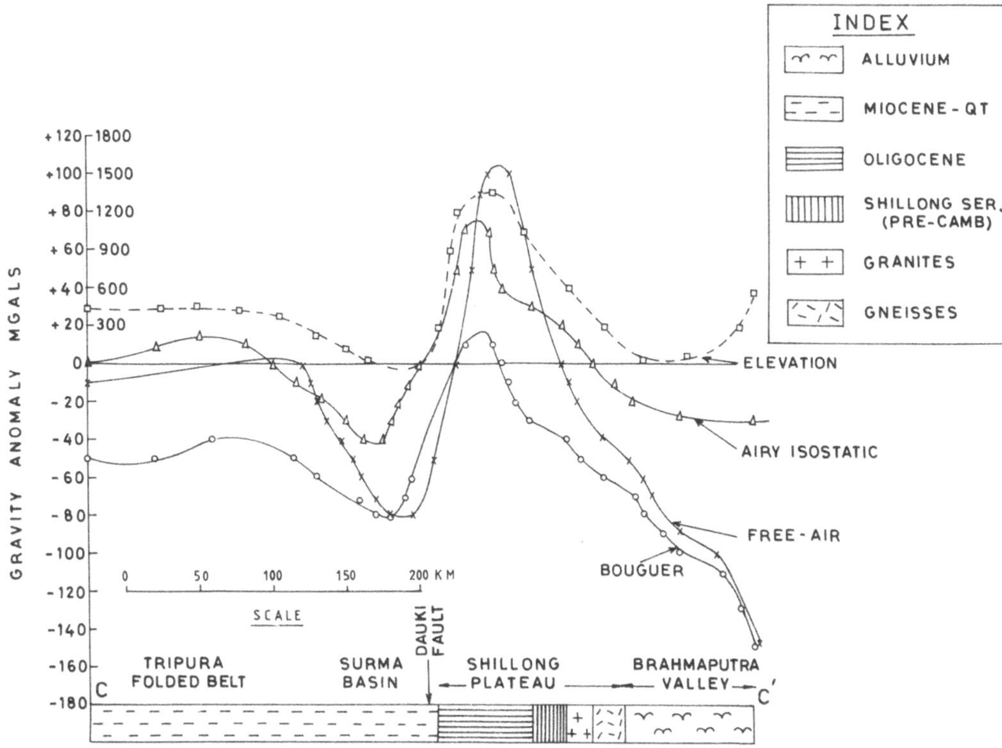

Fig. 11.6. Analysis of gravity anomalies along profile CC′ across Shilliong Plateau. The profile shows that the Shillong Plateau is characterised by highly positive gravityfield.

In general Bouguer anomaly becomes more negative over areas of high elevation on account of root effect, expected on the basis of Airy or Hayford scheme of isostatic compensation. However, this does not happen in the case of Shillong Plateau which has an anomaly-elevation regression coefficient of −28 mgal km^{-1} with elevation (Verma and Mukhopadhyay, 1976). It may be noted that the free-air anomaly is highly positive and generally follows elevation. The nature of free-air, Bouguer and isostatic anomalies suggests that the Shillong Plateau is a horst type structure and is not locally compensated. It is one of the most anomalous areas of India not only from the point of view of gravity but also seismicity (discussed later in this chapter). Taking into account the nature of gravity anomalies and elevation of the plateau, Verma and Mukhopadhyay (1977) have shown that the Shillong Plateau has a higher density (0.12 gm cm^{-3}) than normal.

In contrast to the Shillong Plateau we note that over large parts of Bengal basin the isostatic anomaly is very low (see Figure 11.6). This is in spite of very large thickness of sediments, exceeding 10 km over large areas. This suggests that the Bengal basin is more or less in isostatic equilibrium and is underlain by an intermediate type of crust as suggested by Verma and Mukhopadhyay (1977).

11.5. Seismicity of NE India and N. Burma

Data of historical seismicity shows that all the tectonic zones constituting NE India have been active for the past several hundred years. The area has experienced two devastating earthquakes, with magnitude exceeding 8.0 (on Richter scale), one near the northern edge of the Shillong Plateau in 1897 and the other in Assam Syntaxial Zone in 1950. Besides historical earthquakes with intensity exceeding IX (on R.F. scale) have been observed for earthquakes at Dhubri (1930), near Darjeeling (1899) and near Sylhet (1869).

Seismicity map of NE India and northern Burma prepared from several sources including India Meteorological Department, Oldham (1882), Gutenberg and Richter (1954), National Oceanic and Atmospheric Administration (NOAA) and Lee *et al.* (1976) is shown in Figure 11.7.

Seismicity follows the trend of major tectonic zones in the area. The trend is N–S to NE–SW along Tripura folded belt, Arakan-Yoma folded belt and the Burmese molasse basin, between latitude 22° to 27° N, and longitude 93° to 97° E. Near latitude 27° N the trend changes sharply to NW–SE along the Assam syntaxial zone which further changes to NE–SW near 29° N, 95° E. Seismicity follows the well-known Himalayan thrusts between longitude 95° to 88° E and continues further west along the Nepal Himalaya. The entire area of Shillong Plateau appears to be very active as seen from location of epicenters which are located near its margins as well as the central region. Three events of large magnitude (exceeding 7.0) for years 1869, 1897, and 1930 have been recorded from this plateau. Out of these the event for 1897 was the largest with a magnitude of 8.7 (Gutenberg and Richter, 1954), the intensity reaching a value of up to XII (Oldham, 1899). This earthquake was felt over an area of nearly 4.5 million km^2. Das Gupta and Nandy (1982) have studied the seismicity and its relationship to faults and lineaments. They have noted that all the major lineaments surrounding the Shillong Plateau namely, the Brahmaputra lineament to the north,

Dauki fault to the south, Kopili fault to the east and Yamuna lineament to the west are seismically active. The latter two appear to be deep seated faults with epicenters reaching a depth of nearly 80–100 km.

The Bengal basin which has had a history of sedimentation during whole of Tertiary times is also not inactive. Several lineaments traversing the Bengal basin in different directions appear to be active. These include the Sylhet lineament traversing in NE–SW direction from Dacca towards Sylhet (see Figure 11.7), Tista lineament (mostly along Tista river) traversing in NW–SE direction from Agartala to Siliguri and Padma lineament, also traversing in the same direction. On the eastern side of the Bengal basin high seismic activity is associated with Tripura folded belt and some lineaments traversing in northeasterly direction near Agartala. Chittagong (and its vicinity) located near the southern part of this folded belt is known to have experienced two major earthquakes in 1851 and 1865.

The entire Arakan-Yoma folded belt is characterised by high seismicity with depth of foci gradually increasing towards east. The deepest earthquakes in this belt are

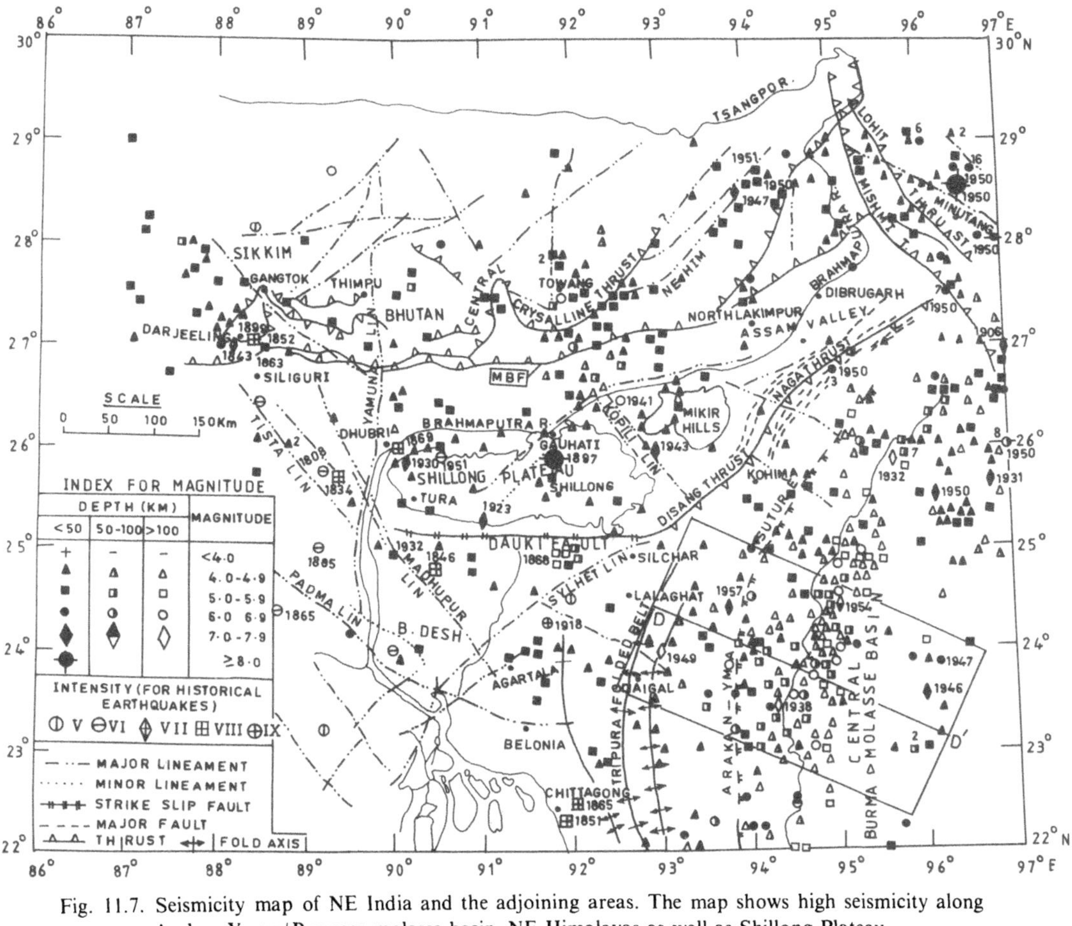

Fig. 11.7. Seismicity map of NE India and the adjoining areas. The map shows high seismicity along Arakan-Yoma / Burmese molasse basin, NE Himalayas as well as Shillong Plateau.

located nearly 250 km, to the east of the Arakan-Yoma mountains (Verma *et al.*, 1976). It may be mentioned here that this seismic belt continues southward along Andaman islands into Sumatra as shown by Verma *et al.* (1978).

Along the Assam syntaxial zone several thrusts (including Lohit and Mishmi thrusts) trending in NW–SE direction appear to be quite active up to 100 km depth. Along Eastern Himalaya and Bhutan epicenters for most of the earthquakes are located appreciably away from Main Boundary Fault (MBF) and genrally close to the central crystalline thrust. Some of the lineaments trending in northeasterly direction also appear to be active. In Sikkim Himalaya (near 27° N, 80–87° E) appreciable seismicity is associated with MBF and other Himalayan thrusts. Quantitative seismicity of the region is discussed in Chapter 12 along with that of the Himalayas.

11.6. Relationship Between Seismicity and Gravity Field in NE India and N. Burma

As discussed in Section 11.5, the Arakan-Yoma folded belt and the Irrawaddy basin are characterised by high seismicity with epicenters reaching up to a depth of nearly 200 km Profile DD′ (Figure 11.8, location shown in Figure 11.7) drawn across this belt in ESE-WNW direction shows the nature of epicentral distribution with depth. For purposes of drawing this profile epicenters within a distance of nearly 100 km were projected on the profile with depth. The figure also shows the nature of Bouguer and isostatic anomalies along the profile along with elevation. It is apparent from the figure that the region of large isostatic anomalies exists to the east of Arakan-Yoma, where the deepest earthquakes are also located. Bouguer anomaly is also minimum in this region.

The relationship between seismicity and anomalous gravity field is also observed in other parts of NE India including Shillong Plateau, Himalaya foothills and Assam Valley. The seismicity as well as gravity field are certainly related to the processes of mountain buidling, subsidence and plateau uplift that have taken place since Tertiary times. These processes appear to be still active.

11.7. Plate Tectonics Model for NE India and N. Burma

Some of the major features of NE India and Burma appear to have been evolved as a result of collision of Indian subcontinent with Eurasia. Desikachar (1974) and Nandy (1976) have explained the evolution of major features of this area in terms of a plate tectonics model. According to them an Andaman-Arakan-Assam geosyncline existed during Mesozoic times between the Burmese landmass and the Indian plate extending from 5° N to 27° N. The Indian plate moved from south to north during Cretaceous-Eocene times as supported by Paleomagnetic studies of Indian rocks (McElhinny, 1973) and oceanic magnetic anomalies (McKenzie and Sclater, 1971). During this period an island arc was formed over the western margin of Burma (The Shan-Tenasserim block). The Indian plate started subducting underneath the Burmese block during Cretaceous-Eocene times. A flysch trough was developed along the margin of the island arc during early stages. As the two plates came closer together the Arakan-Yoma mountains and Naga Hills were formed by thrust movements directed from west towards east in N. Burma and towards northwest in NE India. The existence

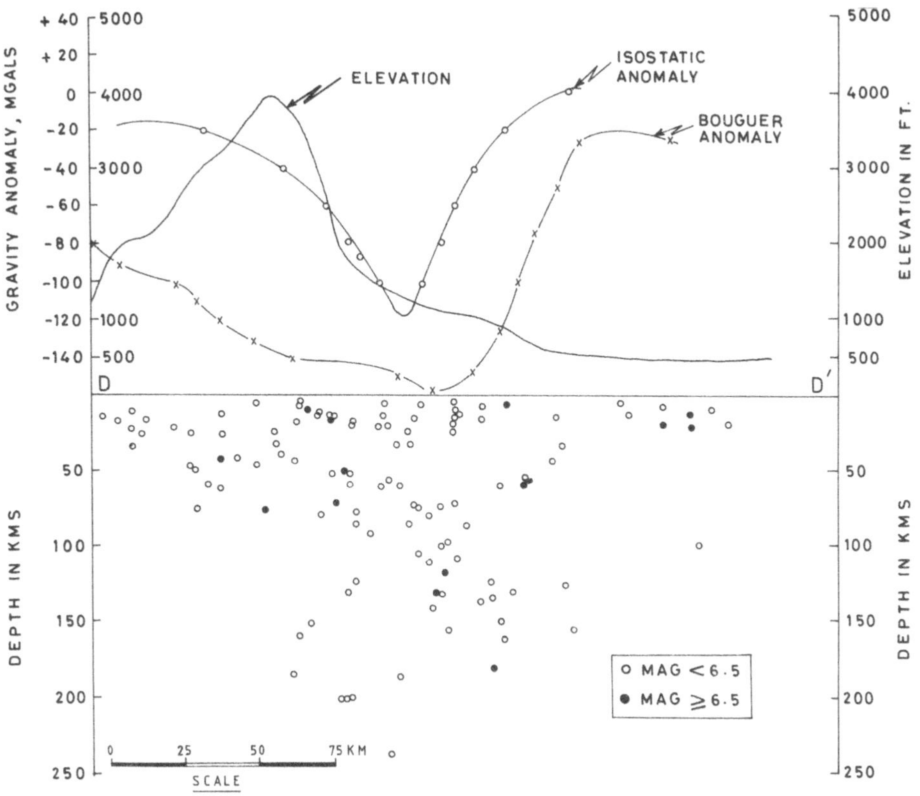

Fig. 11.8. Analysis of gravity and seismicity along profile DD′ from Tripura folded belt across Arakan-Yoma to Burmese plains. The figure shows that isostatic anomaly is appreciably negative where deepest earthquakes are located.

of Ophiolite belt all along the Arakan-Yoma supports this model. Figure 11.9 shows various stages in the development of structures in this area during Cretaceous to Recent times.

Further compressive movements between the two plates during Oligocene to Recent times resulted in the formation of Tripura folded belt to the west of Arakan-Yoma and Burmese molasse basin to its east. The existence of Quaternary volcanoes to the east of Arakan-Yoma supports the subduction model as applicable to island arcs.

The strong correlation between high seismicity, depth of foci and large negative isostatic anomalies to the east of Arakan-Yoma suggests that the subduction processes are still continuing. Watts and Talwani (1974) have shown that flexure of the crust and upper mantle as a result of collision between two plates can cause large negative isostatic anomalies, of the order of few hundred mgals. The large negative isostatic anomaly observed in Burma (see Figure 11.5) could be explained as a result of flexure of the lithosphere in this area. Further information regarding processes taking place at present in India–Burma region is obtained from study of focal mechanism solutions (see Chapter 12) for earthquakes taking place in this area. These show that a considerable amount of thrust faulting is taking place along the Burmese arc at present

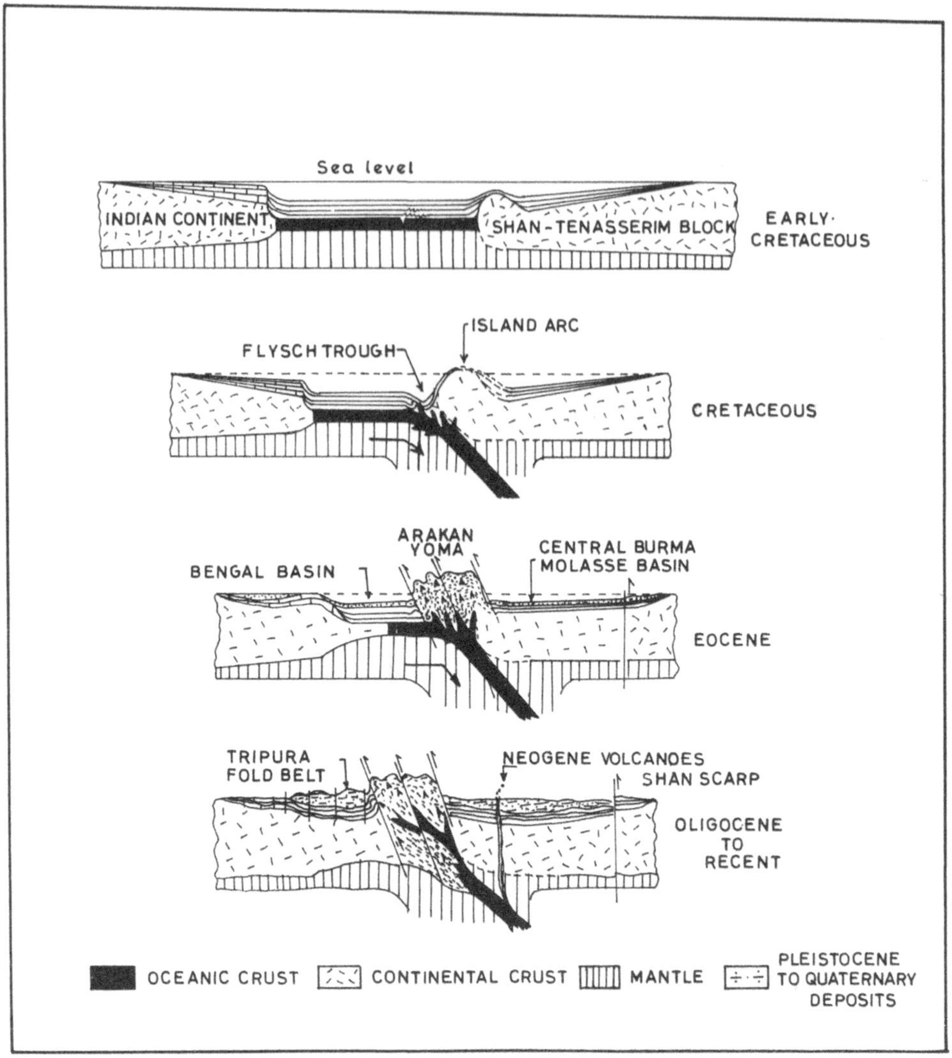

Fig. 11.9. Plate tectonics model for evolution of NE India and the adjoining areas after Nandy (1976). The figure shows different stages in the evolution of the whole area starting from Early Cretaceous times. Figure reproduced with the permission of D. R. Nandy.

with P axis (axis of largest compression) directed nearly in east-west direction. The thrust movements are taking place up to a depth of nearly 200 km at present.

On account of continued movement of the Indian landmass towards the north, the lithosphere underneath the Eastern Himalaya, Assam Valley and Shillong Plateau is still experiencing compression. As a result of this the Shillong Plateau is still rising. These forces are preventing isostatic equilibrium from taking place in the area and are also responsible for its high seismicity.

CHAPTER 12

SEISMICITY OF THE HIMALAYAS AND PLATE TECTONICS

12.1. Introduction

Himalayas, the highest mountains in the world form a well defined arc to the north of Indian Peninsula and extend over a distance of nearly 2500 km from west-northwest (Indus river/Nanga Parbat, 8215 m) to the east-northeast (Brahmaputra river/Namche Barwa, 7755 m). On the northern side of the arc lies the Indus-Tsangpo Valley and to the south the Indo-Gangetic plains. Although it is known that the Himalayas are mostly composed of sedimentary and metamorphic rocks, yet it is not clear how they were sandwiched between the sub-continent of India to the south and the Asian landmass to the north.

The origin of Himalayas appears to be related in some way to the collision between the Indian and the Eurasian plates. McKenzie and Sclater (1971) have carried out an exhaustive study of the Indian ocean and traced the movement of the Indian plate over several thousand kilometers since the Late Cretaceous. They have shown that the rate of movement of Indian plate has been relatively very large, 5–6 cm yr^{-1} during the past 40 m.y. The most likely time of contact between India and Eurasia appears to be sometime during Eocene (Dewey and Bird, 1970; Powell and Conaghan, 1973). Different models have been presented to explain the origin of the Himalayas. Dewey and Birds' model envisages the birth of Himalays as a result of collision between the Indian and the Eurasian plates. The ophiolite suite that follows mostly the Indus and Tsangpo Valley (Indus Suture zone–ISZ, also called Indus Tsangpo Suture, ITS) northeast of the Himalayas (Gansser, 1964, 1980) apparently marks the pre-collision boundary along which the Indian plate was subducted beneath the Asian plate. Subsequently, the low density continental crust and its large thickness prevented further subduction of the Indian lithosphere (Isacks *et al.*, 1968., McKenzie, 1969). In place of subduction other processes such as crustal shortening and mountain building took place resulting in the formation of Himalayas. However, Powell and Conaghan's (1973) model envisages that after the collision the sialic crust of Indian Peninsula subducted below the Asian landmass resulting in abnormal thickness and elevation of the Tibetan plateau. Thus there are major differences between scientists regarding the manner in which the collision between India and Asia took place and what happened later. The present day Indian plate boundary is supposed to lie along the Arakan Yoma-Himalayas and Suleiman/Kirthar ranges of Pakistan. However, analysing the quantitative seismicity of whole of Asia, Kaila and Narain (1976) have propsed that the northern part of the boundary does not lie along the Himalayas but along the Tien Shan-Nan Shan seismic zone.

It is believed that thrusting which took place in the Himalayas south of ISZ during the Oligocene resulted in piling of older rocks of Indian subcontinent on top of one another. With increasing time movements migrated southward towards the Main Central Thrust (MCT) and Main Boundary Fault (MBF) (Molanar *et al.*, 1977). The current tectonic activity in Asia is attributed to continued collision between the Indian and the Eurasian plates (Dewey and Burke, 1973; Molnar and Tapponnier, 1975; Minster *et al.*, 1974). It is envisaged that due to underthrusting of the Indian plate beneath the Eurasian plate stresses are being generated which are being released through earthquakes of various magnitudes.

At present appreciable movements are taking place all along the Himalays. Recent geodetic measurements carried out by Survey of India (Arur, 1982) show that horizontal as well as vertical movements are taking place, practically all along the thrusts of the Himalayas. Movements of the order of 4–13 mm yr^{-1} have been recorded along Beas/Satlitta thrusts, in Punjab and Nahan thrust in U.P. Along Main Central Thrust appreciable movements varying from 8 mm to 40 mm yr^{-1} have also been recorded. In addition to movements along major thrust, several faults and lineaments transverse to the trend of the Himalayas have been found to be active as shown by Valdiya (1976). Before discussing the present day seismicity of the Himalayas and its significance a background of geology and tectonics of the Himalayas is presented.

12.2. Geology and Tectonics of the Himalayas

The Himalayas consist of the following litho-tectonic units from south towards north: (a) Sub-Himalaya, (b) Lesser or Lower Himalaya, (c) Higher Himalaya, (d) Tethys Himalaya, and (e) Indus-Suture. Figure 12.1 shows tectonic map of the Himalayas (after Valdiya, 1973) showing its subdivisions.

(a) The Sub-Himalaya consist of Neogene and later deposits (Siwaliks) which have been deposited in a fore-deep, formed as the Himalayas were rising to the north. They have been folded and thrust by movements, which are continuing to the present day. The Sub-Himalaya are bound to the north by Main Boundary Fault (MBF) which separates the Siwaliks of Mio-Pleistocene age from early Tertiary (Murre-Dharamsala) rocks. In northwestern part of Siwaliks several thrusts have been developed during the Quaternary period. The outermost of these, is the Hoshiarpur thrust, followed by Satlitta, Jawalamukhi and Nahan. The MBF continues all along the Himalayan chain from west to east and is known by different names in different areas. It is Murree thrust in Kashmir, Palampur thrust in Panjab and Nahan/Krol thrust in Uttar Pradesh (U.P.).

(b) The Lower (or Lesser) Himalaya zone is composed of platform type sediments, and metamorphics believed to be in continuation of shield elements, overthrust by southwardly directed nappes and thrust sheets. These Himalaya are delimited to the north by Main Central Thrust (MCT) and to the south by Main Boundary Fault. The metamorphics have been intruded by several granitic bodies and basic volcanics. The granite masses appear to be syntectonically emplaced bodies within the metasediments. In Jammu and Kashmir state, north of Riasi thrust two major thrusts, Murree and Panjal, have been mapped. The former has brought Carboniferous-Eocene rocks and thrust them over the Siwaliks, while the latter has driven the Archaean or Precambrian schists and phyllites with granitic gneisses and thrust them over the

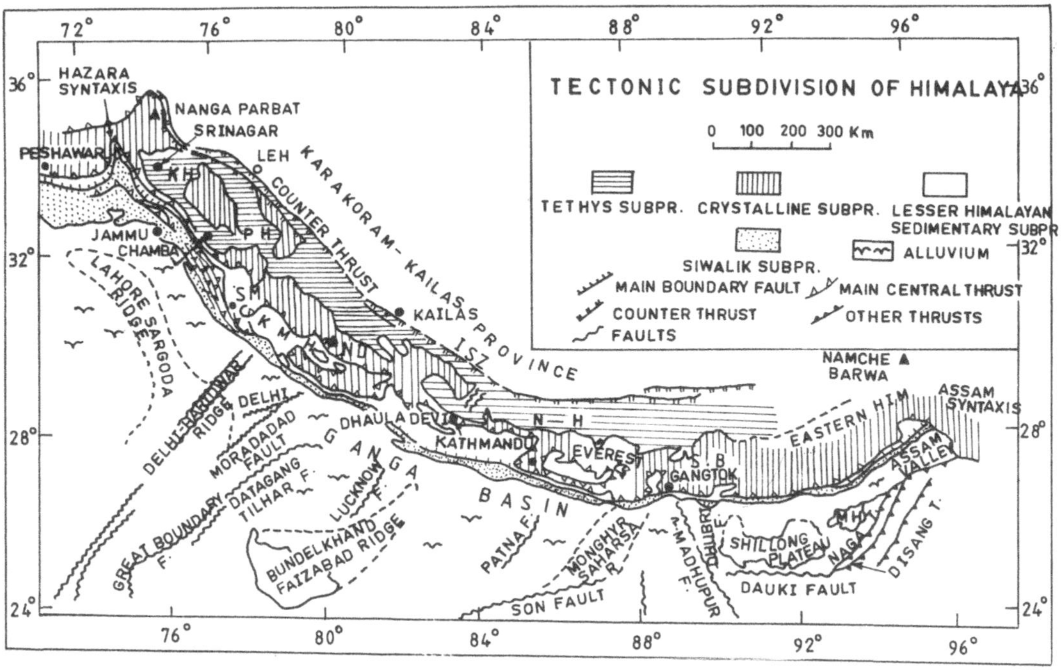

Fig. 12.1. Simplified geologic map of the Himalayas (After Valdiya, 1973) showing various Subprovinces. Abbreviations: (ISZ) Indus Suture Zone, (KH) Kashmir Himalaya, (KMH) Kumaun Himalaya, (MH) Mikir Hills, (NH) Nepal Himalaya, (ND) Nanda Devi, (PH) Panjab Himalaya, (SB) Skikim and Bhutan Himalaya, (SM) Simla, (SP) Shillong Plateau, (U.P.) Uttar Pradesh.

Carboniferous-Eocene rocks.

(c) The Higher Himalaya zone represents the line of greatest uplift and maximum crustal shortening and is the principal metamorphic belt of low grade (mainly of Precambrian age), intruded by granites of later ages, including Tertiary. It is now more or less well established that this metamorphic belt extends from Kashmir in the west to Dihang Valley in Arunachal Pradesh. It has an enormous thickness (15–20 km) and forms the basement upon which the Tethys sediments were deposited. The MCT which marks the southern boundary of this zone has caused southward movement of crystalline rocks over the low grade metamorphic rocks of Lesser Himalayan zone. The intrusive granites invariably bear tourmaline (Jhingaran *et al.*, 1979). The Lower as well as Higher Himalaya are made up of rocks mostly belonging to the Indian shield. They do not represent a geosynclinal origin, which is in quite contrast to the Alpine mountain building events.

(d) The Tethys Himalaya rocks present quite a contrast to those of Lower and Higher Himalaya. The zone comprises fossiliferous sediments ranging in age from Late Precambrian to Upper Cretaceous. The rocks belonging to this zone occur in four separate areas viz., Kashmir, Lahul-Spiti, Kumaun and Nepal. It is believed that the sedimentary pile which was once laid down in Tethys sea must have formed a huge synclinorium at one time. The Tethys zone is delimited to the north along the Upper Indus-Tsangpo Valley by a southward hading counter thrust.

(e) The Indus Suture zone is a structure of fundamental importence north of the Tethys zone and occurs over a length of nearly 2400 km along the Indus and Tsangpo Valleys. It defines a junction between litho-tectonic elements of the Himalaya and Trans-Himalaya. This zone is considered to be a flysch belt which represents a transition from platform to ocean type sediments and becomes increasingly tectonised towards north. The ocean floor is represented by basic rocks alternating with pelagic sediments, 3000–4000 m thick, such as Dras Volcanics of Cretaceous age. A noteworthy feature of this zone is the occurrence of ultrabasic rocks as well as Trans-Himalayan plutons all along its belt. The ultramafic rocks such as Harzburgite give all indication of origin under high temperture and pressure conditions. North of Indus Suture lies the zone of Ladakh batholiths characterised mostly by acidic intrusives, north of which lies another Suture zone known as the Shyok zone (Andrews Speed and Brookfield, 1982).

The Himalayan arc takes a sharp hair pin turn on the western side (near 34° N, 73.5° E) forming a well known structure known as Hazara/Kashmir syntaxis. The Murree and Panjal thrust also take a sharp turn and continue on the western side in Hazara thrusts region. Northeast of Hazara syntaxis lies the Nanga Parbat–Haramosh massif around which the rocks belonging to Indus Sature take a sharp turn. This crystalline axis separates the Kohistan zone on the west from Ladakh zone on the east. On the eastern side (near 28° N, 96° E) the Himalayan arc takes a sharp turn of nearly 90° and meet the structures belonging to the Arakan-Yoma which trend in NNE–SSW to N–S direction. This region is known as Assam syntaxis and is discussed in detail in Chapter 11.

12.3. Seismicity and its Relationship to Faults

For purposes of present discussion, the Himalayas have been longitudinally divided into the following regions.

(1) Northwestern Himalaya, including Kashmir, Panjab and Kumaun Himalaya extending from Hazara Syntaxis (73° E) to Sarda river (80° E) along the Himalaya arc

(2) Central Himalaya, including mostly Nepal, extending from 80° E to 88° E.

(3) Eastern Himalaya extending from 88° E to 97° E.

Relationship of seismicity to faults in the Himalayas is discussed below.

12.3.1. NORTHWESTERN HIMALAYA

Figure 12.2 shows an epicentral distribution map of Northwestern Himalaya. The map is based on the data provided by India Meteorological Dept. (IMD), National Oceanic and Atmospheric Administration (NOAA), Gutenberg and Richter (1954), Oldham (1882), Chandra (1978), and Quittmeyer and Jacob (1979). Generally seismicity follows the NW–SE trend of the Himalayas and often there is cluster of events close to either MCT or the MBF. However, on close scrutiny it is found that the picture is not very simple since most of the faults are thrusts dipping towards north. In Kashmir Himalaya, the Kashmir Valley is found to be very active having experienced several historical earthquakes of intensity VIII and above in 1828, 1885, 1937, and 1947 (Chandra, 1978). The seismicity could be attributed to Panjal

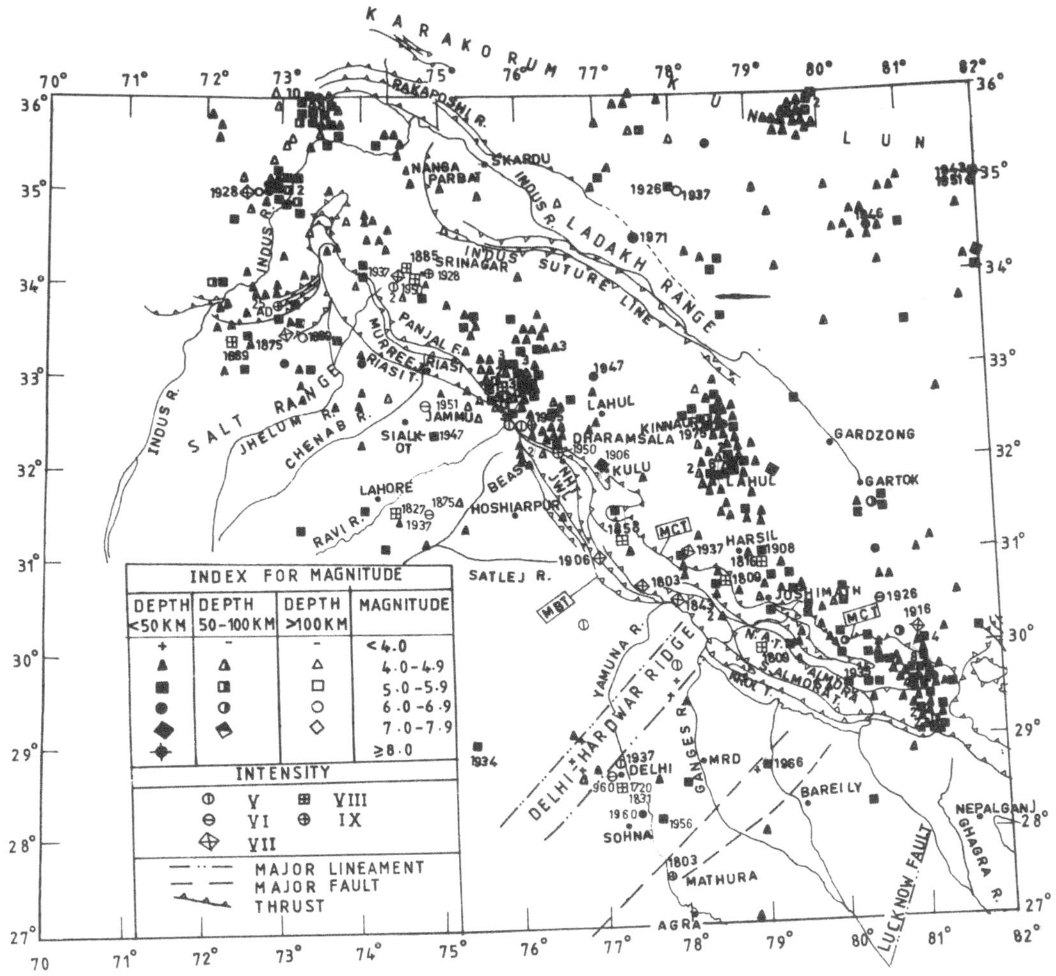

Fig. 12.2. Epicentral distribution map of NW Himalayas and the adjoining areas from historical times to 1980. Location of major Himalayan thrusts is shown in the figure. It may be seen that most of the earthquake epicenters are located north of the Main Boundary Fault (see text). Abbreviations: (JWL) Jawalamukhi thrust, (NAT) North Almora Thrust, (NHT) Nahan Thrust, (MRD) Moradabad.

fault/thrust, Murree or Riasi thrusts lying to the south of Kashmir Valley. The Himalayan trend of seismicity extends northwestward beyond the Hazara syntaxis up to 73° E and has been studied in detail by Armbruster, Seeber and Jacob (1978) from micro-earthquakes. They have named this region as Indus Kohistan Seismic Zone (IKSZ). The Hazara thrusts region and Potwar plateau of Pakistan are found to be fairly active.

Considerable seismicity is observed in Panjab Himalaya where MCT and MBF (locally known as Panjal thrust and Murree thrust) come very close together, near longitude 76° E. Epicenters having depths greater than 50 km are observed close to the above mentioned thrusts. The great Kangra earthquake of 1905 (magnitude, 8.4

according to Gutenberg and Richter) took place very close to the MBF. The earthquake had taken a toll of nearly 19 000 lives and has been studied in detail by Middlemiss (1910). The length of rupture due to this earthquake extended over a distance of nearly 3000 km along the MBF as estimated by Seeber and Armbruster (1981) on the basis of intensity data. It may be mentioned here that in this region a number of thrusts underlie the MBF forming an imbricate structure. These include the Nahan, Satlitta, Jawalamukhi and Hoshiarpur thrusts. Krishnaswamy (1966) after analysing the geology and tectonics of Panjab Himalaya has suggested that the Kangra earthquake was caused by movement along Satilitta thrust which underlies the MBF. Srivastava and Dube (1982) have reported the results of microearthquake studies in NW Himalayas in connection with several dam sites. They have shown that the entire region north of MBF is highly active up to a distance of nearly 100–150 km north of major thrusts. According to them the thrusts located south of MBF are also seismically active.

Between lat. 31° to 33° N and long. 78° to 79° E (MCT and ISZ) a zone of high seismcity is observed to underlie the Tethys Himalaya region. No suitable explanation for such high seismicity is obtainable in terms of activity of major thrusts. It is quite possible that the seismicity is due to some transverse faults in the area as suggested by Kaila and Narain (1976). Between 79° to 80° E most of the seismicity appears to be related to MCT, locally known as Vaikrita thrust. To the south of MBF noticeable seismicity is observed along several features which are oriented transverse to the trend of the Himalayas. The most conspicuous of these is the Delhi-Hardwar ridge (DHR) whose axis extends from 27° N, 75° E to 30° N, 78° E. Several historical earthquakes (1720, 1830, 1831, 1842, 1937) have been reported from the area close to Delhi. The area between Delhi and Sohna appears to be relatively more active. Apart from DHR other transverse faults south of NW Himalayas are: (1) Mathura-Muradabad fault, southeast of DHR, (2) Ravi-Beas fault (located between Ravi and Beas rivers).

12.3.2. CENTRAL HIMALAYA

Epicentral map of Central Himalaya is shown in Figure 12.3. The figure shows that the eastern and the western margins of the area are relatively more active. Between 29° to 30° N, 80–82° E a number of events having magnitude 6.0 and above are located close to MCT. In the area between 82° to 87° E seismicity is fairly scattered. Most of the epicenters in this area lie between MBF and MCT. Some of these could be related to MBF, which is dipping towards north. However, it appears that several transverse faults are also active in this region. These include Dhangi, Andhi, Judi, Thaple and Kathmandu faults. Epicenter for one of the most devastating earthquakes in Indian history, the Bihar earthaquake of 1934 (magnitude 8.3, according to Gutenberg and Richter, 1954) is located fairly south of the MBF. This earthquake had taken a toll of nearly 11 000 lives. The earthquakes have been studied in detail by several officers of Geological Survey of India (Auden and Ghosh, 1934; Dunn *et al.*, 1939). The isoseismals as reported were elongated in E–W direction and had an intensity of X south of the main shock. Singh and Gupta (1980) have obtained a solution for this earthquake using S–Wave polarisation studies. The solution shows thrust cum strike-slip mechanism along an east-west trending, shallow dipping fault plane. The fault length was estimated to be nearly 130 km. The most plausible explanation for this

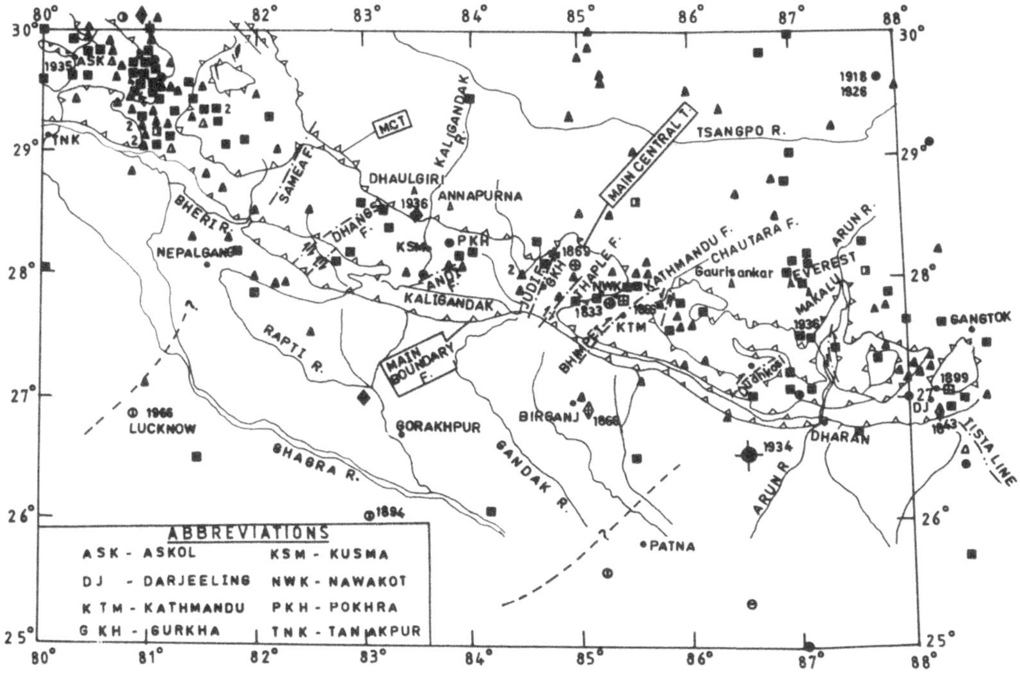

Fig. 12.3. Epicentral distribution map of Central Himalayas (80° to 88° E). Location of several transverse faults in the area and the Himalayan thrusts is shown in the figure. Index for magnitudes as in Figure 12.2.

major earthquake is movement along some thrust underlying the MBF but possibly related to it in some way (see discussion at the end). Seismicity is generally scattered in the Higher Himalaya region north of MCT. The Indus-Tsangpo Suture zone is quite inactive. No well defined seismicity is noted in the Ganga basin south of MBF in this region.

12.3.3. EASTERN HIMALAYA

In Eastern Himalaya between longitude 88–96° E, appreciable seismicity is observed (see Figure 11.7) which appears to be related to the MBF and thrust or lineaments trending in northeasterly direction. It may be mentioned here that two of the most devastating earthquakes in Indian history (1897 over Shillong Plateau and 1950 in Assam syntaxial region) have occurred in northeastern India. The 1897 event had a maximum intensity exceeding X over Shillong Plateau and the surrounding Brahmaputra Valley (see Chapter 11 for discussion). This earthquake is not strictly related to the Himalayas but to the tectonics of Shillong Plateau. The main shock of 1950 earthquake is located beyond the conventional limit of the Himalayas, in Assam syntaxial region, nearly 100 km east of Mishmi thrust (see Figure 11.7, Chapter 11). A large number of aftershocks of this event are located in the Eastern Himalaya between MBF and Tsangpo region, suggesting that large parts of the Himalayas were affected by this event and seismic energy was released by several faults and thrusts. The location of this major event tends to suggest that the Himalayan thrust front may extend eastward into the Assam syntaxial region even as it does north-westward beyond Hazara syntaxis into Indus Kohistan Seismic Zone.

12.4. Quantitative Seismicity

Epicentral distribution map gives some information regarding seismicity and its relationship to tectonics and faults in an area. However, in order to compare the seismicity of one area with another for quantitative purposes a mathematical approach is necessary. The energy released by earthquakes can be defined as tectonic flux which is amount of energy released per unit area per unit time (Ritsema, 1954; St. Amand, 1956). Recently Kaila and Narain (1971) have developed a new method of analysing quantitative seismicity of an area, based on the frequency of occurrence of earthquakes (N) and their magnitude (M). The relationship used by them is,

$$\log_{10} \underset{\text{cum}}{N(M)} = A - bM.$$

Here A represents the area under the incremental curve from $M = 0$ to $M = M_{max}$ on the cumulative regression curve and has been used as a measure of seismicity level of a region. Based on this A and b values for $1° \times 1°$ areas and 30 years time period can be written as,

$$A = 2.695 \frac{\log N_{(det)}}{0.75} + 2.15.$$

$$b = \frac{\log N_{(det)}}{2.36D} + 0.538.$$

Here, N(det) is the number of events detected by seismic observatories during the specified time period and D is the detectability factor taken as 0.75 for Himalaya-Arakan Yoma and surrounding regions (Kaila and Narain, 1971). The A value map using this method for a period of 30 years (1951–1980) is shown in Figure 12.4. The map brings out several interesting features concerning the seismicity of the Himalayas and the surrounding regions.

Along the eastern side the contours have a northeasterly trend with A value reaching a maximum of 8 east of Arakan-Yoma. The trend changes sharply to northwesterly direction along Assam syntaxial region where the value is nearly 6 to 7. The contours have a NE-SW trend over Eastern Himalaya, which changes to E–W and then to NW–SE in accordance with the trend of the major Himalayan thrusts (MBT and MCT). Over large parts of the Himalayas the A value ranges between 5 and 7. Several highs are observed along the Himalayan arc. These are located (from West to East) northwest of Srinagar, near Dalhousie, Kinnaur and Askol with a maximum value of 7, Kathmandu (maximum value, 5), Dzong and north of Lhasa (maximum value, 6). From the A value map it appears that the Kashmir and Panjab Himalayas are relatively more active as compared to Nepal and Eastern Himalaya (barring Assam syntaxial region). At several places the A value contours show trends transverse to the Himalayan strike. For instance along the Delhi-Hardwar ridge (DHR), near Kinnaur, north of Kathmandu and near Lhasa. The trend is attributable to seismicity along

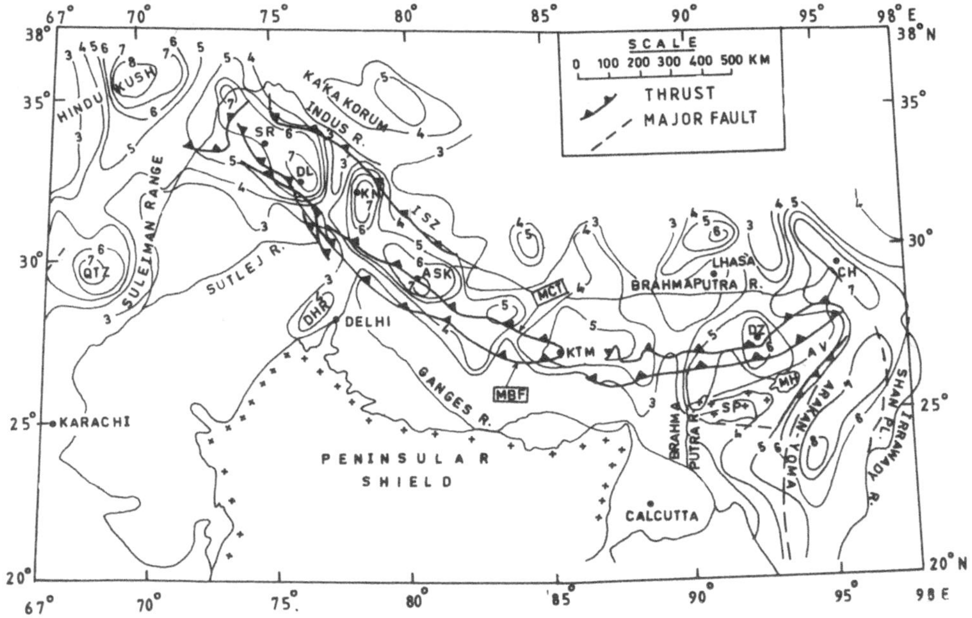

Fig. 12.4. Quantitative seismicity map of the Himalayas and northern part of Arakan-Yoma based on Kaila and Narain's (1971) method, showing *A* values using data for 30 years (1951–1980) period and 1° × 1° areas. The figure shows several highs and lows over the Himalayas and the adjoining areas (see text). Abbreviations: (ASK) Askol, (AV) Assam Valley, (CH) Chimdra, (DHR) Delhi Hardwar ridge, (DL) Dalhousie, (DZ) Dzong, (KN) Kinnaur, (KTM) Kathmandu, (MH) Mikir Hills, (QTZ) Quetta Transverse Zone, (SP) Shillong Plateau, (SR) Srinagar.

transverse features as discussed in the text. Low *A* values are observed at several places e.g. west of Kinnaur (between 32–34° N, 77–78° E) near Gangtok (26–29° N, 89–90° E) and in Assam valley. These areas are seismic gaps and may experience large magnitude earthquakes in the near future. West of Himalayas the contours change sharply to NE–SW direction showing highs near Hindukush and Suleiman Ranges near Quetta (maximum values of 8 and 7 respectively).

12.5. Focal Mechanism Solutions for the Himalayas

12.5.1. INTRODUCTION

The term focal mechanism is used to describe the manner of energy released at the point where the earthquake occurs i.e., at the foucs. Whenever faulting takes place elastic wave energy is released at the focus. The most accepted theory of faulting is that of Reid (1910) who studied the San Francisco earthquake of 1906. According to this theory strain energy builds up in rocks on either side and immediately adjacent to a fault surface. At some point of time the strain builds up to a sufficient degree and unlocks it, resulting in movement of blocks on either side of the fault plane in opposite direction and generation of seismic waves. This theory assumes that a single couple acts along the fault plane. Seismologists now favour double couple type of focal mechanism which is consistent with the requirement of static equilibrium that exists before failure, such that net unbalanced forces are equal to zero. This type of mechanism also produces compressions and dilatations in four quadrants. The compressions and dilatations generated at distant stations (away from the focus) are picked up by seismic instruments from arrival of *P* waves. Generally compression is indicated by upward motion of a recorder and dilatation by a downward motion. Knowing the distance of a recording station from the epicenter and the nature of movement recorded for a particular event, the first motions are plotted on the lower hemisphere of a focal sphere using stereographic projection and i vs Δ tables of Nuttli (1969).

The compartments containing compressions and dilatations plotted on the focal sphere can be well demarcated by means of two planes (one of which is the actual fault plane), if the number of observations recording an event are sufficient and the observatories are equipped with modern instruments. The points where the two nodal planes meet is known as the null (*B*) axis. Poles for 1st and 2nd nodal planes are located 90° away from the B axis along great circle paths. Axes of maximum compressive stress (*P*) and tensile stress (*T*) are located 45° away from the Poles in compartments of dilatations and compressions, respectively. Excellent reviews of methods of focal mechanism determination are given by Stauder (1962) and Brumbaugh (1979). General nature of normal, thrust and strike-slip solutions is shown in Figure 12.5.

12.5.2. DISCUSSION OF FOCAL MECHANISM SOLUTIONS

Study of focal mechanism solutions for the Himalayas throws a significant light on the nature of processes taking place at present. Several workers including Chandra (1978), Chaudhury and Srivastava (1976), Das Gupta *et al.* (1982), Fitch (1970), Molnar *et al.* (1973), Rastogi *et al.* (1973), Rastogi (1974, 1976), Tandon and Srivastava (1975a, b) have contributed appreciably to this study. For some events a number of workers have obtained different solutions. These are generally based on the data of P wave first arrivals as reported by ISS Bulletins or the data obtained from World-Wide Standarised Seismograph Network (WWSSN). In case of multiple solutions, the ones obtained from WWSSN records have been chosen for presentation. Only those solutions for which well defined compartments of compressions and dilatation have been obtained as reported in literature are listed in Table 12.1 along with their solution parameters. For purposes of discussion the Himalayas have been divided into Northwestern, Central and Northeastern parts (including N. Burma).

12.5.2.1. *Northwest Himalaya*

Figure 12.5 shows focal mechanisms for this area. In northwestern part, west of Nanga Parbat–Haramosh massif a number of solutions (C–11 (1), C–12 (2), C–12 and C–14) have been reported. In this area a number of thrusts of crystalline rocks striking in NW–SE or E–W direction and dipping towards NE have been mapped (Gansser, 1964). First three solutions support thrust faulting along northwesterly and northeasterly trending nodal planes, while solution C–14 suggests normal faulting along one of the nodal planes. This event is probably due to some local feature (fault or a lineament) trending in north-easterly direction. Solution PN–1 is due to Pattan earthquake of December 28, 1974, which is located in north-western part of Indus Kohistan Seismic Zone (as discussed earlier) and shows thrust faulting. The solution

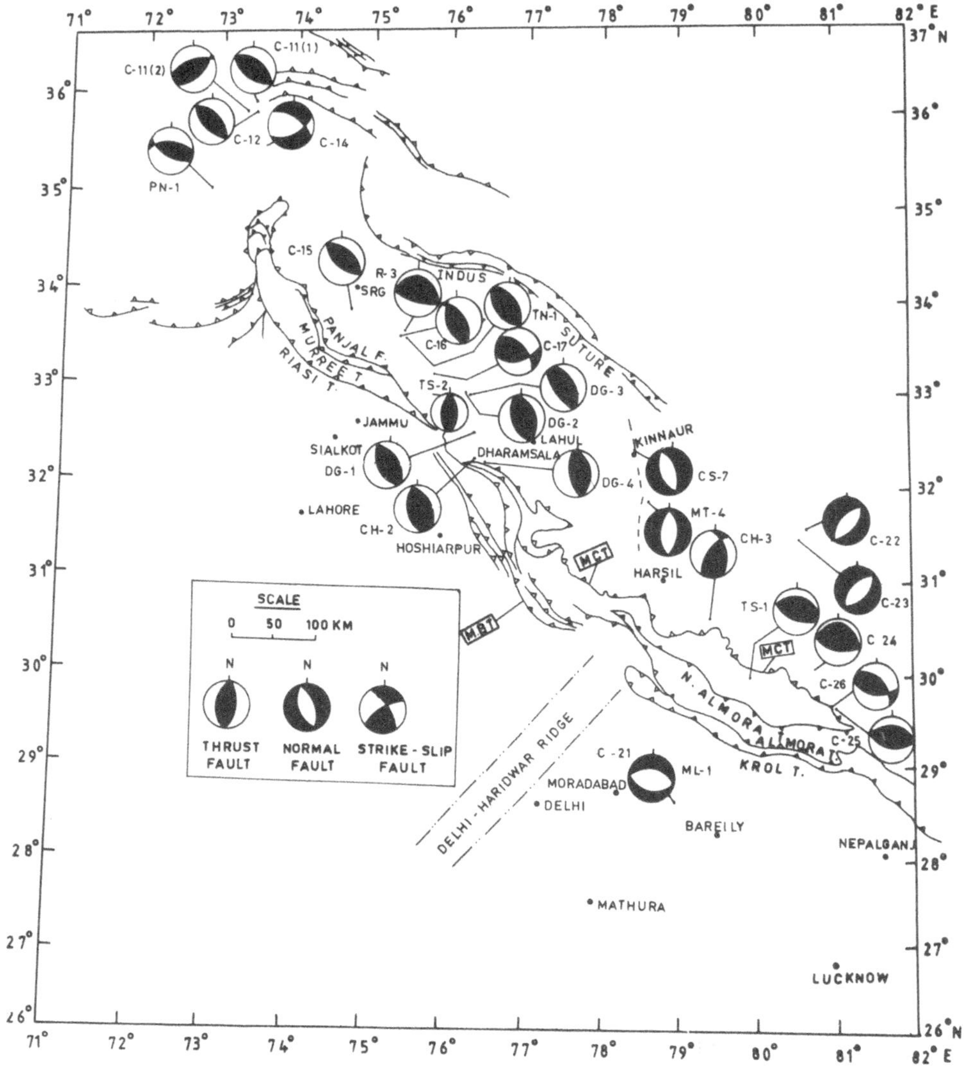

Fig. 12.5. Focal mechanism solutions reported for NW Himalayas. It may be noted that while most of the solutions show thrust faulting along the Himalayan belt, a few events located along transverse features, show normal mechanism. Parameters for solutions are given in Table 12.1.

supports extension of Himalayan thrusts into Kohistan zone, beyond the Hazara syntaxis.

A large number of focal mechanisms for the main Himalayan arc (between MBF and ISZ) show thrust faulting along NW–SE oriented nodal planes. These include C–15, C–16, C–17, C–24, C–25, C–26, CH–3, DG–1, DG–2, DG–3, DG–4, R–3, TN–1, TS–2, and TS–13. Focal depth for these events varies from 20 to 50 km. Since most of the seismicity in NW Himalaya is attributable to MBF and MCT (including Riasi, Krol, Murree and Panjal thrusts), these solutions suggest that a considerable amount of energy is being released through thrust faulting along the Himalayas at present. In

Tethys Himalaya of Panjab solutions for Kinnaur earthquake of July 19, 1975 and its aftershocks (MT-4 and CS-7) have been obtained by Molnar and Tapponnier (1978) and Chaudhury and Sirvastava (1976). Both the solutions show normal faulting along N-S or NNW-SSE oriented nodal planes. Singh *et al.* (1976) have studied the effects of this earthquake and concluded that this event was caused by N-S trending Kaurik-Chango fault. In this region seismicity also has a north-south trend which is probably due to several N-S trending faults in the area. It appears that this region may be in tension and vertical movements are taking place along some of the transverse features at present.

Solutions C-22 and C-23 are for two events located close to ISZ and show normal faulting along NE-SW oriented nodal planes. The two events lie in continuation of the Aravalli trend (Delhi-Hardwar ridge). It appears that the Aravallis may extend northeastward across the Himalayas up to ISZ and a portion of these may have underthrust the Eurasian plate as suggested by Chandra (1978). Tension along NW-SE direction may have caused normal faulting as shown by these two events.

In the Ganga basin of Uttar Pradesh Molnar *et al.* (1973) and Chandra (1978) have obtained identical solutions for an event located in the Ganga basin. The solutions (C-21 and ML-1) show normal faulting along ENE-WSW trending planes. There are no known faults with this trend in this area. Known faults (such as Moradabad fault) have a NE-SW strike direction. It appears that the earthquake has resulted from tension which is prevailing in this area as a result of bending of lithosphere due to continental subduction (or collision) in the Himalayas (Chandra, 1978).

12.5.2.2. *Central Himalaya*

In Central Himalaya a number of focal mechanism solutions (C-26, C-27, F-1, ML-2, ML-6, MT-6, MT-2, R-5) for events located close to MCT show thrust faulting as shown in Figure 12.6. However, a few show strike-slip faulting as well. Valdiya (1976) has pointed out that several faults transverse to the trend of the Himalayas have been very active since Late Tertiary times. Seismic map of the Central Himalaya supports this inference to some extent. The events for solutions MM-1 and TS-7 show strike-slip faulting and are located very close to Judi fault and Chautara fault respectively, in Nepal Himalaya. The solutions suggest left lateral motion along these transverse features.

12.5.2.3. *Eastern Himalaya and Northern Burma*

East of Nepal Himalaya, the MBF as well as the MCT continue into Sikkim, Bhutan and NEFA Himalayas. Although the MBF is well mapped geologically, MCT is not. Focal mechanism solutions obtained for the entire NE India and N. Burma are shown in Figure 12.7. It is apparent from the Figure that a number of solutions (C-28, C-30, C-31, C-32, C-33, C-34, C-35, F-5, ML-4, MT-1 and R-10) suggest thrust faulting along E-W or NE-SW oriented nodal planes, which is the predominant trend of MBF and other Himalayan thrusts in the area. Only two solutions C-29 and R-8 whose epicenters lie close to the Central Crystalline thrust show thrust faulting along NW-SE trending fault planes. Local strike-slip faulting is suggested by solution C-36 along one of the northeasterly trending lineaments.

East of 95° E the Himalayan trend takes a sharp turn from NE-SW to NW-SE,

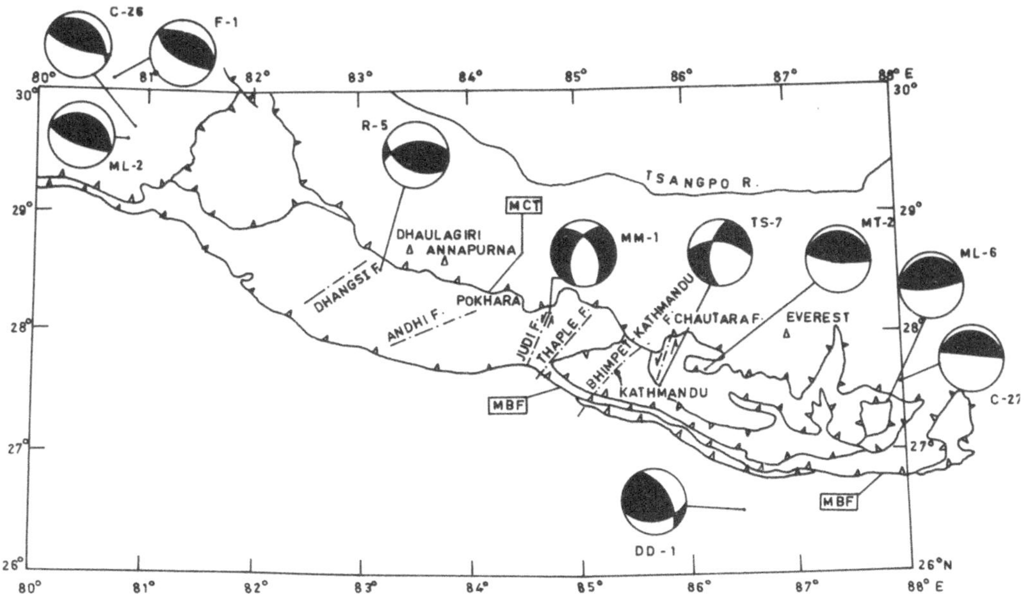

Fig. 12.6. Focal mechanisms for the Central Himalayas, (80° −88° E). It may be noted that most of the solutions show thrust faulting, while two solutions (MM−1 and TS−7) show strike-slip faulting. These are attributable to movement along some of the transverse faults in the Himalayas. Parameters for the solutions are given in Table 12.1.

where Lohit and Mishmi thrusts form an important part of Assam syntaxial region. The great Assam earthquake of August 15, 1950 (magnitude 8.7) had taken place in this region. According to Poddar (1953) isoseismals of this major event are elongated in southwesterly direction, which suggests that the causative fault for this earthquake could be a northeasterly trending fault. Ben Menahem *et al.* (1974) have determined a solution (BM−1) for this event which suggests predominant strike-slip faulting along NW−SE or NE−SW trending planes. The northwesterly trend agrees with the isoseismal trend. However, Chen and Molnar (1977) have shown that a shallow dipping thrust plane can also fit the data very well. On this basis Seeber and Armbruster (1981) have suggested that this earthquake has been caused by the Himalayan 'detachment', which according to them is the locus of all great Himalayan earthquakes.

The Assam Valley is found to be seismically rather inactive (see Figure 11.7). One focal mechanism (TS−6) has been obtained for an event located in this area. The solution suggests thrust faulting, probably related to the Brahmaputra lineament.

As seen in Figure 11.7 the Arakan-Yoma and Burmese mollase basin are seismically very active. A number of solutions have been obtained by several workers for events located from this area as shown in Figure 12.7. It may be noticed that solutions C−44, C−45, C−46, CS−4, R−12, R−13, TS−8 and V−2 show predominantly thrust faulting, while C−35, C−39, C−43, CS−1, CS−4, TS−11 show thrust faulting with appreciable strike-slip components. Events for C−41, C−42, TS−12, V−3 and V−4 show large strike-slip motion. In these solutions the nodal plane oriented in NNE−SSW, N−S or NE−SW

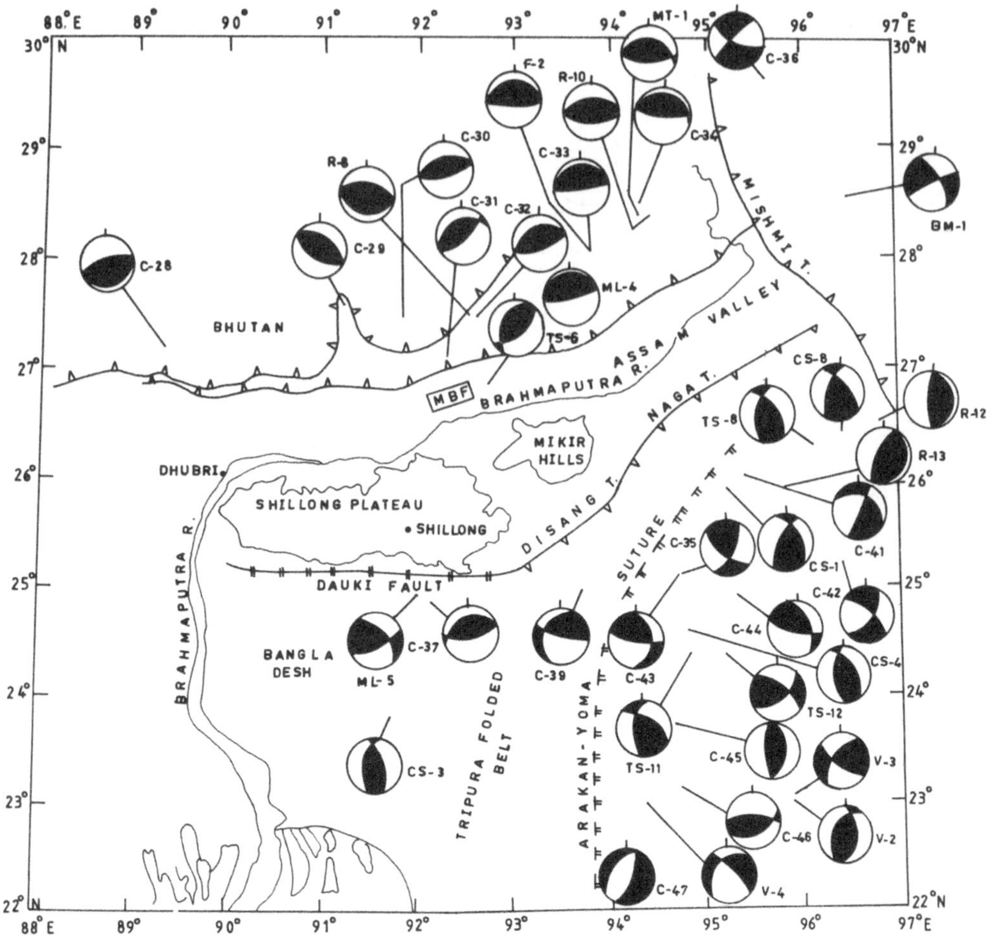

Fig. 12.7. Focal mechanism reported for NE Himalayas, Bengal basin and Arakan-Yoma seismic belt. It is interesting to note that practically all the solutions up to 95° E longitude along the Himalayan belt show thrust faulting. A number of solutions to the east of Arakan-Yoma show thrust faulting with appreciable strike-slip components. Parameters for the solutions are given in Table 12.1.

direction is most likely the fault plane in accordance with the trend of the Arakan-Yoma seismic belt. The sense of motion in most of the thrust-with-strike-slip solutions is right lateral.

From the analysis of gravity data and seismicity in this region (see Chapter 11) it has been brought out that the Indian plate is underthrusting the Burmese landmass to the east. The focal mechanism solutions however suggest that in addition to underthrusting the Indian Plate is also moving northward (slightly) relative to the Burmese landmass.

Four mechanisms, C–41, C–42, C–47 and V-4 show normal faulting on the western side of Arakan-Yoma. These can be explained in terms of tensions prevaling in some parts as a result of underthrusting of the lithosphere to the east. To the west of Arakan-Yoma lies the Tripura fold belt where solution CS–3 suggests compression along E–W direction. In the northern part of Bengal basin focal mechanism ML–5 shows strike-slip movement close to the Dauki fault, which is probably related to faulting along NE–SW trending Sylhet lineament in Bengal basin (see Fig. 12.7).

Summarising the results we may say that the most predominant mode of energy release in the entire Himalayan belt and the Arakan-Yoma is by thrust faulting which is attributable to activity along all the major thrusts. However, to a minor extent strike-slip and normal faulting is also taking place along the transverse features. Away from the major thrusts, on the continental side of the lithosphere tensions may prevail and seismic energy is released through normal faulting.

12.6. Orientation of P Axes

From a focal mechanism solution we can determine the azimuth and dip of principal compressive stress axis (*P*) and tension axis (*T*). Thrust faulting is generally caused by compressive stress prevailing in an area, while normal faulting is caused by tension. The directions of *P* axes prevailing along the Himalaya-Arakan-Yoma belt is shown in Figure 12.8. It is apparent from the figure that compressive stress is generally acting in E–W to NE–SW direction along the Arakan-Yoma. The direction slightly changes to NNW–SSE in the northern part. Along the Himalayan belt the predominant direction is N–S to NE–SW, nearly at right angles to the trend of the Himalayas. On the western side of the Kashmir syntaxis the azimuth changes again to NW–SE. It is evident that practically on all sides of the Indian Peninsula compressive stress is acting at right angles to the trend of the major mountain ranges.

12.7. Models for the Evolution of the Himalayas

Several models have been proposed to explain the evolution of the Himalayas. All these envisage the drift of the Indian subcontinent from southern to the northern latitudes as suggested from Paleomagnetic evidence (McElhinny, 1973) and pattern of magnetic anomalies in the Indian Ocean (McKenzie and Sclater, 1971). Relative position of the Indian landmass with respect to Eurasia at different times as inferred from magnetic anomalies is shown in Figure 12.9 (after, Molnar and Tapponnier, 1975).

The most likely contact between India and Eurasia appears to be sometime in Eocene (Dewey and Bird, 1970; Dewey and Burke, 1973). What happened after this collision is mostly controversial.

It is generally agreed that the Himalayas were formed as a result of major thrust movements directed from north towards south. According to Gansser (1966) thrusting within the Himalayas began in post-Eocene times. A giant underthrusting of the Indian crustal block along a shallow dipping fault zone resulting in doubling of the crust beneath the Tibetan landmass has been envisaged by Powell and Conaghan (1973) as shown in Figure 12.10a. Molnar and Tapponnier (1975), Tapponnier and Molnar (1976) have on the other hand suggested a more diffuse deformation (after the collision) over a broad zone involving crustal thickening through thrusting, magmatism and folding. They depict indentation of Indian subcontinent into Asia resulting in displacement of several lithospheric fragments along major strike-slip

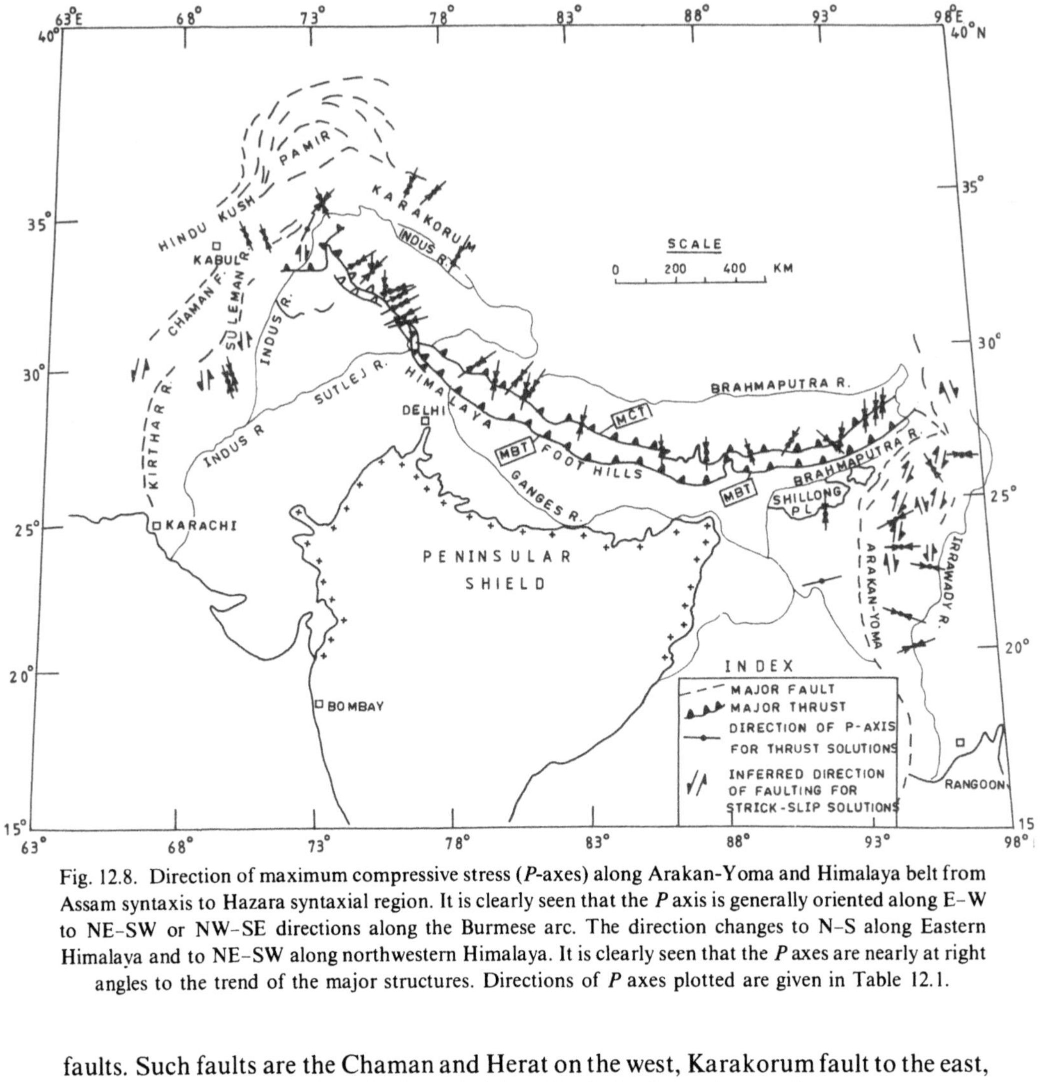

Fig. 12.8. Direction of maximum compressive stress (*P*-axes) along Arakan-Yoma and Himalaya belt from Assam syntaxis to Hazara syntaxial region. It is clearly seen that the *P* axis is generally oriented along E–W to NE–SW or NW–SE directions along the Burmese arc. The direction changes to N–S along Eastern Himalaya and to NE–SW along northwestern Himalaya. It is clearly seen that the *P* axes are nearly at right angles to the trend of the major structures. Directions of *P* axes plotted are given in Table 12.1.

faults. Such faults are the Chaman and Herat on the west, Karakorum fault to the east, Kang Ting fault, Kun Lun fault and Altyn Tagh fault to the north.

The origin of the Himalayas is undoubtedly related to that of the Tibetan Plateau. In order to explain the evolution of this plateau Chang and Cheng (1973) have suggested that the plateau has been evolved as a result of successive underthrusting of several plates during five tectonic cycles ranging from Caledonian (Early Paleozoic) to Himalayan (Cenozoic). The successive cycles are younger in age from north towards south as shown schematically in Figure 12.10b. The model highlights the relative antiquity of deformation processes and implies that the Tibetan Plateau developed throughout Phanerozoic times by successive accretion of meridional belts. The Himalayas occupy the last phase in this scheme. An interesting feature of the Asian tectonics is that the deformation of the lithosphere occurs over a vast area stretching from the Himalayas up to Tien-Shan and even beyond.

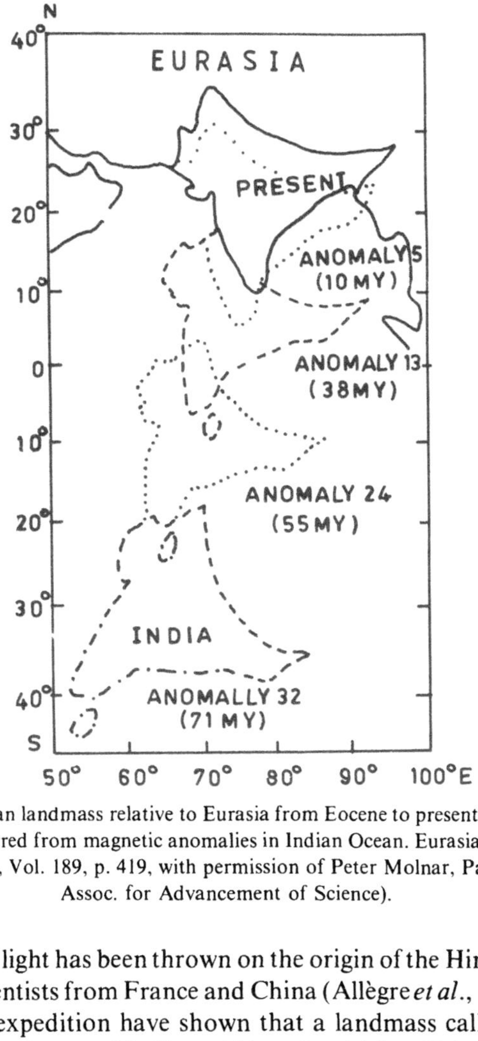

Fig. 12.9. Position of Indian landmass relative to Eurasia from Eocene to present times (after Molnar and Tapponnier, 1975), as interred from magnetic anomalies in Indian Ocean. Eurasia is kept arbitrarily fixed. (Reproduced from *Science*, Vol. 189, p. 419, with permission of Peter Molnar, Paul Tapponnier and Am. Assoc. for Advancement of Science).

Recently some new light has been thrown on the origin of the Himalayas through the joint efforts of geoscientists from France and China (Allègre *et al.*, 1984). The results of this French-Chinese expedition have shown that a landmass called the Lhasa block existed between the landmass of India and Eurasia which collided with the Quantang block of China around 120–140 m.y. ago, forming a suture named as Bangong-Nujiang Suture (BNS). The Indian landmass which drifted from the southern hemisphere collided with the Lhasa block around 40 m.y. ago at Indus-Tsangpo Suture (ITS). As a result of this, part of the Indian lithosphere could have subducted beneath the Lhasa block. Also as a result of melting of the existing crust major granites were intruded between the MCT and ITS. The DSS experiments carried out by this expedition show that the Moho has thrust contact with different blocks at several places, jumping from nearly 70 to 50 km (as shown in Figure 12.11a). According to these scientists there have been at least six continental thrusts between the Main Boundary Fault (MBF) (also called Main Boundary Thrust – MBT) and the BNS (shown schematically in Figure 12.11b). Considerable thickening of the crust appears to have taken place as a result of underthrusting of the continental blocks. All these thrusts must have been very active

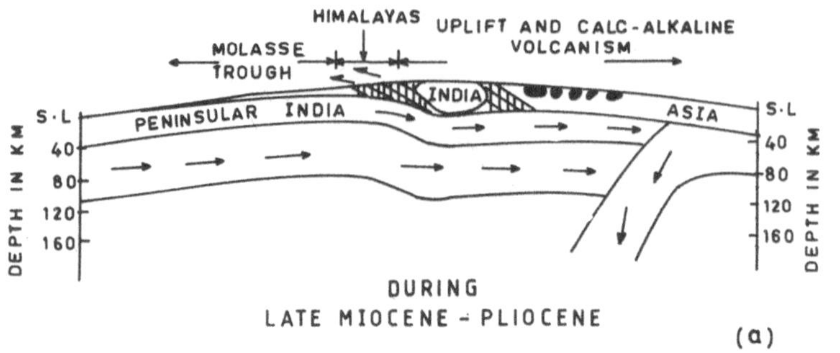

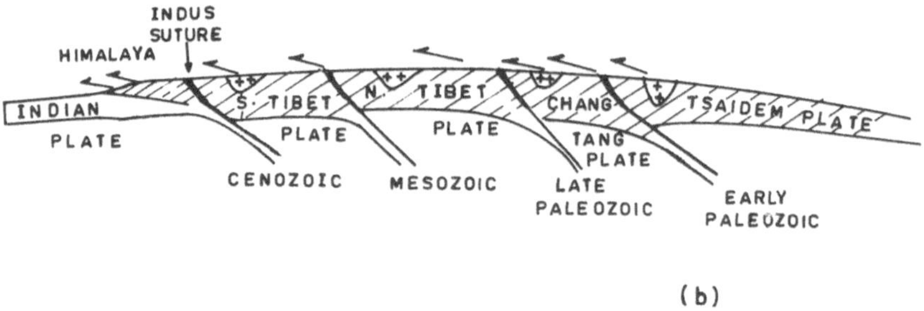

Fig. 12.10 (a). Underthrusting of Indian landmass beneath Asia during Miocene to Pliocene times as envisaged by Powell and Conaghan (1973). (b) Thrusting of northern blocks towards south from Caledonian (Early Paleozoic) to Cretaceous times as suggested by Chang and Cheng (1973) in order to explain the evolution of Tibet. (Reproduced from *EPSL*, Vol. 20, p. 7 with permission of C. M. Powell, N. J. Conaghan and Elsevier Publ. Co., Copyright, 1973).

during geological times. However, at present the major activity appears to have migrated towards MBF (Molnar *et al.*, 1977; Seeber *et al.*, 1981).

A noteworthy feature of the Himalayan seismicity is that great earthquakes (having magniitude 8.0 and above) have occurred south of the MBF. This includes the Kangra earthquake of 1905, Bihar earthquake of 1934, the Assam earthquakes of 1897 and 1950. Analysing these great earthquakes Seeber and Armbruster (1981) have suggested a model (shown in Figure 12.12) to explain the present day seismicity of the Himalayas. According to this model great ruptures have taken place in the Himalayan arc along a 'detachment' which separates the overlying sedimentary wedge from the subducting basement. The 'detachment' dips at shallow angles towards south and is aseismic between great earthquakes.

The subducting Indian Shield meets the overlying Tethys slab at the Basement Thrust Front (BTF) which lies close to MCT. It is noticed that a number of thrust earthquakes occur close to MCT. Topography north of BTF includes the Higher and Tethys Himalayas and parts of Tibetan Plateau, which are in static equilibrium with the crust thickened by continental thrusts as suggested by Allègre *et al.* (1984) or continental subduction as suggested by Powell and Conaghan (1973). According to Seeber and Armbruster (1981) the Himalayan front has ruptured 6–8 times since 1800. Great earthquakes are likely to occur after every 30 years along the Himalayan arc. The probable locations for such major rupturing in future are Kashmir, Kumaun Himalaya and Assam Valley, which appears to be a seismic gap (Khatri and Tyagi, 1983).

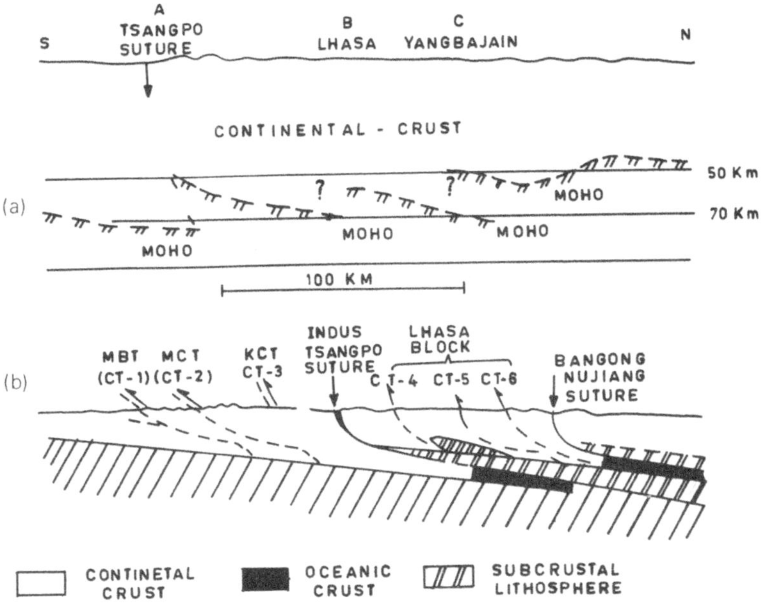

Fig. 12.11a. Moho configuration as obtained from wide angle refraction profile in parts of Tibet (Allègre *et al.*, 1984). The Moho shows a complex topography with jumps of the order of 20 km., particularly near Tsangpo Suture Lhasa and Yangbajain. (b) A model for the structure of lithosphere in Tibet-Himalaya region. The model shows that the Tibetan Plateau and the Himalayas have been evolved as a result of several continental thrusts (CT_1 to CT6) and obduction along Indus-Tsangpo Suture (ITS) and Bangong-Nujiang suture. (MBT) Main Boundary Thrust, (MCT) Main Central Thrust, (KCT) Kangmar Continental Thrust. (Figures reproduced from *Nature*, Vol. 307, pp. 17–22 with permission of C. J. Allègre and McMillan Journals Ltd., Copyright, 1984.)

Analysis of seismicity and focal mechanisms in the Himalayas shows that forces which gave birth to the Himalayas are still very active and the entire Himalayas as well as the Ganga-Brahmaputra foredeep are experiencing the effects of these forces.

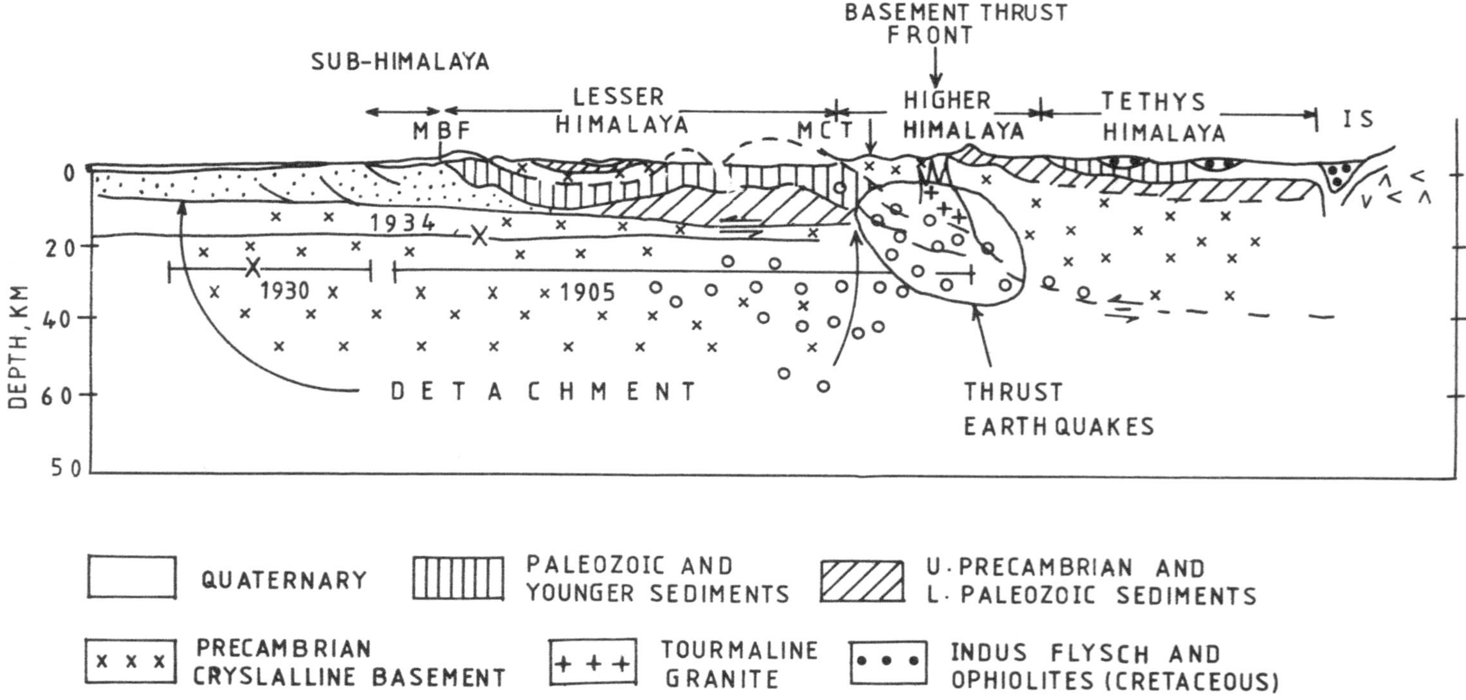

Fig. 12.12. Model for the Himalayas after Seeber and Armbruster (1981). The model suggests that the Himalayas are getting detached from the Precambrian crystalline basement along a detachment which is underthrusting the Tibetan landmass. Most of the large magnitude earthquakes along the Himalayan belt have occurred along the upper surface of the detachment, south of the MBF. (Reproduced from *AGU Monograph, Geodynamics Ser.*, Vol. 3 with permission of L Seeber, J. G. Armbruster and Am. Geophy. Union, Copyright, 1981).

TABLE 12.1

Parameters for focal mechanism solutions discussed in text

Sl. No.	Date D M Y	Origin time H M S	Lat. °N	Long. °E	Depth (km)	Magnitude	Pole of 1st Nodal plane		Pole of 2nd Nodal plane		B. Axis		P. Axis		T. Axis		Reference
							Az. (deg)	Pl. (deg)	Az. (deg)	Pl. (deg)	Az. (deg)	Pl. (deg)	Az. (deg)	Pl. (deg)	Az. (deg)	Pl. (deg)	
1	2	3	4	5	6	7	8	9	10	11	12	13	14	15	16	17	18
BM-1	15. 8.1950	14:09:30	28.5	97.0	25	8.7	246	30	337	2	70	62	200	18	296	22	Ben Menahem, 1974
C-1	1. 8.1966	21:03:01	30.1	68.6	33	6.0	187	82	278	83	229	79	322	1	53	11	Chandra, 1978
C-1	24. 1.1966	07:23:10	29.9	69.70	26	5.6	300	27	171	72	254	20	155	24	19	58	,,
C-3	7. 2.1966	04:26:11	29.9	69.70	10	6.0	327	45	147	45	57	0	147	0	237	90	,,
C-4	7. 2.1966	23:06:34	30.3	69.90	11	5.8	182	55	321	43	256	21	163	6	58	68	,,
C-5	28.12.1972	16:57:46	34.7	70.4	63	5.3	328	3	148	87	58	0	148	42	328	48	,,
C-6	15. 5.1969	20:39:49	34.6	70.9	22	5.6	324	16	165	75	254	5	160	30	353	60	,,
C-10	28.12.1974	12:11:44	35.1	72.9	22	6.0	334	24	138	60	243	5	151	23	345	67	,,
C-11 (1)	3. 9.1972	16:48:29	36.0	73.4	36	6.3	218	30	38	60	308	0	38	15	218	75	,,
C-11 (2)	3. 9.1972	23:03:52	35.9	73.3	33	5.6	324	40	165	50	62	11	153	6	271	77	,,
C-12	4. 9.1972	13:42:18	35.9	73.4	33	5.8	217	35	37	55	307	0	37	10	217	80	,,
C-14	29. 1.1965	20:06:24	35.6	73.6	33	5.7	341	56	220	20	120	25	253	26	19	22	,,
C-15	2. 9.1963	01:34:31.6	33.9	74.7	44	5.1	235	70	55	20	325	0	235	25	55	65	,,
C-16	20. 2.1967	15:18:39	33.7	75.3	18	5.6	227	45	45	46	322	7	46	0	145	85	,,
C-17	16. 1.1973	21:31:26	33.2	75.7	42	5.1	352	48	230	35	117	31	4	7	280	69	,,
C-20	22. 6.1965	05:49:28	36.2	77.6	107	5.7	219	3	39	87	129	0	219	48	39	42	,,
C-21	15. 8.1966	02:15:34	28.7	78.9	53	5.6	191	13	11	60	101	0	191	75	11	15	,,
C-22	6. 3.1966	02:10:54	31.6	80.6	12	5.7	315	45	135	45	45	0	315	90	195	0	,,
C-23	6. 3.1966	02:15:57	31.5	80.5	50	6.0	315	55	135	35	45	0	135	80	315	10	,,
C-24	26. 9.1964	0:46:2′8	30.1	80.7	50	6.2	211	72	24	17	115	3	207	28	19	62	,,
C-25	16.12.1966	20:52:16	29.7	80.7	15	5.8	190	51	10	39	280	0	190	6	10	84	,,
C-26	27.16.1966	10:41:8′6	29.7	80.9	37	6.1	223	62	26	20	120	7	212	18	9	71	,,
C-27	12. 1.1965	13:32:24	27.6	88.0	23	6.1	181	78	1	12	271	0	181	33	1	57	,,
C-28	27. 3.1964	23:03:42	27.2	89.3	32	6.3	346	71	166	19	256	0	346	26	166	64	,,
C-29	18. 2.1964	03:48:36	27.5	91.1	30	5.6	212	50	32	40	302	0	212	5	32	85	,,

Table 12.1 (*Continued*)

1	2	3	4	5	6	7	8	9	10	11	12	13	14	15	16	17	18
C-30	15. 9.1967	10:32:49	27.4	91.8	57	5.8	353	30	173	60	83	0	173	15	153	75	,,
C-31	1. 9.1964	13:22:36	27.2	92.3	33	5.7	314	20	134	70	44	0	134	25	314	65	,,
C-32	26. 9.1966	05:10:56	27.5	92.6	19	5.5	353	20	163	70	253	0	163	25	343	65	,,
C-33	21.10.1964	23:09:19	28.1	93.8	37	5.9	354	3	174	87	84	0	174	42	354	48	Chandra, 1978
C-34	14. 3.1967	06:58:03	28.5	94.3	12	5.8	6	6	186	84	96	0	186	39	6	51	,,
C-35	19. 2.1970	22:12:02	25.1	94.7	33	5.5	38	38	290	22	176	42	77	10	337	46	,,
C-36	15. 6.1965	07:59:19	29.6	95.6	30	5.6	313	0	42	20	221	70	88	14	354	14	,,
C-37	21. 6.1963	15:26:29	24.9	92.1	53	5.7	354	12	124	74	262	8	154	31	5	55	,,
C-38	19. 6.1963	10:47:24	25.0	92.1	51	5.9	351	25	171	65	81	0	171	20	351	70	,,
C-39	18. 6.1965	08:17:40	25.0	93.7	66	5.8	132	68	22	5	290	20	180	32	53	44	,,
C-40	28. 9.1966	14:00:22	27.5	100.1	27	6.1	230	50	50	40	320	0	50	85	230	5	,,
C-41	29. 7.1970	10:16:19	26.0	95.4	59	6.5	200	60	109	0	20	28	265	39	135	38	,,
C-42	30. 5.1971	15:44:16	25.2	96.4	15	5.8	308	10	214	4	97	78	201	11	352	3	,,
C-43	29.12.1971	22:27:02	25.1	94.7	33	5.5	107	10	254	57	107	30	214	26	337	48	,,
C-44	12. 7.1964	20:15:59	24.9	95.3	155	6.7	12	20	250	60	110	24	211	24	340	55	,,
C-45	13. 7.1964	10:58:48	23.7	94.7	117	6.5	88	30	274	60	181	5	272	14	71	75	,,
C-46	17.10.1969	01:25:12	23.1	94.7	134	6.0	327	74	184	14	91	10	348	30	197	57	,,
C-47	22. 1.1964	15:58:46	22.4	93.6	88	6.1	288	22	140	64	24	14	275	66	120	23	,,
C-48	27. 2.1964	15:10:49	21.7	94.4	102	6.4	254	70	88	17	357	4	258	26	95	63	,,
CH-1	16. 1.1974	21:31:26	33.2	75.7	42	5.5	119	34	242	52	17	28	252	45	174	62	Choudhury *et al.*,
CH-2	5.11.1968	02:02:45	32.3	76.5	33	4.9	235	66	74	30	336	5	242	22	86	67	1974
CH-3	22. 6.1969	01:33:24	30.6	79.4	19	5.4	64	72	227	20	320	6	56	27	222	63	,,
CS-1	28. 4.1969	12:50:15	25.9	95.2	50	5.2	126	13	220	62	35	24	286	27	150	55	Choudhury and
CS-2	19. 2.1970	07:10:02	27.4	94.0	18	5.5	258	42	15	30	130	35	220	6	321	55	Srivastava, 1976
CS-3	2. 2.1971	07:59:56	23.8	91.76	37	5.4	100	45	260	45	0	10	263	0	180	89	,,
CS-4	26. 6.1971	02:16:37	24.6	94.8	76	5.2	84	36	236	48	342	16	250	4	144	76	,,
CS-5	17. 7.1971	15:00:56	26.4	93.2	50	5.3	39	43	288	18	184	40	74	15	333	45	,,
CS-6	1.11.1972	21:53:46	26.4	96.4	97	5.3	225	40	122	16	14	45	270	14	167	40	,,
CS-7	19. 1.1975	08:00:24	32.45	78.43	60	5.3	63	54	252	35	157	5	267	80	63	10	,,
DD-1	15. 1.1934	08:43:18	26.50	86.50	20	8.4	244	19	10	60	145	22	46	24	275	58	Singh and Gupta, 1980
DG-1	25. 2.1971	04:30:24	32.60	76.3	150	—	247	56	60	34	153	3	243	12	48	80	DasGupta *et al.*, 1982
DG-2	11.12.1975	10:09:50	33.0	76.17	42	5.0	76	16	256	74	164	0	256	27	75	63	,,

Table 12.1 (*Continued*)

1	2	3	4	5	6	7	8	9	10	11	12	13	14	15	16	17	18
DG-3	1. 1.1976	00:24:52	32.97	76.12	40	5.3	60	34	248	56	152	30	243	12	48	80	,,
DG-4	14. 6.1978	16:12:05	32.24	76.61	7	5.0	78	18	258	72	348	0	258	26	78	64	,,
F-1	26. 9.1964	00:46:03	30.1	80.7	50	6.2	40	30	220	60	135	0	220	15	40	75	Fitch, 1970
F-2	14. 3.1967	06:58:03	28.41	94.29	12	5.8	3	10	183	80	93	0	183	35	3	55	Fitch, 1970
ML-1	15. 8.1966	02:15:34	28.7	78.93	53	5.6	42	59	222	31	132	0	222	76	42	14	Molnar *et al.*, 1973
ML-2	16.12.1966	20:52:16	29.62	80.79	15	5.8	190	66	10	24	100	0	190	21	10	69	,,
ML-3	20. 2.1967	15:18:39	33.63	75.33	18	5.6	168	25	42	50	272	28	10	14	123	57	Molnar *et al.* 1973
ML-4	19. 2.1970	07:10:02	27.40	94.0	18	5.5	347	5	167	85	77	0	167	40	347	50	,,
ML-5	19. 6.1963	15:26:29	24.97	92.06	53	5.7	235	20	330	16	91	60	192	5	287	24	,,
ML-6	12. 1.1965	13:55:20	27.4	87.84	23	6.1	0	15	180	75	90	0	180	30	0	60	,,
MM-1	31. 1.1975	12:38:52	28.1	84.72	33	5.4	213	10	119	25	324	63	168	28	74	10	Mukhopadhyay, 1984
MM-2	17. 7.1971	15:00:55	26.46	93.23	49	5.3	191	2	97	56	280	32	161	38	39	34	,,
MT-1	14. 3.1967	06:58:03	28.4	94.29	12	5.8	183	80	183	35	93	0	183	35	3	55	Molnar and Tapponier
MT-2	24. 3.1974	14:16:03	27.73	86.11	33	5.7	7	2	187	88	97	0	187	43	7	47	1978
MT-3	4. 6.1975	02:24:33	35.87	79.85	33	5.7	90	26	198	30	327	48	140	42	233	3	,,
MT-4	19. 7.1975	06:10:55	31.92	78.61	40	5.3	90	40	270	50	0	0	90	85	270	5	,,
PN-1	28.12.1974	12:11:47	35.1	72.90	12	5.0	200	52	38	38	295	15	205	8	76	80	Pennington, 1979
R-1	26. 6.1963	14:09:20	36.4	76.6	89	5.3	25	88	205	2	203	2	25	43	205	47	Rastogi, 1973, 1974,
R-2	22. 6.1965	05:49:28	36.2	77.6	107	5.7	48	56	228	34	318	0	48	11	228	79	1976
R-3	21. 2.1967	12:37:46	33.7	75.4	41	5.1	190	60	10	30	280	0	190	15	10	75	,,
R-4	3. 9.1972	16:48:29	36.0	73.4	36	6.3	182	64	2	26	274	0	182	19	2	71	,,
R-5	1. 6.1965	07:52:24	28.5	83.2	20	5.3	177	50	33	34	302	19	197	9	84	70	,,
R-6	12. 1.1965	13:32:24	27.6	88.0	23	6.1	189	88	9	2	278	0	189	43	9	47	,,
R-8	9.12.1965	20:26:05	27.4	92.5	29	5.3	191	26	11	64	282	0	11	19	191	71	,,
R-9	26. 9.1966	05:10:56	27.5	92.6	19	5.5	204	72	24	18	114	0	204	27	24	63	,,
R-10	11. 3.1967	16:56:50	28.4	94.4	12	5.3	183	60	3	30	272	0	183	15	3	75	,,
R-11	19. 6.1963	10:47:24	25.0	92.1	51	5.9	183	80	3	10	93	0	183	35	3	55	,,
R-12	22. 9.1962	06:51:28	26.5	96.8	33	6.3	273	74	93	16	183	0	273	29	93	61	,,
R-13	3. 6.1964	02:49:15	25.9	95.8	100	5.5	293	84	113	6	203	0	293	39	113	51	,,
TN-1	20. 2.1967	15:18:40	33.6	75.4	25	5.1	66	33	240	57	150	3	247	13	66	78	Tandon, 1972
TS-1	10. 3.1972	14:36:16	33.9	72.2	45	4.9	172	29	12	58	268	10	140	70	3	14	Tandon and
TS-2	24.10.1973	05:23:57	33.1	75.9	33	5.4	78	54	276	34	175	4	82	10	304	78	Srivastva, 1975a
TS-3	11. 2.1968	20:38:27	34.2	78.7	24	5.1	245	5	352	76	128	10	34	38	238	56	,,

Table 12.1. (*Continued*)

1	2	3	4	5	6	7	8	9	10	11	12	13	14	15	16	17	18
TS-4	12. 6.1968	04:29:22	24.9	91.9	44	5.3	324	27	80	40	210	38	116	8	15	55	,,
TS-5	29.12.1971	22:27:10	25.2	94.7	46	5.0	206	40	125	34	3	38	246	0	180	60	,,
TS-6	30. 6.1969	08:51:54	26.9	92.7	44	5.0	338	41	118	42	230	22	138	2	48	70	,,
TS-7	26. 2.1970	19:30:15	27.6	85.7	96	5.0	197	14	97	39	303	50	158	40	54	15	,,
TS-8	29. 8.1969	10:02:50	26.3	96.1	72	5.0	99	36	230	46	346	24	238	5	146	72	Tandon and
TS-9	10.10.1971	18:25:17	23.0	95.9	48	4.9	53	22	140	46	302	38	98	50	202	13	Srivastava, 1975a
TS-10	14.10.1971	12:55:22	23.1	95.9	47	5.1	348	9	262	27	102	63	307	27	212	10	,,
TS-11	2.10.1966	04:31:49	24.4	94.8	75	4.9	204	21	98	38	318	47	58	11	158	35	,,
TS-12	22. 3.1954	23:42:12	24.4	95.2	180	7.5	217	28	322	28	90	46	348	1	266	43	,,
TS-13	28.12.1958	05:34:42	30.0	79.9	30	6.2	12	36	192	54	282	0	192	44	12	80	,,
V-1	18. 2.1965	04:26:35	25.0	94.2	45	5.4	89	40	308	40	200	20	22	67	288	1	Verma *et al.*, 1976
V-2	10.10.1971	18:25:17	23.0	95.9	46	4.9	118	54	274	33	11	14	104	10	233	74	,,
V-3	14.10.1971	12:55:22	23.06	95.9	47	5.1	133	58	32	8	298	32	187	31	63	43	,,
V-4	22.10.1966	03:03:24	23.04	94.3	72	5.1	338	48	232	14	130	40	271	44	23	21	Verma *et al.*, 1980

BIBLIOGRAPHY

Acharya, S.: 1964, *Stratigraphy of Banded Iron Formation of Diateri* - Tomaka area, Cuttack Dist., Orissa, India. Proc. 22nd. Int. Geol. Congr. Pt. 10, pp. 13-21.

Agarwal, R. C. and Agarwal, P. N.: 1982, 'Monitoring of Contemporary Seismic Activity for Project Sites', *Proc. 4th. Int. Assoc. of Engg. Geol.* **8**, 19-31.

Airy, G. B.: 1955, *Phil. Trans. Roy Soc. Lond.* **145**, 101-104.

Allègre, C. J., Courtillot, V., Tapponnier, P., Hirn, A., Mattauer, M., Coulon, C., Jaeger, J. J., Achache, J., Schärer, U., Marcoux, J., Beng, J. P., Girardeau, J., Arimjo, R., Gariepy, C., Gopel, C., Tindong. Li, Xuchang, Xiao, Chenfa, Chang., Tonglin, Hau., Xibin, Wang., Wamming, Den., Haibin, Sheng., Yougong, Cao., Zhou Ji., Hongrong, Qiu., Peisheng, Bao., Songchan, Wang., Bixiang, Wang., Yaoxiu, Z., and Roughna Xu: 1984, 'Structure and Evolution of Himalaya-Tibet Orogenic Belt', *Nature* **307**, 17-22.

Andrews-Speed, C.P. and Brookfield, M.E.: 1982, 'Middle Paleozoic to Cenozoic Geology and Tectonic Evolution of the Northwest Himalaya', *Tectonophysics* **82**, 253-275.

Anhaeusser, C. R.: 1973, 'The Evolution of the Early Precambrian Crust of S. Africa', *Phil. Trans. Roy Soc. Lond. Ser. A.* **273**, 359-388.

Armbruster, J., Seeber, L., and Jacob, K. A.: 1978, 'The Northwestern Termination of Himalayan Mountain Front, Active Tectonics from Microearthquakes', *J. Geophy. Res.* **83**, 269-283.

Arur, M. G.: 1982, 'Results of Crustal Movement Studies in India', Proc. 4th. Int. Assoc. of Engg. Geol., N. Delhi, pp. 11-18.

Aswathanarayana, U.: 1964, 'Isotopic Ages from the Eastern Ghats and Cuddapahs of India', *J. Geophy. Res.* **69** (16), 3479-3486.

Aswathanarayana, U.: 1968, 'Metamorphic Chronology of the Precambrian Provinces of S. India', *Can. J. Earth Sci.* **5**, 591-600.

Auden, J. B. and Ghosh, A. M. N.: 1934, 'Preliminary Account of Earthquake of 15 Jan. 1934 in Bihar-Nepal Region', *Rec. Geol. Surv. India* **68**(2), 117-239.

Bala Subramaniyam, M. N. and Chatterji, P. K.: 1975, 'Some Aspects of the Geology of Eastern Ghats', *Geol. Surv. Ind. Misc. Pub.* **23**, 223-232.

Bandopadhyay, T. K.: 1981, 'Geophysical Studies over Raniganj Coalfield Using Gravity and Resistivity Method', Ph. D. thesis, Indian School of Mines, Dhanbad.

Banerji, P. K.: 1968, 'Revision of the Stratigraphy, Structure and Metamorphic History of Gangpur Series, Sundargarh Dist', *Rec. Geol. Surv. India* **95**(2), 327-348.

Banks, R. J., Parkar, R. L., and Huestis, S. P.: 1977, 'Isostatic Compensation on a Continental Scale, Local vs Regional. Mechanism', *Geophy. J. Roy. Astr. Soc.* **51**, 431-452.

Basu, A. R., Ray, S. L., Saha, A. K., and Sarkar, S. N.: 1981, '3800 m.y. Old Crust in Eastern India and Early Mantle Differentiation', *Science* **212**, 1502-1506.

Beckinsale, R. D., Drury, S. A., and Holt, R. N.: 1980, '3360 m.y. Old Gneisses from the South Indian Craton', *Science* **283**, 469-470.

Ben Menahem, A., Aboodi, E., and Schild, R.: 1974, 'The Source of Great Assam Earthquake, an Interplate Wedge Motion', *Phy. Earth and Plan. Inter.* **9**, 265-289.

Berry, M. F. and Fuchs, F.: 1973, 'Crustal Structure of the Superior and Grenville Provinces of the NE Canadian Shield', *Bull. Seis. Soc. Am.* **63**, 1393-1432.

Bhanger, A. R.: 1976, 'Mechanism Solution of Kinnaur (Himachal) Earthquake of January 19, 1975', *Tectonophysics* **31**, T5-T11.

Bowie, Wm.: 1917, *Investigation of Gravity and Isostasy*, USCGS Spl. Pub. No. 40.

Bowie, Wm.: 1924, *Isostatic Investigations and Data Since 1915*, USCGS Spl. Pub. No. 99.

Brock, B. B.: 1952, *An Approach to the Rift Valley Problem*, Congress Geologique International, Algiers, Section 3, pp. 225-239.

Brooks, M.: 1966, 'Regional Gravity Anomalies Attributable to Basic Instrusions in Orogenic Belts', *Geophy. J. Roy. Astr. Soc.* **12**, 29-31.

Brooks, M.: 1970, 'Positive Bouguer Anomalies in Some Orogenic Belts', *Geol. Mag.* **107**(4), 399-400.

Brumbaugh, D. S.: 1979, 'Classical Focal Mechanism Techniques for Body Waves', *Geophy. Surv.* **3**, 279-329.

Brunschweiler, R. O.: 1966, 'On the Geology of Indo-Burman Ranges (Arakan Coast and Yoma, Chin Hills, Naga Hills)', *J. Geol. Soc. Austr.* **13**, 137-194.

Bullard, E.C.: 1956, 'Gravity Measurements in East Africa', *Trans. Roy. Soc. (London), Ser. A.* **235**, (757), 445-531.

Burrard, S. G.: 1901, *Survey of India*, Prof. Pap., No. 5.

Burrard, S.G.: 1912, *On the Origin of Himalaya Mountains*, Surv. of India. Prof. Pap.. No. 12.

Burrard, S. G.: 1918, *Investigation of Isostasy in Himalayan and Neighbouring Regions*, Surv. of India, Prof. Pap., No. 17.

Chandra, U.: 1975a, 'Fault Plane Solution and Tectonic Implication of Pattan Earthquake of December 28, 1974', *Tectonophysics* **28**, T19-T24.

Chandra, U.: 1975b, 'Seismicity, Earthquake Mechanism and Techtonics of Burma, 20° N-28° N', *Geophy. J. Roy. Astr. Soc.* **40**, 367-381.

Chandra, U.: 1978, 'Seismicity, Earthquake Mechanisms and Tectonics along the Himalayan Mountain Range and Vicinity', *Phy. Earth and Plan. Inter.* **16**, 109-131.

Chang, Ch. F. and Cheng, H. S. L.: 1973, 'Some Tectonic Features of Mt. Jolmo Lungma Area, Southern Tibet, China.' *Scientia Sinica* **16**, 257-265.

Chatterji, G. C. and Ghosh, P. K.: 1970, 'Tectonic Framework of Peninsular Gondwanas in India', *Rec. Geol. Surv. India* **98** (Pt. 2), 1-15.

Chaudhury, H. M., Srivastava, H. N., and Subba Rao, J., 1974, 'Seismotectonic Investigations of the Himalayas', *Himalayan Geol.* **4**, 481-491.

Chaudhury, H. M. and Srivastava, H. N.: 976, 'Seismicity and Focal Mechanisms of Some Recent Earthquakes in NE India', *Annali di Geophysica* **29** (1-2), 41-56.

Chen, W. P. and Molnar, P.: 1977, 'Seismic Movements of Major Earthquakes and Average Rate of Slip in Central Asia', *J. Geophy. Res.* **82**, 2945-2969.

Choudhury, S. K.: 1975, 'Gravity and Crustal Thickness in the Indo-Gangetic Plains and Himalayan Region, India', *Geophy. J. Roy. Astr. Soc.* **40**, 441-452.

Chugh, R. S.: 1974, *Study of Recent Crustal Movements in India and Future Programs*, Presented at International Symposium on Recent Crustal Movements, Zurich.

Chugh, R. S.: 1977, '*A Geophysical Study of Aravalli Range Deformation*', International Symposium on Recent Crustal Movements. Surv. of India, unpublished report.

Chugh, R. S.: 1978, *Geotectonics of Kashmir, Himachal, Garhwal and Kumaun Himalayas and Ladakh Region*, Surv. of India, unpublished report.

Chugh, R. S. and Bhattacharji, J. C.: 1975, *Study of Isostasy in the Himalayan Region*, Seminar on Himalayan Geology, held at Nainital, November 1973.

Cordell, L. and Henderson, R. G.: 1968. 'Iterative Three Dimensional Solution of Gravity Data Using a Digital Computer', *Geophysics* **33** (4), 596-601.

Crawford, A. R. and Compston, W.: 1967, *New Rb-Sr. Age Determination for the Precambrian of S. India* (Abst.), Conference on Geochronology of Stratified Rocks, Alberta, Canada, p. 33.

Crawford, A. R.: 1969a, 'India, Ceylon and Pakistan; New Age Data and Comparison with Australia', *Nature, London* **223**, 280-284.

Crawford, A. R.: 1969b, 'Reconnaissance Rb-Sr Dating of the Precambrian Rocks of Southern Peninsular India', *J. Geol. Soc. India* **10**, 117-166.

Crawford, A. R.: 1974, 'Indo-American Gondwanaland and the Distorsion of the Granulite Belt', *Tectonophysics* **22**, 141-157.

Crawford, A. R. and Compston, W.: 1973, 'The Age of Cuddapah and Kurnool System S. India', *J. Geol. Soc. Australia* **19**, 453-464.

Crookshank, H.: 1938, 'The Western Margin of the Eastern Ghats in Southern Jeypore', *Rec. Geol. Surv. India* **73** (Pt. 3), 398-433.

Crookshank, H.: 1963, 'Geology of Southern Bastar and Jeypore from the Bailadila Range of Eastern Ghats', *Mem. Geol. Surv. India* **87**, 1–149.

Daly, R.: 1969, *Strength and Structure of the Earth*. Hafner Pub. Co., N.Y., pp. 30–45.

Das Gupta, A., Srivastava, H.N., and Basu Mallick, S.: 1982, 'Source Mechanism of Earthquakes in Kangra-Chamba Region of Himachal Pradesh, India', *Bull. Ind. Soc. Earthq. Tech.* **19**, 102–116.

De, A.: 1969, 'Anorthosites of Eastern Ghats', in Y. W. Isachsen (ed), *Origin of Anorthosites and Related Rocks*, N. Y. State Mus. Sci. Serv. Mem., **18**, 425–434.

Deb, S.: 1966, *Contributions to the Geology of Singhbhum*, Pub. by Jadavpur Univ., Calcutta.

Desikachar, S. V.: 1974, 'A Review of Tectonic and Geological History of Eastern India in Terms of Plate Tectonics Theory', *J. Geol. Soc. India* **15**, 137–149.

Dewey, J.F. and Bird, J. M.: 1970, 'Mountain Belts and New Global Tectonics', *J. Geophy. Res.* **75**, 2625–2647.

Dewey, J. F. and Burke, K. C. A.: 1973, 'Tibetan, Variscan and Precambrian Basement Reactivation, Product of Continental Collisional Orogeny', *J. Geol.* **81**, 683–692.

Dorman, L. M. and Lewis, B. T. R.: 1970, 'Experimental Isostasy I; Theory of Determination of Isostatic Response to Concentrated Load', *J. Geophy. Res.* **75**, 3357–3365.

Dorman, L. M. and Lewis, B. T. R.: 1972, 'Experimental Isostasy 3. Inversion of Isostatic Green Function and Lateral Density Changes', *J. Geophy. Res.* **77**, 3068–3077.

Dunn, J. A.: 1929, 'Geology of Northern Singhbhum, Including Parts of Ranchi and Manbhum Districts', *Mem. Geol. Surv. India,* V. 54.

Dunn, J. A.: 1937, 'Mineral Deposits of Eastern Singhbhum and the Surrounding Area,' *Mem. Geol. Surv. India*, **69**, (Pt. 1).

Dunn, J. A.: 1940, 'Stratigraphy of South Singhbhum,' *Mem. Geol. Surv. India* **63** (Pt. 3).

Dunn, J. A. and Dey, A. K.: 1942, 'Geology and Petrology of Eastern Singhbhum', *Mem. Geol. Surv. India* **69** (Pt. 2).

Dunn, J. A., Auden, J. B., Ghosh, A. M. N., Roy, S. C., and Wadia D. N.: 1939, 'The Bihar-Nepal Earthquake of 1934', *Mem. Geol. Surv. India* **73**.

Evans, P.: 1964, 'The Tectonic Framework of Assam', *J. Geol. Soc. India* **5**, 80–86.

Evans, P. and Crompton, W.: 1946, 'Geological Factors in Gravity Interpretation by Evidence from India and Burma', *Q. J. Geol. Soc. London* **102**, 211–249.

Fermor, L. L.: 1919, Proc. Ind. Sci. Congr., p. 125.

Fermor, L. L.: 1930, 'On the Age of Aravalli Range', *Rec. Geol. Surv. India* **62**, 391–409.

Fermor, L. L., 1931, 'An Attempt at Correlation of Ancient Schistose Rocks of Peninsular India', *Mem. Geol. Surv. India* **70**, (Pt. 1), pp. 1–217.

Fermor, L. L.: 1936, 'An Attempt at Correlation of Ancient Schistose Formations of Peninsular India', *Mem. Geol. Surv. India* **70** (Pt. 1), p. 55; (Pt 2), pp, 53–217.

Fermor, L. L.: 1936–40. 'An Attempt at Correlation of Ancient Schistose Formations of Peninsular India', *Mem. Geol. Surv. India* **70** (Pt. 1), p.55; (Pt. 2), p. 53–324.

Fitch, T. J.: 1970, 'Earthquake Mechanisms of the Himalayan-Burmese and Andaman Regions and Continental Tectonics in Central Asia', *J. Geophy. Res.* **75**, 2699–2709.

Fox, C. S.: 1931, 'Gondwana System and Related Formations', *Mem. Geol. Surv. India* **58**, 1–241.

Fyfe, W. S.: 1970, 'Some Thoughts on the Granite Magma', in G. Newall and N. Rast (eds.), *Mechanism of Igneous Intrusion*, Liverpool, Gallery Press, pp. 201–216.

Fyfe, W. S. and Leonardos, Jr. O. H.: 1977, 'Speculations on the Causes of Crustal Rifting and Subduction, with Application to Atlantic Margins of Brazil', *Tectonophysics* **42**(1), 29-36.

Gansser, A.: 1964, *Geology of the Himalayas*, Interscience Publishers, London, pp. 235–245.

Gansser, A.: 1966, 'The Indian Ocean and the Himalayas, a Geophysical Interpretation', *Ecl. Geol. Helv.* **59**, 831–848.

Gansser, A.: 1980, 'The Significance of Himalayan Suture Zone', *Tectonophysics* **62**, 37–52.

Gaposchkin, E. M.: 1973, *Smithsonian Standard Earth (III)*, Smithsonian Astrophysical Observatory, Spec. Rep. 353.

Gaposchkin, E. M.: 1974, 'Earth's Gravity Field to Eighteenth Degree and Geocentric Coordinates for 104 Stations from Satellite and Terrestrial Data', *J. Geophy. Res.* **79**, 5377–5411.

Gaposchkin, E. M. and Lambeck, K.: 1970, *Smithsonian Standard Earth (II)* Smithsonian Astrophy. Obs. Spec. Rep. No. 315.

Gaposchkin E. M. and Lambeck, K.: 1971, 'The Earth's Gravity Field to Sixteenth Degree and Station Coordinates from Satellite and Terrestrial Data', *J Geophy. Res.* **76**, 4855–4883.

Garland, G. D.: 1965, *The Earth's Shape and Gravity*, Pergamon Press, London, pp. 93–95.

Gee, E. R.: 1932, 'Geology and Coal Resources of the Raniganj Coalfield', *Mem. Geol. Surv. India* **61**.

Geos-3: 1979, *J. Geophy. Res.* V. **84**, No. 88, special issue.

Ghosh, P. K.: 1941, 'The Charnockite Series of Bastar State and Western Jeypore', Rec. Geol. Surv. India, 75; Prof. Pap. 15, pp. 3–53.

Gibb, R. A.: 1968, 'The Densities of Precambrian Rocks from Manitoba', *Can. J. Earth. Sci.* **5**, 433–438.

Gibb, R. A. and Halliday, D. W.: 1974 *Gravity Measurements in Southern Dist. of Keewatin and Southeastern Dist. of McKenzie*, Grav. Map Ser., Earth Phy. Br., Dept. of Mines, Ottawa, Canada, No. 124–131, p. 136.

Gibb, R. A. and Thomas, M. D.: 1976, 'Gravity Signature of Fossil Plate Boundaries in the Canadian Shield', *Nature* **262**(5365), 199–200.

Girdler, R. W.: 1964, 'Geophysical Studies of Rift Valleys', *Physics and Chemistry of the Earth* **5**, 121–156.

Girdler, R. W., Fairhead, J. D., Searle, R. C., and Sowerbutts, T. C.: 1969, 'The Evolution of Rifting in Africa', *Nature* **229**, 66–71.

Glikson, A. Y. and Lambert, I. B.: 1976, 'Vertical Zonation and Petrogenesis of the Early Precambrian Crust in Western Australia', *Tectonophysics* **30**, 55–90.

Goguel, J.: 1949, 'La structure des fosses Africans et la Gravimetric', *Ann. Geophys. Fr.* **5**, 174–175.

Grady, J. C.: 1971, 'Deep Main Faults in South India', *J. Geol. Soc. India* **12**(1), 56–62.

Gulatee, B. L.: 1956, *Gravity Data in India*, Surv. of India, Tech. Pap. 10, pp. 1–95.

Gutenberg, B. and Richter, C. F.: 1954, *Seismicity of the Earth and Associated Phenomenon*, Princeton Univ. Press, Princeton, N. J.

Hammer, S. I.: 1939, 'Terrain Correction for Gravity Stations', *Geophysics* **4**, 184–194.

Hayford, J. F.: 1909, *The Figure of the Earth and Isostasy from Measurements in U.S.*, U.S. Coast and Geodetic Survey.

Heiskanen, W. A.: 1938. *New Isostatic Tables for the Reduction of Gravity Values Calculated on the Basis of Airy's Hypothesis*', Pub. No. 2. Isostatic Institute (IAG), Helsinki.

Heiskanen, W. A. and Vening Meinesz, F. A.: 1958, *The Earth and its Gravity Field*, McGraw Hill Book Co., N.Y., pp. 159–161, 187–220.

Hirn, A., Lepine, J. C., Jobert, G., Sapin, M., Wittlinger, G. Xin, X. Z., Yuan, G. E., Jing, W. X. Wen, T. J., Bai, X. S., Pandey, M. R., and Tater, J. M.: 1984, 'Crustal Structure and Variability of the Himalayan Border of Tibet', *Nature* **307**, 23–25.

Holland, T. H.: 1900, 'The Charnockite Series: A Group of Archaean Hypersthene Rocks in Peninsular India', *Mem. Geol. Surv. India* **88** (Pt. 2), 117–149.

Holmes, A.: 1949, 'The Age of Uraninite and Monzonite from Post Delhi Pegmatites of Rajasthan', *Geol. Mag.* **86**, 288–302.

Holmes, A.: 1950, 'The Age of Uraninite from a Pegmatite near Singar, Gaya Dist India', *Am. Mineralogist* **35**, 19.

Holmes, A.: 1955, 'Dating of Precambrians of Peninsular India and Ceylon', *Proc. Geol. Assoc. Canada* **7** (Pt. 2), 81–106.

Holmes, A.: 1965, *Principles of Physical Geology*, Nelson, London, pp. 1288.

Illies, F. H.: 1970, 'Graben Tectonics as Related to Crust–Mantle Interactions', *Graben Problems, International Rift Symposium*, F. H. Illies and K. Fuchs (eds.).

I. M. D.: 1982, India Meteorological Dept., *Report* on *Seismic Activity Around Pong, Pandoh and Salal Dams for the Period 1981–82* (unpublished).

Iyengar, S. V. P.: 1971, 'Structure, Stratigraphy and Correlation of the Dharwar Super Group Rocks of Mysore and the Adjoining Areas', *Seminar on Precambrian Geology of Peninsular Shield,* Geol. Surv. India, pp. 29–31.

Iyengar, S. V. P. and Anand Alwar, M. A.: 1965, *The Dhanjori Geosyncline and its Bearing on Stratigraphy of the Singhbhum, Keonjhar and Mayurbhanj districts*, Min. Geol. Met. Inst. India, D. N. Wadia commemorative volume, pp. 138–162.

Iyengar, S. V. P. and Banerjee, S.: 1964, *Magmatic Phases Associated with the Precambrian Tectonics of Mayurbhanj dist., Orissa, India,* Proc. 22nd. Int. Geol. Congr. Sec. 10, pp. 510–538.

Isacks, B., Oliver, J., and Sykes, L. R.: 1968, 'Seismology and the New Global Tectonics', *J. Geophy. Res.* **73**, 5855.
IUGG.: 1967, *International Union of Geodesy and Geophysics*, General Assembly, held in Switzerland.
Jeffreys, H. 1959. *The Earth*, Cambridge University Press.
Jhingaran, A. G., Thakur, V. C., and Tandon, S. K.: 1979, 'Structure and Tectonics of the Himalaya', *Himalayan Geol. Seminar*, Sect. II A, Misc. Pub. 41, Geol. Surv. India, pp. 1–37.
Jones, H. C.: 1934, 'Iron Ores of Bihar and Orissa', *Mem. Geol. Surv. India* **63**, (Pt. 3).
Kahle, H. G. and Talwani, M.: 1973, 'Gravimetric Indian Ocean Geoid', *Zeitschrift für Geophysik* **39**, 167–187.
Kahle, H. G., Chapman, M., and Talwani, M.: 1978, 'Detailed 1° × 1° Gravimetric Indian Ocean Geoid and Comparison with GEOS-3 Radar Altimeter Profiles', *Geophy. J. Roy. Astr. Soc.* **55**, 703–720.
Kaila, K. L.: 1981, 'Structure and Seismotectonics of the Himalaya - Pamir - Hindukush Region, and the Indian Plate Boundary', in H. K. Gupta and F. M. Delany (eds), *Zagros–Hindukush–Himalaya Geodynamic Evolution*, Pub. Am. Geophy. Un., Geodynamics Ser. 3..
Kaila, K. L.: 1982, 'Deep Seismic Sounding Studies in India', *Geophy. Res. Bull.* **20** (3), 309–328.
Kaila, K. L. and Narain, H.: 1971, 'A New Approach for the Preparation of Quantitative Seismicity Maps as Applied to Alpide Belt, Sunda Arc and Adjoining Areas', *Bull. Seis. Soc. Am.* **61**, 1275–1291.
Kaila, K. L., Gaur, V. K., and Narain, H.: 1972, 'Quantitative Seismicity Maps of India', *Bull. Seis. Soc. Am.* **62**, 1119–1132.
Kaila, K. L. and Narain, H.: 1976, *Evolution of the Himalaya Based on Seismotectonics and Deep Seismic Soundings*, Proc. Himalayan Geol. Sem. New Delhi, Sec. IIB., Geol. Surv. India, Misc. Pub. No. 41, pp. 1–41.
Kaila, K. L., Krishna, V. G., Roy Chowdhury, K., and Narain, H. 1978, 'Structure of the Kashmir Himalaya from Deep Seismic Studies', *J. Geol. Soc. India* **19**, 1–20.
Kaila, K. L., Roy Chowdhury, K., Reddy, P. R. Krishna, V. G., Narain, H., Subotin, S. I., Sollugib, V. G., Chekurov, A. V., Kharetchko, G. E., Lazarenko, M. A., and Ilchenko, T. V.: 1979, 'Crustal Structure along Kaveli-Udipi Profile in Indian Peninsular Shield from Deep Seismic Sounding', *J. Geol. Soc. India* **20**, 307–333.
Kaila, K. L., Reddy, P. R., Krishna, V. G., Roy Chowdhury, K., Tiwari, H. C., Murthy, P. R. K., and Tripathi, K. M.: 1979, 'Crustal Investigations in India from Deep Seismic Soundings', *Geophy. Res. Bull.* **17**, 273–292.
Kaila, K. L. and Bhatia, S. C.: 1981, 'Gravity Study along the Kavali - Udipi Deep Seismic Sounding Profile in Indian Peninsular Shield: Some Inferences About the Origin of Anorthosites and Eastern Ghats Orogeny', *Tectonophysics* **79**, 129–143.
Kailasam, L. N. and Qureshy, M. N.: 1964, *On Some Anomalous Gravity Anomalies in India*, Adv. Frontiers in Geology and Geophysics. Pub. Ind. Geophy. Un., pp. 135–146.
Kanungo, D. N.: 1972, 'Metamorphism in the Eastern Part of Gangpur', *J. Geol. Soc. Ind.* **13**(2), 120–130.
Kanungo, D. N. and Mallick, N. K.: 1967, *Structure and Stratigraphic Position of Gangpur Ser. in Archaeans of Peninsular India*, Proc. 1st Upp. Mantle Proj. Symposium, N.G.R.I., pp. 458–478.
Karunakaran, C.: 1970, *Ultramafic and Related Rocks in Southern Peninsular India*, Proc. 2nd Upper Mantle Proj. Symposium, N.G.R.I., Hyderabad, pp. 361–376.
Katz, M.B.: 1978, 'Tectonic Evolution of the Archaean Granulite Facies Belt of Srilanka, S. India', *J. Geol. Soc. India* **19**, 185–205.
Khan, A. H. and Azad, J.: 1963, *The Geology of Pakistan Gas Field*, Proc. 2nd Symp. Developments of Petroleum Resources, ECAFE, Min. Res. Dev. Ser. 10, pp. 275–282.
Khatri, K. N. and Tyagi, A. K.: 1983, 'Seismicity Pattern in the Himalayan Plate Boundary and Identification of Areas of High Seismic Potential', *Tectonophysics* **96**, 281–297.
King, L. C.: 1980. *Mem. Geol. Surv. India* **18**, 209.
Kono, M.: 1974, 'Gravity Anomalies in Eastern Nepal and Their Implications to the Crustal Structure of the Himalayas', *Geophy. J. Roy. Astr. Soc.* **39**, 283–299.
Krishnan, M. S.: 1937, 'Geology of Gangpur State', *Mem. Geol. Surv. India* **71**.
Krishnan, M. S.: 1960, *Geology of India and Burma*, Higgin Bothams, Madras, India.
Krishnan, M. S.: 1960, 'Precambrian Stratigraphy of India', *Report Int. Geol. Cong. Norden* **9**, 95–107.
Krishnabrahmam, N. and Kanungo, D. N.: 1976, 'Inference of Granitic Batholiths by Gravity Studies in South India', *J. Geol. Soc. India* **17**(1), 45–53.

Krishnaswamy, V. S.: 1966, 'Seismological Considerations Relating to the Planning and Design of Salal Hydal Project in *Jammu and Kashmir State*', Proc. 3rd Symp. on Earthq. Engg. Roorkee University, Roorkee, pp. 44–66.

Larki, P., Kivioja, L., and Heiskanen, W. A.: 1961, *Topographic-Isostatic Reduction Maps for the World for Hayford Zones, 18-1, Airy-Heiskanen System, T = 30 km*, Pub. Isostatic Institute of International Association of Geodesy, Helsinki.

Lee, W. H. K., Wu, F. T., and Jacobson, C.: 1976, 'A Catalogue of Historical Earthquakes in China, Compiled from Recent Publications', *Bull. Seis. Soc. Am.* **66**, 2003–2016.

Le Fort, P.: 1975, 'Himalayas, the Collided Range; Present Knowledge of Continental Arc.' *Am. J. Sci.* **275** (A), 1–44.

Le Pichon, X.: 1968, 'Sea Floor Spreading and Continental Drift'. *J. Geophy. Res.* **73**, 3661–3698.

Lewis, B. T. R. and Dorman, L. M.: 1970, 'An Isostatic Model for U.S.A. Derived from Gravity and topographic Data', *J. Geophy.* Res. **75**, 3567–3586.

Lerch, F. J.: 1983, 'Status of the Geopotential', *Rev. of Geophysics and Space Physics* **21**(3), 560–565.

Lerch, F. J. Klosko, S. M., Laubscher, R. E., and Wagner, C. A.: 1979, 'Gravity Model Improvement Using GEOS-3', *J. Geophy. Res.* **84**, 3897–3915.

Marussi, A.: 1964, *Italian Expedition to the Karakorum (K^2) and Hindukush*. Sci. Rep. II. Geophysics of the Karakorum, Pub. E. J. Brill, Leiden.

Mathur, S. P.: 1974, 'Crustal Structure in Southwestern Australia from Seismic and Gravity Data', *Tectonophysics* **24**, 154–182.

Mathur, L. P. and Evans, P.: 1964, *Oil in India*, Proc. 22nd Int. Geol. Congr., New Delhi, pp. 86.

McElhinny, M. W.: 1973, *Paleomagnetism and Plate Tectonics*, Cambridge University Press, England.

McKenzie, D.P.: 1969, 'Speculations on the Causes and Consequences of Plate Motions', *Geophy. J. Roy. Astr. Soc.* **18**, 1–32.

McKenzie, D. P. and Sclater, J. G.: 1971, 'Evolution of the Indian Ocean since and Late Cretaceous', *Geophy. J. Roy. Astr. Soc.* **24**, 437–528.

Middlemiss, C. S.: 1910. 'The Kangra Earthquake of 4th April, 1905', *Mem. Geol. Surv. India* **38**.

Minister, J.B., Jordan, T.H., Molnar, P., and Haines, E., 1974, 'Numerical Modelling of Instantaneous Plate Tectonics', *Geophy. J. Roy. Astr. Soc.* **36**, 541–576.

Miyashiro, A.: 1961, 'Evolution of Metamorphic Belts', *J. Petrology* **2**, 277–311.

Molnar, P. and Tapponnier, P.: 1975, 'Cenozoic Tectonics of Asia: Effects of a Continental Collision', *Science* **189**, 419–426.

Molnar, P. and Tapponnier, P.: 1978, 'Active Tectonics of Tibet' *J. Geophy. Res.* **83**, 5361–5369.

Molnar, P., Fitch, T. J., and Wu, F. T.: 1973, 'Fault Plane Solutions of Shallow Earthquakes, Contemporary Tectonics of Asia', *Earth and Planet. Sci. Lett.* **19**, 101–112.

Molnar, P., Chen, W. P., Fitch, T. J., Tapponnier, P., Warsi, W. E. K., and Wu, F. T.: 1977, *Structure and Tectonics of the Himalaya: A Brief Summary of Relevant Geophysical Observations*, Colloques Internaux du CNRS, No. 268, Ecologie et Geologie de L' Himalaya, pp. 269–293.

Mueller, St. and Rybach, L.: 1974, *Crustal Dynamics in the Central Part of Rhine Graben, Approaches to Tophrogenesis*, Proc. International Rift Symposium, F. H. Illies and K. Fuchs (eds.)

Mukhopadhyay, Manoj: 1984, 'Seismotectonics of Transverse Lineaments Across the Eastern Himalaya and Its Foredeep', *Tectonophysics* (communicated).

Murthy, M. V. N.: 1970, *Tectonic and Mafic Igneous activity in Northeast India in Relation to Upper Mantle*, Proc. 2nd Upper Mantle Proj. Symposium, N.G.R.I., Hyderabad, pp. 287–304.

Murthy, M. V. N., Viswanathan, T. V., and Roy Chowdury, S.: 1971, 'The Eastern Ghats Group', *Records Geol. Surv. India* **101**, 15–42.

Naha, A. K.: 1965, 'Metamorphism in Relation to Stratigraphy, Structure and Movements in Parts of Eastern Singhbhum, Bihar', *Quart. J. Min. Met. Soc. India* **37**, 41–88

Nandy, D. R.: 1976, *Geological Set-up of the Eastern Himalaya and the Patkoi-Naga-Arakan-Yoma* (Indo-Burman) Hill Ranges in Relation to the Indian Plate Movement, Proc. Himalayan Geol. Seminar, Sec. II A, Geol. Surv. India, Misc. Pub. No. 41, 205–213.

Nandy, D. R.: 1980, 'Tectonic Patterns in Northeastern India', *Ind. J. Earth Sci.* **7**, 103–107.

Narayanswami, S.: 1970, *Tectonic Setting and Manifestation of the Upper Mantlke in the Precambrain Rocks of South India*, Proc. 2nd. Upper Mantle Project Symposium, N.G.R.I., Hyderabad, India, pp. 378–403.

Narayanswami, S.: 1975, *Proposal for Charnockite-Khondalite System in the Archean Shield of Peninsular India*, Geol Surv. India, Misc. Pub. No. 23, pp. 1–26.
Nautiyal, S. P.: 1966, *Precambrian of Mysore Plateau*, Presidential address, Indian Science Congress, Geological Section.
Nuttli, O. W.: 1969, 'Tables of Angles of Incidence of P Waves at Focus, Calculated from 1968 P-Tables', *Earthquake notes*, 40, 21–25.
Oldham, R. D.: 1917, 'The Structure of the Himalayas and the Gangetic Plains as Elucidated by Geodetic Observations in India', *Mem. Geol. Surv. India* **49**(11), 149–301.
Oldham, R. D.: 1882, 'A Catalogue of Indian Earthquakes from the Earliest to the End of 1869', *Mem. Geol. Surv. India* **19**, 163–215.
Oldham, R. D.: 1899, 'Report on Great Earthquake of 12th June, 1897', *Mem. Geol. Surv. India* **29**, 1–319.
Pakiser, L. C. and Steinhart, J. S.: 1964, 'Explosion Seismology in Western Hemisphere' in H. Odishaw (ed.), *Research in Geophysics*, pp. 123–147.
Pascoe, E.: 1950, *A Manual of Geology of India and Burma*, Govt. of India Pub. pp. 483–485.
Pennington, W. D.: 1979, 'A Summary of Field and Seismic Observations of Pattan Earthquake', in Abdul Farah and De Jong (eds.), *Geodynamics of Pakistan*, pp. 143–147.
Pichamuthu, C. S.: 1953, *The Charnockite Problem*, Spl. Pub. by Mysore Geol. Assoc.
Pichamuthu, C. S.: 1962, 'Some Observations on the Structure, Metamorphism and Geological Evolution of Peninsular India', *J. Geol. Soc. India* **3**, 106–118.
Pichamuthu, C. S.: 1965, 'Regional Metamorphism and Charnockitization in Mysore State, India', *Indian Mineralogist* **6**, 119–126.
Pichamuthu, C. S.: 1967, 'The Precambrians of India', in K. Rankama (ed.), *The Geological System, The Precambrians V. 3*, John Wiley and Sons, pp. 1–90.
Pichamuthu, C. S.: 1971, 'Precambrian Geochronology of Peninsular India', *J. Geol. Soc. India* **12**, 262–273.
Poddar, M. C.: 1953, *A Short Note on the Assam Earthquake of 15th August 1950*. Pub. I. Central Board of Geophysics, Govt of India.
Powell, C. M. and Conaghan, N. J.: 1973, 'Plate Tectonics and the Himalayas', *Earth and Planet Sci. Lett.* **20**, 1–20.
Prasada Rao, G. H. S. V., Murthy, Y. G. K., and Deekshitulu, M. N.: 1964, *Precambrian Structure and Ignerous Activity in Parts of Balasore, Keonjhar, Cuttack, Dhenkanal and Sundergarh Districts, Orissa*. Proc. 22nd. Int. Geol. Cong. Sec. 4, 390–401.
Pratt, J. H.: 1855, 'On the Attraction of Himalayan Mountains and the Elevated Regions on the Plumb Line in India', *Phil. Trans. Roy. Soc. Lond.* **195**, 53–55.
Pratt, J. H.: 1859. *Phil. Trans. Roy. Soc. Lond.* **149**, 747–763.
Pretorius, D. A.: 1974, *The Structural Boundary Between the Keapvaal and Sonama Crustal Provinces*, Econ. Geol. Res. Unit., Univ. Witwatersrand, Johannesberg, S. Africa, Inf. Circ., 88, 27.
Quittmeyer, R. C. and Jacob, K. H.: 1979, 'Historical and Modern Seismicity of Pakistan, Afghanistan and SE Iran. Bull', *Seis. Soc. Am.* **69**, 773–823.
Quittmeyer, R. C., Farah, A., and Jacob, K.H.: 1979, 'Seismicity of Pakistan and its Relationship to Surface Faults' in Abdul Farah and De Jong (ed.), *Geodynamics of Pakistan*, pp. 271–284.
Qureshy, M. N.: 1969, 'Thickenning of Basaltic Layer as Possible Cause for the Uplift of Himalaya – A Suggestion Based on Gravity Data', *Tectonophysics* **7**(2), 137–157.
Qureshy, M. N.: 1971, 'Relation of Gravity to Elevation and Rejuvenation of Blocks in India', *J. Geophy. Res.* **76**, 545–557.
Qureshy, M. N., Krishna Brahmam, N., Garde, S. C., and Mathur, B. K., 1968, 'Gravity Anomalies and Godavari Rift, India', *Geol. Soc. Am. Bull.* **79**, 1221–1230.
Qureshy, M. N., Bhatia, S. C., and Subba Rao, D. V.: 1972, 'Preliminary Results of Some Gravity Survey in Singhbhum Area of Orissa', *J. Geol. Soc. India* **13**, 238–246.
Qureshy, M. N., Krishnabrahmam, N., and Aravamadhu, P. S.: 1969, 'Gravity Surveys on Some Granitic Bodies in S. India', *J. Ind. Geoscience Assoc.* **11**, 171–181.
Qureshy, M. N., Venkatachelam, S. V., and Subrahmanayam, C.: 1974, 'Vertical Tectonics in Middle Himalaya: An Appraisal from Recent Gravity Data', *Bull. Geol. Soc. Am.* **85**, 921–926.
Radhakrishna, B. P.: 1956, *The Closepet Granite of Mysore State*, Bangalore, Pub. Mysore Geologists Association.
Radhakrishna, B. P.: 1967, *The Present Status of Geological and Geochemical Studies in the Dharwar*

Region, Proc. 1st Symposium Upper Mantle Proj., N. G. R. I., Hyderabad (India), pp. 191–210.

Radhakrishna, B. P. and Vasudeva, V. N.: 1977, 'The Early Precambrian of the Southern Indian Shield', *J. Geol. Soc. India* **18**(10), 525–541.

Ramamurthy, V. and Sadasivaiah, M. S.: 1967, 'Rb-Sr Ages of Charnockites and Other Rocks from the Precambrian Complex of Mysore India', in *Conference on Geochronology of Precambrian stratified rocks (Abs)*, Edmonton, p. 73.

Rama Rao, B.: 1945a, 'The Archean Complex of Mysore' *Mysore Geol. Dept. Bull.* **17**.

Rama Rao, B.: 1945b, 'The Charnockite Rocks of Mysore (S. India)', *Mysore Geol. Dept. Bull.* **18**.

Rama Rao, B.: 1962, *Handbook of Geology of Mysore State*, S. India, Bangalore Printing and Publ. Co., p. 264.

Rao, G. Venkateshwar: 1976, 'Heat Flow and Heat Generation Relationship in Gondwana basins of Damodar and Mahanadi Valleys', Ph.D. thesis, Osmania University, Hyderabad.

Rapp, R. H.: 1978, *A Global 1° × 1° Anomaly Field Combining Satellite GEOS-3 Altimeter and Terrestrial Anomaly Data*, Rep. 278, Dept. of Geodetic Science, Ohio State University, Columbus, U.S.A.

Rapp, R. H.: 1981, *The Earth's Gravity Field Using SEASAT Altimeter Data, Terrestrial Data and Other Data*, Rep. 32, Dept. of Geodetic Science, Ohio State University, Columbus, Ohio, U.S.A.

Rapp, R. H.: 1983, 'The Determination of Geoid Undulation and Gravity Anomalies from SEASAT Altimeter Data', *J. Geophy Res.* **88**, 1552–1562.

Rastogi, B. K.: 1974, 'Earthquake Mechanism and Plate Tectonics in Himalayan Region', *Tectonophysics* **21**, 47–56.

Rastogi, B. K.: 1976, 'Source Mechanism Studies and Contemporary Tectonics in Himalayan and Nearby Regions', *Bull. Int. Inst. Seis.* **14**, 99–134.

Rastogi, B. K., Singh, J., and Verma, R. K.: 1973, 'Earthquake Mechanisms and Tectonics of Assam-Burma Region', *Tectonophysics* **18**, 355–366.

Ray, S.: 1976, *Metamorphism and Metasomatic Rocks of India*, K. P. Basu Publ. Co., Calcutta, pp. 114–126.

Reeves, C. V. and Hutchins, D. G.: 1975, 'Crustal Structure in Central Southern Africa', *Nature* **254**, 408–409.

Reid, H. F.: 1910, *The California Earthquake of April 18, 1906*, Report of State Earthquake Investigatigation Committee, 11, Carnegie Institute of Washington, Pub. No. 87.

Ritsema, A. R.: 1954, 'The Seismicity of Sunda Arc in Space and Time', *Indonesia Madjalah Ilmu Alam* **110**, 41–49.

Saha, A. K.: 1964a. *Emplacement of the Northern Part of Singhbhum Granite, E. India–A study in Precambrian Magmatism and Granitisation*, Proc. 22nd. Int. Geol. Congr. Sect. 10, pp. 366–385.

Saha, A. K.: 1964b, 'On the Stratigraphic Position of Basic Lavas South of Jagannathpur, Singhbhum Dist', Bull. Geol. Soc. India **1**, 1–65.

Saha, A. K.: 1966, *Magmatectonics of the Northern Part of Singhbhum Granite Complex, Eastern India*, Proc. 24th. Int. Geol. Congr. Sec. 2, pp. 147–155.

Saha, A. K., 1972, *Petrogenetic and Structural Evolution of the Singhbhum Granite Complex, Eastern India*, Rep. 24th Int. Geol. Congr. Sec. 2, pp. 149–157.

Saha, A. K., 1975, 'The Mayurbahanj Granite - Precambrian batholith in Eastern India - A Review', *Ind. J. Earth Sci.* **7**, 12–26.

Sarkar, S. N.: 1957, 'Stratigraphy and Tectonics of the Dongargarh System', *J. Sci. and Engg. Res.* **1**(2), 237–268.

Sarkar, S. N.: 1958. 'Stratigraphy and Tectonics of the Dongargarh System', *J. Sci. and Engg. Res.* **2**(2), 145–160.

Sarkar, S.N.: 1968, *Precambrian Stratigraphy and Geochronology of Peninsular India*, Dhanbad Publishers, India.

Sarkar, S. N.: 1980, 'Present Status of Precambrian Stratigraphy and Geochronology of Peninsular India - A review', *Ind. J. Earth Sci.* **7**, 12–16.

Sarkar, S. N.: 1983, 'Present Status of Precambrian Stratigraphy and Geochronology of Peninsular India: A Synopsis', *Ind. J. Earth Sci.* **10**, 104–106

Sarkar, S. N., Polkanov, A. A., Gerling, E. K., and Chukrov, F. V.: 1964, *Precambrian Geochronology of Peninsular India*, Proc 22nd. Int. Geol. Congr. Pt. 10, p. 102–130.

Sarkar, S. N. and Saha, A. K.: 1966, 'On the Stratigraphic Relationship of the Metamorphic Rocks of

Singhbhum - Gangpur Region', in S. Deb (ed.), *Contributions to the Geology of Singhbhum*, pp. 22–31.

Sarkar, S. N., Gerling, E. K., Polkanov. A. A., and Chukov, F. V., 1967, 'Precambrian Geochronology of Peninsular India: A Synopsis', *Science and Culture* **30**, 527–537.

Sarkar, S. N., Saha, A. K., and Miller, J. A., 1969. 'Geochronology of the Precambrian Rocks of Singhbhum and Adjacent Regions', *E. India. Geol. Mag.* **106**(1), 15–45.

Sarkar, S. N. and Saha, A. K.: 1977, 'Present Status of the Precambrian Stratigraphy, Tectonics and Geochronology of Singhbhum, Keonjhar, Mayurbhanj Region, Eastern India', *Ind. J. Earth Sci.*, S. Ray Volume, pp. 37–55.

Sarkar, S. N., Gopalan, K., and Trivedi, J. R.: 1981, 'New Data on the Geochronology of the Precambrians of Bhandara - Drug, C. India', *Ind. J. Earth Sci.* **8**, 131–151.

Sarkar, S. N. and Saha, A. K.: 1983, 'Structure and tectonics of the Singhbhum - Orissa Iron - Ore Craton', *Recent Researches in Geology* **10**, 1-25.

Sarma, A. U.S., 1980. 'Study and Analysis of Gravity Field over Singhbhum and Adjoining Area', Ph. D. thesis, Indian School of Mines, Dhanbad.

Searle, R. C.: 1970, 'Evidence from Gravity Anomalies for Thinning of Lithosphere Beneath the Rift Valleys in Kenya', *Geophy. J. Roy. Astr. Soc.* **21**, 13–31.

Seasat: 1982, *SEASAT Project, J. Geophy. Res.* **87**, No. C 5 (Special issue).

Seeber, L. and Armbruster, J. G.: 1981, *Great Detachment Earthquakes Along the Himalayan Arc and Long Term Forecasting*, Earthquake prediction, an international review. Maurice Ewing Ser. 4, Pub. Am. Geophy. Un. pp. 259–277.

Seeber, L., Armbruster, J. G., and Quittmeyer, R. C.: 1981, 'Seismicity and Continental Subduction in Himalayan Arc', in H.K. Gupta and F.M. Delany, (eds.), *Zagros-Hindukush-Himalaya, Geodynamic Evolution*. Geodynamics ser., Pub. Am. Geophy. Un. pp. 215–242.

Shackleton, R. M.: 1976, 'Shallow and Deep Level Exposures in Archaean Crust in India and Africa', in B. F. Windley (ed.), *The Early History of the Earth,* Wiley-Inter Science Pub., pp. 317–322.

Singh, S., Jain, A. K., Sinha, P., Singh, V. N., and Srivastava, L. S.: 1976, 'The Kinnaur Earthquake of January 19, 1975: a Field Report', *Bull. Seis. Soc. Am.*, **66**, 887–901.

Singh, D. D. and Gupta, H. K.: 1980, 'Source Dynamics of Two Great Earthquakes of the Indian Subcontinent: The Bihar - Nepal Earthquake of January 15, 1934 and the Quetta Earthquake of May 30, 1935', *Bull. Seis. Soc. Am.* **70**, 757–773.

Sreenivasan, R. and Sreenivas, B. L.: 1975, *A Reappraisal of Some Aspects of the Geology of Dharwars of Mysore State, India*, Geol. Surv. Ind. Misc. Pub. No. 23, pp. 49–70.

Sreenivasan, R. and Sreenivas, B. L.: 1977, 'Some New Geological Features from the Landsat Imagery of Karnataka', *J. Geol. Soc. India* **18**, 589–597.

Srivastava, L. S. and Somayajulu, J. G.: 1966, *The Seismicity of the Area Around Delhi*, Proc. Third Symposium on Earthquake Engg. pp. 417–422.

Srivastava, L. S. and Dube, R. K.: 1982, 'Seismicity Studies of Some Important Dams in India', *Proc. Int. Assoc. of Engg. Geol.* **8**, 219–227.

St. Amand, P.: 1956, 'Two Proposed measurements of Seismicity', *Bull. Seis. Soc. Am.* **46**, 217–218.

Stauder, W.: 1962, 'Focal Mechanism of Earthquakes' Adv. in *Geophysics,* **9**, 1–71, Academic Press.

Subba Rao, D. V.: 1977, 'Gravity Studies over Mahanadi and South Rewa Gondwana Basins of India', Ph.D. thesis, Andhra University, Waltair.

Subramaniam, A. P.: 1960, *Petrology of Charnockite Suite of Rocks Around St. Thomas Mount and Pallavaram near Madras City, India,* Rept. 21st. Int. Geol. Cong. Session 8, p. 394–403.

Subramaniam, A. P.: 1967, 'Charnockites and Granulites of Southern India - A review', *Medd. Fra. Dansk Geol. Forening* **17**, 473–493.

Subrahmanyam, C.: 1978a., 'Relationship of Gravity Field to Structural Provinces of South India', Ph. D. thesis, Indian School of Mines, Dhanbad.

Subrahmanyam, C.: 1978b, 'On the Relation of Gravity Anomalies to Geotectonics of the Precambrian Terrains of Southern Indian Shield', *J. Geol. Soc. India* **19**, 251–263.

Subrahmanyam, C. and Verma, R. K.: 1982, 'Gravity Interpretation of the Dharwar Greenstone Gneiss - Granite terrain in Southern India Shield and Its Geological Implications, *Tectonophysics* **84**, 225–245.

Swiminath, J., Ramakrishnan, M., and Viswanatha, M. N.: 1976, 'Dharwar Stratigraphic Model and Karnataka Craton Evolution', *Rec. Geol. Surv. India* **107**, (2), 149–175.

Takin, M. L.: 1966, 'An Interpretation of Positive Gravity Anomaly over Bombay on the West Coast of India', *Geophy. J. Roy. Astr. Soc.* **11**, 527–537.

Talwani, M., Worzel, J. L., and Landisman, M.: 1959, 'Rapid Gravity Computation of Two Dimensional Bodies with Application to Mendocino Submarine Fracture Zone', *J. Geophy. Res.* **64**, 49–59.

Tandon, A. N.: 1954, 'A Study of Assam Earthquake of August 1950 and it After-Shocks', *Ind. J. Meteor. and Geophy.* **50**, 95–137.

Tandon, A. N.: 1972, 'The Anantnag Earthquake', *Ind. J. Meteor. and Geophy* **23**, 491–502.

Tandon, A. N. and Srivastava, H. N.: 1975a, 'Focal Mechanism of Some Recent Himalayan Earthquakes and Regional Plate Tectonics', *Bull. Seis. Soc. Am.* **65**, 963–969.

Tandon, A.N. and Srivastava, H.N. 1975b, 'Fault Plane Solutions as Related to Faults in and Around India', *Ann. di. Geofisica* **13**, 27

Tanner, J. G.: 1969, 'A Geophysical Study of Structural Boundaries in the Eastern Canadian Shield', Ph.D. thesis, Univ. Durham, England (unpublished).

Tapponnier, P. and Molnar, P.: 1976, 'Slip-Line Field theory and Large-Scale Continental Tectonics', *Nature* **264**, 319–324.

Telford, W. M., Geldart, L. P., Sheriff, R. E., and Key, D.A.: 1976, *Applied Geophysics*, Cambridge University Press, England, p. 21.

Thomas, M. D.: 1974., *The Correlation of Gravity and Geology in Southeastern Quebec and Southern Labrador*, Gravity map ser., Earth Phy. Br., Dept. of Energy, Mines and Resources, Ottawa, Canada, p. 49.

Thomas, M. D. and Tanner, J. G.: 1975, 'Cryptic Suture in the Eastern Grenville Province', *Nature* **256**, 392–394.

Vaidyanadhan, R.: 1967, *An Outline of Geomorphic History of India South of Lat. 18° N.*, Proc. Symp. On Geomorphological Studies in India, Center of Adv. Studies, Univ. Sagar, pp. 121–130.

Valdiya, K. S.: 1973, 'Tectonic Framework of India, a review and Interpretation of Recent Structural and Tectonic Studies', *Geophy. Res. Bull.* **11**, 79–114.

Valdiya, K. S.: 1976, 'Himalayan Transverse Faults and Folds and Their Parallelism with Subsurface Structures of North Indian Planes', *Tectonophysics* **32**, 353–386.

Vening Meinesz, F. A.: 1931. 'Une nouvelle methode pour la reduction isostatique regionale de l'intensite du la pesanteur', *Bull. Geod.* No. 29.

Vening Meinesz, F. A.: 1941a. *Gravity over Hawaiian Archipelago and over the Maderia Area*, Proc. Netherland Acad. Westensia, pp. 44.

Vening Meinesz, F. A.: 1941b, *Tables for Regional and Local Isostatic Reduction (Airy System) for Gravity Values*, Pub. Netherland Geodetic Commission, Waltman, Delft, Holland.

Vening Meinesz, F. A. and Heiskanen, W. A.: 1958, *The Earth and Its Gravity Field*, McGraw Hill, N. Y., pp. 470.

Venkatasubramanian, V. S. and Narayanaswamy, R.: 1974a. 'Rb–Sr. Whole Rock Isochron Studies on Granitic Rocks of Chitradurga and N. Mysore', *Jour. Geol. Soc. Ind.* **15**, 77–81.

Venkatasubramanian, V. S. and Narayanaswamy, R.: 1974b 'Primary and Metamorphic Rb–Sr Chronology in Some Areas of South Mysore', *Jour. Geol. Soc. Ind.* **15**, 200–205.

Verma, R. K. and Debabrata Ghosh: 1974, 'Gravity Survey over Jharia Coalfield, India', *Geophy. Res. Bull.* **12**, 165–175.

Verma, R. K. and Mukhopadhyay, M.: 1976, 'Tectonic Significance of Anomaly – Elevation Relationship in Northeastern India', *Tectonophysics* **34**, 117–133.

Verma, R. K. and Ghosh, D.: 1977, 'Gravity Field, Structure and Tectonics of Some Gondwana basins of Damodar Valley', Geophysical Case Histories of India, 1, 97–112, AEG, Hyderabad.

Verma, R. K. and Mukhopadhyay, M.: 1977, 'An Analysis of Gravity Field in Northeastern India', *Tectonophysics* **42**, 283–317.

Verma, R. K., Mazumdar, R., Ghosh, D. Ashish Ghosh, and Gupta, N. C.: 1976, 'Results of Gravity Survey over Raniganj Coalfied, India', *Geophy. Prosp.* **24**, 19–30.

Verma, R. K., Ahluwalia, M. S., and Mukhopadhyay, M.: 1976a., 'Seismicity, Gravity and Tectonics of Northeast India and N. Burma', *Bull. Seis. Soc. Am.* **66**, 1683–1694.

Verma, R. K., Mukhopadhyay, M., and Ahluwalia, M. S.: 1976b. 'Earthquake Mechanisms and Tectonic Features of Northern Burma', *Tectonophysics*, **32**, 382–399.

Verma, R. K., Mukhopadhyay, M., and Roy, B. N.: 1977, 'Seismotectonics of Himalaya and Continental

Plate Convergence', *Tectonophysics* **42**, 319–335.

Verma, R. K., Mukhopadhyay, M., and Bhuin, N. C.: 1978a 'Seismicity, Gravity and Tectonics in Andaman Sea', *J. Phy. Earth,* **26** (Suppl.), 5233–5248.

Verma, R. K., Mukhopadhyay, M., Roy, S. K., and Sinha, R. P. P.: 1978b, 'An Analysis of Gravity Field over Northern Singhbhum', *Tectonophysics* **44**, 41–63.

Verma, R. K., Bhuin, N. C., and Mukhopadhyay, M.: 1979, 'Geology, Structure and Tectonics of Jharia Coalfied, India–a Three Dimensional Model', *Geoexploration* **17**, 305–324.

Verma, R. K., Bandopadhyay, T. K., and Mukhopadhyay, M.: 1980a, 'Gravity Field, Structure and Tectonics of Raniganj Coalfield–A Three Dimensional Model' *J. Geol. Soc. India* **21**, 117–127.

Verma, R. K., Mukhopadhyay, M., and Nag, A. K.: 1980b. 'Seismicity and Tectonics in S. China and Burma', *Tectonophysics* **64**, 84–96.

Verma, R. K., Sarma, A. U. S., and Mukhopadhyay, M.: 1984. 'Gravity Field over Singhbhum, Its Relationship to Geology and Tectonic History', *Tectonophysics,* **106**, 87–107.

Vidyadharan, K. T., Sukamaran, P. V., and Nair, M. M.: 1977, 'A Note on the Occurrence of Anorthosite Near Perinthatta, Talipar Amba Taluk, Cannanore Dist., Kerala', *J. Geol. Soc. India*, **18**(a), 519–520.

Vinogradov, A. P. and Tugarinov. A. I., Zhykov, C., A. I., Zhykov, C., Staphikova N., Bibikova, E., and Khorre, K: 1964, *Geochronology of the Indian Precambrians*, Report 22nd. Int. Geol. Congr. Pt. 10, pp. 553–567.

Vinogradov, A. P. and Tugarinov, A. I.: 1968, *Geochronology of the Indian Precambrian* Report 22nd. Int. Geol. Congr. Sec. 10.

Viswanathan, M. M. and Tareen, J. A. K.: 1970, 'The Nilgiri Charnockite', *Indian Mirteralogist* **11**, 78–86.

Viswanatha, M. N. and Ramakrishnan, M.. 1975, 'The PRe-Dharwar Supercrustal Rocks of Sargur Schist Complex in Southern Karnataka and Their Tectonic Metamorphic Significance', *Indian Mineralogist* **16** 48–65.

Wadia, D. N.: 1942, *The Making of India*, Proc. 29th. Indian Science Congress, Pt. 2, p. 3–23.

Wadia, D. N.: 1949, 1953, *Geology of India*, McMillian and Co., London.

Wadia, D. N.: 1975, *Geology of India*, Tata McGraw Hill Pub. Co., New Delhi.

Walcott, R. I.: 1970, 'Flexural Rigidity, Thickness and Viscosity of the Lithosphere', *J. Geophy. Res.* **75**, 3941.

Warsi, W. E. K. and Molnar, P.: 1977, *Gravity Anomalies and Plate Tectonics in the Himalaya*, Proc. Colloques Internationaux du CNRS, Ecologie et Geologie de l' Himalaya, Paris, pp. 463–478.

Watts, A. B. and Cochran, J. R.: 1974, 'Gravity Anomalies and Flexure of Lithosphere along the Hewaiian–Emperor Seamount Chain', *Geophy. J. Roy. Astr. Soc.* **38**, 119–141.

Watts, A. B. and Talwani, M.: 1974, 'Gravity Anomalies Seaward of Deep-Sea Trenches and Their Tectonic Implications', *Geophy. J. Roy. Astr. Soc.* **36**, 57–90.

Wellman, P.: 1978, 'Gravity Evidence for Abrupt Changes in Mean Crustal Density at the Junction of Australian Crustal Blocks', *BMR Jour. Aust. Geol. and Geophy* **3**, 153–162.

Wellman, P. and McElhinny, M. W.: 1970., 'K–Ar Age of the Deccan Traps, India', *Nature,* **227**, 595–596.

Wilcox, L. E.: 1974, *An Analysis of Gravity Prediction Methods for Continental Areas*. Defence Mapping Agency, Aerospace Center, Pub. No. 74–001.

Windley, B. F.: 1973, 'Crustal Development in the Precambrian'. *Phil. Trans. Roy. Soc. Lond. Ser.* A. **273**, 321–341.

Windley, B. F. and Bridgwater, D.: 1971, 'The Evolution of the Archean Low and High Grade Terrains', *Geol. Soc. Austr. Spl. Pub.* **3**, 33–36.

Windley, B. F. and Selvan, T. A.: 1975, 'Anorthosites and Associated Rocks of Tamil Nadu, S. India' *J. Geol. Soc. Ind.* **16**, 209–214.

Woollard, G. P.: 1959, 'Crustal Structure from Gravity and Seismic Measurements', *J. Geophy. Res.* **64** (10), 1521–1544.

Woollard, G. P.: 1962, *The Relation of Gravity Anomalies to Surface Elevation, Crustal Structure and Geology*, Univ. Wisconsin, Geophy, Polar Res. Center, Rept., 62–9.

Woollard, G. P.: 1966. *Regional Isostatic Relations in U.S.A.*, Geophy. Un. Monograph. No. 10, Am. Geophy. Un. pp. 557–594.

Woollard. G. P.: 1969, *Regional Variations* in *Gravity*, Am. Geophy. Un. Monograph No. 13, pp. 320–343.

Worze, J. L. and Shurbet, G. L.. 1965, *Gravity Interpretation from Standard Oceanic and Continental Crustal Sections*, Geol. Soc. Am. Spl. Pap. 62, pp. 87–100.

AUTHOR INDEX

SUBJECT INDEX

GPSR Compliance

The European Union's (EU) General Product Safety Regulation (GPSR) is a set of rules that requires consumer products to be safe and our obligations to ensure this.

If you have any concerns about our products, you can contact us on ProductSafety@springernature.com

In case Publisher is established outside the EU, the EU authorized representative is:

Springer Nature Customer Service Center GmbH
Europaplatz 3
69115 Heidelberg, Germany

Batch number: 10755414

Printed by Printforce, the Netherlands